Growth and reproductive strategies of freshwater phytoplankton

Growth and reproductive strategies of freshwater phytoplankton

Edited by

CRAIG D. SANDGREN

The University of Wisconsin – Milwaukee

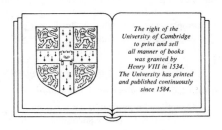

The right of the
University of Cambridge
to print and sell
all manner of books
was granted by
Henry VIII in 1534.
The University has printed
and published continuously
since 1584.

CAMBRIDGE UNIVERSITY PRESS

Cambridge

New York   Port Chester   Melbourne   Sydney

Published by the Press Syndicate of the University of Cambridge
The Pitt Building, Trumpington Street, Cambridge CB2 1RP
40 West 20th Street, New York, NY 10011-4211, USA
10 Stamford Road, Oakleigh, Victoria 3166, Australia

First published 1988
First paperback edition 1991

Printed in the United States of America

Library of Congress Cataloging-in-Publication Data

Growth and reproductive strategies of freshwater phytoplankton /
[edited by] Craig D. Sandgren.
p. cm.
Contents: General introduction ; The ecology of chrysophyte
flagellates / Craig D. Sandgren – Ecology of the Cryptomonadida, a
first review / Dag Klaveness – Freshwater armored
dinoflagellates / Utsa Pollingher – Ecology of freshwater
planktonic green algae / Christine M. Happey-Wood – Growth and
survival strategies of planktonic diatoms / Ulrich Sommer – Growth
and reproductive strategies of freshwater blue-green algae
(Cyanobacteria) / Hans W. Paerl – Physiological mechanisms in
phytoplankton resource competition / David H. Turpin – Selective
herbivory and its role in the evolution of phytoplankton growth
strategies / John T. Lehman – Functional morphology and the
adaptive strategies of freshwater phytoplankton / Colin S. Reynolds.
ISBN 0-521-32722-9
1. Freshwater phytoplankton – Growth. 2. Freshwater phytoplankton –
Reproduction. I. Sandgren, Craig D.
QK935.G76 1988
589.4 – dc19   87–30543

British Library Cataloguing in Publication Data

Growth and reproductive strategies of
freshwater phytoplankton.
1. Freshwater phytoplankton
I. Sandgren, Craig D.
589.4   QK935

ISBN 0-521-32722-9 hardback
ISBN 0-521-42910-2 paperback

CONTENTS

CONTRIBUTORS

Dr. Christine M. Happey-
Wood
School of Plant Biology
University College of North
Wales
Bangor Gwynedd LL57 2UW
Great Britain

Dr. Dag Klaveness
Department of Biology
University of Oslo
P.O. Box 1027
N-0315 Blindern
Oslo 3, Norway

Dr. John T. Lehman
Department of Ecology and
Evolutionary Biology
University of Michigan
Ann Arbor, MI 48109 USA

Dr. Hans W. Paerl
Institute of Marine Sciences
University of North Carolina
3407 Arendale St.
Morehead City, NC 28557
USA

Dr. Utsa Pollingher
Israel Oceanographic and
Limnological Research
Tel Shikmona
P.O. Box 8030
Haifa 31080, Israel

Dr. Colin S. Reynolds
Freshwater Biological
Association
The Ferry House
Ambleside, Cumbria
LA22 0LP
Great Britain

Dr. Craig D. Sandgren
Department of Biological
Sciences & Center for
Great Lakes Studies
University of Wisconsin –
Milwaukee
P.O. Box 413
Milwaukee, WI 53201 USA

Dr. Ulrich Sommer
Max-Planck-Institut für
Limnologie
August Thienemann Strasse 2
Postfach 165
D-2320 Plön
Federal Republic of Germany

Dr. David H. Turpin
Department of Biology
Queen's University
Kingston, Ontario K7L 3N6
Canada

Chapter 1

GENERAL INTRODUCTION

CRAIG D. SANDGREN

There is hardly a lack of published information concerning the ecology of freshwater phytoplankton. The growth in numbers of journals, journal articles, and published symposia in this research area since the appearance of Hutchinson's seminal thesis (Hutchinson 1967) and Lund's review paper (Lund 1965) is impressive. A number of recent works have attempted to synthesize the topic from various viewpoints (e.g., Morris 1980; Moss 1980; Platt 1981; Round 1981; Meyers & Strickler 1984; Reynolds 1984a; Harris 1986; Sommer et al. 1986; Munawar & Talling 1986).

One important trend during this recent surge of interest in freshwater phytoplankton ecology has been the simultaneous emphasis on *both* growth and loss processes as mediated through resource supply, trophic exchange, and physical mixing of the system (Crumpton & Wetzel 1982; Reynolds et al. 1982; Lehman & Sandgren 1985). Related topics of particular interest have been the role of limiting nutrient supply ratios in competitive interactions (Rhee & Gotham 1980; Tilman 1982; Tilman, Kilham & Kilham 1982; Smith 1983; Kilham & Kilham 1984; Terry, Laws, & Burns 1985); mixotrophic nutrition of algae and heterotrophic flagellates, with associated food web complexities (Paerl 1982; Hobbie & Williams 1984; Porter et al. 1985; Bird & Kalff 1986,

This volume is an outgrowth of a symposium on the ecology of freshwater phytoplankton held at the Second International Phycological Symposium, Copenhagen, Denmark, in August 1985. The editor would like to thank the organizing committee of the congress and, particularly, Drs. Ø. Moestrup and J. Kristiansen for the opportunity to organize the symposium. The editor would also like to thank all the participants in the symposium for their timely and scholarly contributions to this volume. Special thanks are extended to Drs. D. Klaveness and H. Paerl who did not participate in the original symposium, but have contributed excellent chapters to this volume and have substantially added to the balance of coverage of important freshwater phytoplankton groups. Travel to the congress by the editor was sponsored by NSF grant BSR-85-96027.

1987; Fenchel 1986a,b; Sherr & Sherr 1987); zooplankton feeding mechanisms and nutrient remineralization (Porter 1976, 1977; Lehman 1980a,b; Lehman & Scavia 1982a,b; Koehl 1984; Strickler 1984; Sterner 1986; Fryer 1987); the role of turbulence in phytoplankton periodicity (Knoechel & Kalff 1975; Reynolds 1980, 1984b; Reynolds, Wiseman, & Clark 1984; Sommer 1984; Paerl, Chapter 7, this volume); organism morphology and size as scaling factors (Lewis 1976; Margalef 1978; Reynolds 1980, 1984b; Werner & Gilliam 1984); and "cascading" effects of changes within individual trophic levels on whole system structure and production (Carpenter & Kitchell 1984; Neill 1984; Bergquist, Carpenter, & Latino 1985; Carpenter, Kitchell, & Hodgson 1985; Drenner, Threkeld, & McCracken 1986; Mills, Furney, & Wagner 1987; Siegfriegld 1987; Threkeld 1987). One outgrowth of these many studies has been increasing emphasis on the physiological and behavioral attributes of the *individual organisms* – the algae, protozoa, zooplankton, and fish species – and renewed recognition of the importance of the actual species mix for determining plankton dynamics and response to perturbations. It has become increasingly clear that individual species may constitute important keystone predators or may act as critical "bottlenecks" to the flow of information and resources through planktonic food webs. Planktonic species are certainly more than bits of carbon and chlorophyll that may be freely substituted one for another without affecting planktonic food web dynamics and productivity.

Much insight into the dynamics of planktonic ecosystems has been gained from careful study of the reproductive and feeding ecology of individual fish and zooplankton species (e.g., King 1977; Dumont & Green 1980; Kerfoot 1980; Zaret 1980; Werner & Gilliam 1984; Lampert 1985; Kerfoot & Sih 1987). However, understanding of the ecology or "life history strategies" of phytoplankton species remains a formidable task because of the comparatively great diversity of species and the frustratingly great interannual and interlake variability in seasonal distribution patterns. The problem has been intensified in the modern literature by a tendency to emphasize short-term data sets and only dominant species in publication. Despite these inconsistencies and shortcomings in the data, limnologists have long recognized general trends suggesting that taxonomic groups of phytoplankton and morphological types of cells characteristically predominate the plankton of lakes at only certain times of the year. As common examples, we can cite the spring diatom blooms and the summer succession of coccoid green algae, cyanobacteria, and dinoflagellates so typical of moderately productive dimictic lakes. These general distributional trends likely represent strong interactions between species and both biotic and abiotic components of aquatic ecosystems. Attempts at

A. Typical Life History for Seasonally-Restricted Planktonic Micro-algae:

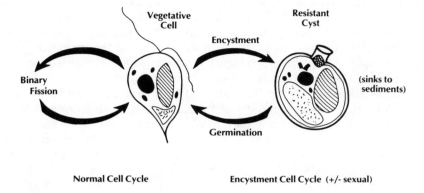

Normal Cell Cycle   Encystment Cell Cycle (+/- sexual)

B. Planktonic Cells Have TWO Sets of Life History Adaptations:

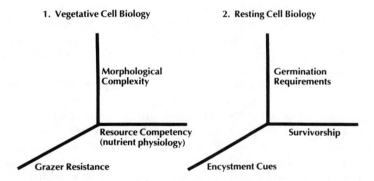

Fig. 1-1. Interacting aspects of phytoplankton population and reproductive ecology that compose the elements of a planktonic alga's life history strategy. (A) Typical life history for many phytoplankton species exhibiting an alternation between vegetative cells and resting cells (resting cyst in this example). (B) Axes representing the characteristics by which microalgae diversify and adapt for persistence in stressful, heterogeneous planktonic habitats.

understanding the nature of these strong interactions have occupied much of the modern autecological literature on freshwater phytoplankton.

The chapters in this book contribute to an understanding of the mechanisms generating phytoplankton distributional patterns by summarizing (1) the morphological, nutritional, and physiological characteristics of the algae, and (2) the competitive and trophic interactions affecting members of the major taxonomic classes of freshwater phytoplankton. A relatively unique feature of these contributions is that the authors have also attempted to integrate information about the reproductive biology of species with the ecology of the vegetative cell populations. Adopting this integrative approach has permitted a great deal of both limnological and phycological literature to be synthesized, in some instances for the first time. Such integration is an important consideration because, though some planktonic algae seemingly perennate via small refuge populations of vegetative cells in the plankton (e.g., araphid pennate diatoms such as *Asterionella*), the majority of freshwater phytoplankton incorporate resistant benthic resting stages into their life history strategies (Fig. 1-1a). Attempts have also been made here to recognize distinctive "packaging plans" of physiological, morphological, and reproductive features among freshwater phytoplankton that represent evolutionary solutions to the problems of persistence in stressful and periodic planktonic habitats (Fig. 1-1b). This evolutionary consideration is a second relatively unique feature of the book. It is hoped that the combination of organismal and evolutionary approaches to freshwater phytoplankton ecology emphasized here will provide new insights into the interplay of both proximate (short-term) and ultimate (evolutionary) factors influencing phytoplankton dynamics.

The book is organized into two sections. Chapters 2–7 concern the ecology of important taxonomic classes of phytoplankton. Authors were asked to consider the validity and utility of the interactions outlined in Fig. 1-1 for the group of algae in which they have special interest and expertise. Each of the last three chapters (8–10) constitutes a more detailed discussion of one of the axes in the left-hand panel of Fig. 1-1b in regard to its effects on phytoplankton in general.

REFERENCES

Bergquist, A. M., Carpenter, S. R., and Latino, J. C. (1985). Shifts in phytoplankton size structure and community composition during grazing by contrasting zooplankton assemblages. *Limnol. Oceanogr.,* 30, 1037–45.

Bird, D. F. and Kalff, J. (1986). Bacterial grazing by planktonic lake algae. *Science,* 231, 493–5.

(1987). Algal phagotrophy: regulating factors and importance relative to photosynthesis in *Dinobryon* (Chrysophyceae). *Limnol. Oceanogr.,* 32, 277–84.

Carpenter, S. R. and Kitchell, J. F. (1984). Plankton community structure and limnetic primary production. *Am. Nat.,* 124, 159–72.

Carpenter, S. R., Kitchell, J. R., and Hodgson (1985). Cascading trophic interactions and lake productivity. *Bioscience,* 35, 634–9.

Crumpton, W. G. and Wetzel, R. G. (1982). Effects of differential growth and mortality in the seasonal succession of phytoplankton populations in Lawrence Lake, Michigan. *Ecology,* 63, 1729–39.

Drenner, R. W., Threkeld, S. T., and McCracken, M. D. (1986). Experimental analysis of the direct and indirect effects of an omnivorous filter-feeding culpeid on plankton community structure. *Can. J. Fish. Aquat. Sci.,* 43, 1935–45.

Dumont, H. J. and Green, J., eds. (1980). Rotatoria: Proceedings of the Second International Rotifer Symposium. *Hydrobiologia, 73.* The Hague: Dr. W. Junk.

Fenchel, T. (1986a). The ecology of heterotrophic microflagellates. In: *Advances in Microbial Ecology,* vol. 9, ed. K. C. Marshall, pp. 57–97. New York: Plenum Press.

(1986b). *Ecology of Protozoa: The Biology of Freeliving Phagotrophic Protists.* Madison, WI: Science Tech.

Fryer, G. (1987). The feeding mechanisms of the Daphniidae (Crustacea: Cladocera): recent suggestions and neglected considerations. *J. Plank. Res.,* 9, 419–32.

Harris, G. P. (1986). *Phytoplankton Ecology: Structure, Function and Fluctuation.* New York: Chapman and Hall.

Hobbie, J. H. and Williams, P. J. LeB., eds. (1984). *Heterotrophic Activity in the Sea.* NATO Conf. Ser. 4, Mar. Sci. V. 15. New York: Plenum Press.

Hutchinson, G. E. (1967). *A Treatise on Limnology. Vol. 2. Introduction to Lake Biology and the Limnoplankton.* New York: Wiley.

Kerfoot, W. C., ed. (1980). *Evolution and Ecology of Zooplankton Communities.* Hanover, NH: University Press of New England.

Kerfoot, W. C. and Sih, A., eds. (1987). *Predation: Direct and Indirect Impacts on Aquatic Communities.* Hanover, NH: University Press of New England.

Kilham, S. S. and Kilham, P. (1984). The importance of resource supply rates in determining phytoplankton community structure. In: *Trophic Interactions within Aquatic Ecosystems,* eds. D. G. Meyers and J. R. Strickler, pp. 7–27. AAAS Selected Symposium 85, American Association for the Advancement of Science. Boulder, CO: Westview Press.

King, C. E., ed. (1977). *Proceedings of the First International Rotifer Symposium. Archiv. Hydrobiol. Beih. 8.*

Knoechel, R. and Kalff, J. (1975). Algal sedimentation: the cause of diatom blue-green succession. *Verh. Internat. Verein. Limnol.,* 19, 745–54.

Koehl, M. A. R. (1984). Mechanisms of particle capture by copepods at low Reynolds numbers: possible modes of selective feeding. In: *Trophic Interactions within Aquatic Ecosystems,* eds. D. G. Meyers and J. R. Strickler, pp. 135–66. AAAS Selected Symposium 85. Boulder, CO: Westview Press.

Lampert, W., ed. (1985). Food limitation and the structure of zooplankton communities. *Ergeb. Limnol.*, 24 *(Arch. Hydrobiol.)*.

Lehman, J. T. (1980a). Release and cycling of nutrients between planktonic algae and herbivores. *Limnol. Oceanogr.*, 25, 620–32.

(1980b). Nutrient recycling as an interface between algae and grazers in freshwater communities. In: *Evolution and Ecology of Zooplankton Communities*, ed. W. C. Kerfoot, pp. 251–63. Hanover, NH: University Press of New England.

Lehman, J. T. and Sandgren, C. D. (1985). Species-specific rates of growth and grazing loss among freshwater algae. *Limnol. Oceanogr.*, 30, 34–46.

Lehman, J. T. and Scavia, D. (1982a). Microscale patchiness of nutrients in plankton communities. *Science*, 216, 729–30.

(1982b). Microscale nutrient patches produced by zooplankton. *Proc. Nat. Acad. Sci. USA*, 79, 5001–5.

Lewis, W. M. (1976). Surface-volume ratio: implication for phytoplankton morphology. *Science*, 192, 885–7.

Lund, J. W. G. (1965). The ecology of freshwater phytoplankton. *Biol. Rev.*, 40, 231–93.

Margalef, R. (1978). Life-forms of phytoplankton as survival alternatives in an unstable environment. *Oceanol. Acta*, 1, 493–509.

Meyers, D. G. and Strickler, J. R. (1984). *Trophic Interactions within Aquatic Ecosystems*. AAAS Selected Symposium 85. Boulder, CO: Westview Press.

Mills, E. L., Forney, J. L., and Wagner, K. J. (1987). Fish predation and its cascading effects on the Oneida Lake food chain. In: *Predation: Direct and Indirect Impacts on Aquatic Communities*, eds. W. C. Kerfoot and A. Sih, pp. 118–31. Hanover, NH: University Press of New England.

Morris, I. (1980). *The Physiological Ecology of Phytoplankton, Studies in Ecology*, vol. 7. London: Blackwell Scientific.

Moss, B. (1980). *Ecology of Fresh Waters*. Oxford: Blackwell Scientific.

Munawar, M. and Talling, J. R., eds. (1986). The seasonality of freshwater phytoplankton: a global perspective. *Hydrobiologia*, 138. The Hague: Dr. W. Junk.

Neill, W. E. (1984). Regulation of rotifer densities by crustacean zooplankton in an oligotrophic montane lake in British Columbia. *Oecologia*, 48, 164–77.

Paerl, H. W. (1982). Interactions with bacteria. In: *The Biology of Cyanobacteria*, eds. N. G. Carr and B. A. Whitton, pp. 441–61. Oxford: Blackwell Scientific.

Platt, T. (1981). *Physiological Bases of Phytoplankton Ecology. Can. Bull. Fish. Aquat. Sci.* 210. Ottawa: Fisheries and Oceans Canada.

Porter, K. G. (1976). Enhancement of algal growth and productivity by grazing zooplankton. *Science*, 192, 1332–4.

(1977). The plant-animal interface in freshwater ecosystems. *Amer. Sci.*, 65, 159–70.

Porter, K. G., Sherr, E. B., Sherr, B. F., Pace, M., and Sanders, R. W. (1985). Protozoa in planktonic food webs. *J. Protozool.*, 32, 409–15.

Reynolds, C. S. (1980). Phytoplankton assemblages and their periodicity in stratifying lake systems. *Holarct. Ecol.,* 3, 141–59.

(1984a). *The Ecology of Freshwater Phytoplankton.* Cambridge: Cambridge University Press.

(1984b). Phytoplankton periodicity: the interactions of form, function and environmental variability. *Freshwat. Biol.,* 14, 111–42.

Reynolds, C. S., Thompson, J. M., Ferguson, A. J. D., and Wiseman, S. W. (1982). Loss processes in the population dynamics of phytoplankton maintained in closed systems. *J. Plank. Res.* 4: 561–600.

Reynolds, C. S., Wiseman, S. W., and Clark, M. J. O. (1984). Growth and loss-rate responses of phytoplankton to intermittent artificial mixing and their potential application to the control of planktonic algal biomass. *J. Appl. Ecol.,* 21, 11–39.

Rhee, G-Y. and Gotham, I. J. (1980). Optimum N:P ratios and coexistence of planktonic algae. *J. Phycol.,* 16, 486–9.

Round, F. E. (1981). *The Ecology of Algae.* Cambridge: Cambridge University Press.

Sherr, E. B. and Sherr, B. F. (1987). High rates of consumption of bacteria by pelagic ciliates. *Nature,* 325, 710–11.

Siegfried, C. A. (1987). Large-bodied crustacea and rainbow smelt in Lake George, New York: trophic interactions and phytoplankton community composition. *J. Plank. Res.,* 9, 27–39.

Smith, V. H. (1983). Low nitrogen to phosphorus ratios favor dominance by bluegreen algae in lake phytoplankton. *Science,* 221, 669–71.

Sommer, U. (1984). Sedimentation of principal phytoplankton species in Lake Constance. *J. Plank. Res.,* 6, 1–14.

Sommer, U., Gliwicz, Z. M., Lampert, W., and Duncan, A. (1986). The PEG-model of seasonal succession of planktonic events in fresh waters. *Arch. Hydrobiol.,* 106, 433–71.

Sterner, R. W. (1986). Herbivore's direct and indirect effects on algal populations. *Science,* 231, 605–7.

Strickler, J. R. (1984). Sticky water: a selective force in copepod evolution. In: *Trophic Interactions within Aquatic Ecosystems,* eds. D. G. Meyers and J. R. Strickler, pp. 135–66. AAAS Selected Symposium 85. Boulder, CO: Westview Press.

Terry, K. L., Laws, E. A., and Burns, D. J. (1985). Growth rate variation in the N:P requirement ratio of phytoplankton. *J. Phycol.,* 21, 323–9.

Threkeld, S. T. (1987). Experimental evaluation of trophic-cascade and nutrient-mediated effects of planktivorous fish on plankton community structure. In: *Predation: Direct and Indirect Impacts on Aquatic Communities,* eds. W. C. Kerfoot and A. Sih. Hanover, NH: University Press of New England.

Tilman, D. (1982). *Resource Competition and Community Structure.* Princeton: Princeton University Press.

Tilman, D., Kilham, S. S., and Kilham, P. (1982). Phytoplankton community ecology: the role of limiting nutrients. *Ann. Rev. Ecol. Syst.,* 13, 349–72.

Werner, E. E. and Gilliam, J. F. (1984). The ontogenetic niche and species

interactions in size-structured populations. *Ann. Rev. Ecol. Syst.,* 15,
393–425.

Zaret, T. M. (1980). *Predation and Freshwater Communities.* New Haven, CT:
Yale University Press.

Chapter 2

THE ECOLOGY OF CHRYSOPHYTE FLAGELLATES: THEIR GROWTH AND PERENNATION STRATEGIES AS FRESHWATER PHYTOPLANKTON

CRAIG D. SANDGREN

INTRODUCTION

The goal of this volume as stated in the general introduction is to integrate information about growth and reproduction of groups of phytoplankton into identifiable suites of characteristics that can be considered successful adaptive "strategies" for survival. Chrysophytes are a most intriguing group of freshwater algae for which to consider sets of adaptations for persistence in heterogeneous planktonic habitats. Most members of the class Chrysophyceae are unicellular and colonial flagellates that have apparently evolved in, and are restricted to, freshwater planktonic habitats. Among common species there exists a broad diversity in the size, shape, and surface ornamentation of cells as well as in the morphology and complexity of colonies. These algae are nutritionally opportunistic, many species apparently being capable of switching among autotrophic, heterotrophic, and phagotrophic mechanisms for energy acquisition in response to changes in the planktonic environment. They commonly exhibit markedly seasonal

Original research in this paper relating to resting cyst survivorship and germination was supported by NSF grant BSR-85-96027. I would like to thank Dr. M. Boraas for his critical reading of the manuscript, and my wife, Maria Terres, for continual assistance and tolerance during manuscript preparation.

growth cycles in lakes with one or two restricted population maxima observed during a year, thus suggesting strong interactions with periodic physical, chemical, and biotic components of lakes. Such demographic patterns would appear to require effective strategies for successful perennation from one growth season to the next, and chrysophyte populations are indeed often observed to produce resistant resting cysts during conditions of intolerable stress for the vegetative cells. As a group, then, chrysophyceean algae would seem to be particularly well adapted for survival in the plankton of lakes. Combinations of characteristics relating to algal form, reproduction, nutrition, and biotic or abiotic interactions with the environment that can be summarized as life history strategies may be most easily documented for these algae.

Chrysophytes are among the most poorly known freshwater phytoplankton with regard to their reproductive biology, nutrition, and ecology. Although a number of genera are very common planktonic constituents of low or moderately productive lakes, chrysophytes, for a number of reasons, have seldom been the subject of limnological or laboratory study. First, they do not typically form dense blooms or create the water quality problems that have stimulated research on freshwater cyanobacteria and diatoms. Second, chrysophyte importance has undoubtedly been underestimated in many lake studies owing to difficulties in adequately preserving cells by standard limnological sampling methodology as well as the specialized techniques required for accurate identification of species in even common genera. Finally, difficulties experienced in culturing these algae have long discouraged attempts at critical in vitro study of life history, physiology, or nutritional requirements for all but a few "weed" species. To a large extent, then, chrysophytes have been ignored during the growth of quantitative phytoplankton ecology that has occurred in the last 25 years. Reviews concerning freshwater phytoplankton have often continued to rely on classical observational and correlative studies for chrysophyte information. A specific review of the ecology of freshwater chrysophytes has never been published, although Kristiansen has published a review of chrysophytes as environmental indicators (Kristiansen 1986a,b).

There is currently increased interest in planktonic chrysophytes stimulated primarily by three recent developments. The first is the tremendous increase since approximately 1970 in limnological studies of the myriad small, softwater, and largely oligotrophic lakes of the north temperate regions of North America and Scandinavia. These lakes, frequently dominated by flagellate chrysophytes and cryptomonads, have proven to be very sensitive to acidification by atmospheric pollutants, and an understanding of their biota and chemistry has been a primary focus of recent funding in lake science. A second development that has increased limnological awareness of chrysophyte algae is that chrysophycean microfossils (siliceous scales and cysts) are well preserved in sediments and are becoming increasingly important as paleolimnological indicators, particularly, again, in studies concerned with historical and geographical documentation of the spread of acid precipitation (Smol 1980; Adam & Mahood 1981; Carney & Sandgren 1983; Smol, Charles, & Whitehead 1984; Cronberg 1986). As this use has increased, the need to more thoroughly understand the ecology and physiological limitations of the vegetative cells producing these indicator fossils has become more profound. Finally, the ability of chrysophytes, as well as other small flagellate protozoans, to heterotrophically and phagotrophically cycle particulate or dissolved organic carbon has begun to receive extensive critical study. These flagellates are being demonstrated to play a previously unsuspected large role in aquatic food webs (Bird & Kalff 1986; Porter in press).

This chapter reviews the available literature concerning the distribution, demography, nutrition, ecological interactions, and reproductive biology of planktonic chrysophytes. As suggested in the general introduction, the emphasis will be on distinguishing supportable generalizations rather than documenting examples of apparent inconsistencies or concentrating on the ecology of individual species. An attempt will be made to summarize this information in such a manner that algal form, reproduction, and physiological characteristics may be considered as interactive adaptations in response to specific environmental forces in periodic, stressful planktonic habitats. The analyses will concentrate on species of the most common genera, primarily *Dinobryon, Chromulina, Ochromonas, Chrysococcus, Spiniferomonas* (= *Chromophysomonas*), *Mallomonas, Synura, Chrysosphaerella*, and *Uroglena*.

MORPHOLOGICAL DIVERSITY AND BEHAVIOR

The fundamental importance of phytoplankton morphology in contributing to the distribution and seasonality of individual species has been emphasized particularly by Margalef (1978) and Reynolds (1980, 1984a,b). A schematic summary of the potential interaction between lake characteristics and algal morphology is presented in Fig. 2-1. Basic characteristics of the algal class Chrysophyceae and cytological features typical of chrysophyte cells can be found in general phycological texts (Bold & Wynne 1985) or in reviews cited therein. Genera within the class are grossly distinguished on the basis of general growth form, presence and number of flagella, and nature of the cell covering.

**The Role of Phytoplankton Cell Morphology in Determining
Gross Aspects of Species' Biogeographical and Seasonal Distribution**

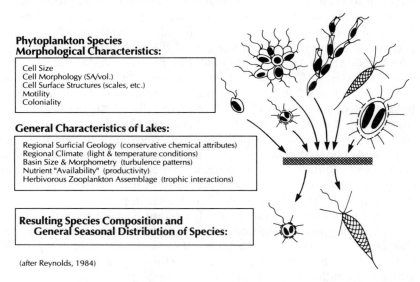

**Phytoplankton Species
Morphological Characteristics:**

Cell Size
Cell Morphology (SA/vol.)
Cell Surface Structures (scales, etc.)
Motility
Coloniality

General Characteristics of Lakes:

Regional Surficial Geology (conservative chemical attributes)
Regional Climate (light & temperature conditions)
Basin Size & Morphometry (turbulence patterns)
Nutrient "Availability" (productivity)
Herbivorous Zooplankton Assemblage (trophic interactions)

**Resulting Species Composition and
General Seasonal Distribution of Species:**

(after Reynolds, 1984)

Fig. 2-1. Generalized schematic of relationship between morphological characteristics of planktonic algae and features of lakes. The "filtering effect" of lake environments on the potential morphological species pool is emphasized. Summarized primarily from discussions in Reynolds (1984a).

In lake plankton, chrysophytes are predominantly represented by *flagellate* genera, although a few coccoid and rhizopodial forms are occasionally important (e.g., *Chrysidiastrum, Chrysocapsa, Stichogloea*). Both unicellular and colonial growth forms are common (Fig. 2-2), and the individual cells may be covered by the cell membrane alone ("naked"; e.g., *Chromulina, Ochromonas, Uroglena*), by an organic lorica that is sometimes secondarily mineralized (e.g., *Chrysococcus, Dinobryon*), or by layers of often elaborately ornamented siliceous scales and bristles (e.g., *Paraphysomonas, Spiniferomonas, Mallomonas, Synura, Chrysosphaerella*). There are occasional reports of transitory palmelloid or temporary sessile stages for some chrysophyte flagellates (see Fenchel 1986; Anderson 1987), but it appears safe to presume that such morphological alterations are exceptional and not of broad ecological importance. Formation of resistant resting cysts is a much more common phenomenon, however, and is therefore of demographic and strategic importance. The biology of chrysophyte cysts will be considered under the topic of reproductive strategies.

Morphological Diversity Among Some Common Planktonic Chrysophytes

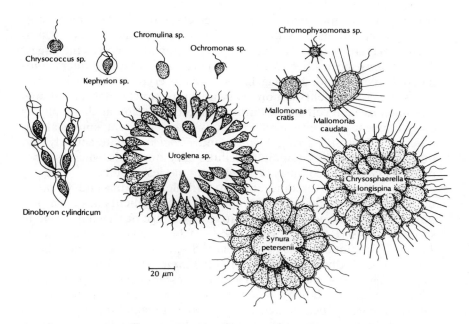

Fig. 2-2. Some morphological diversity of common planktonic chrysophyte flagellates. Both unicellular and colonial genera are depicted with the three common types of cell coverings: *center*, membrane only ("naked"); *left*, organic lorica; *right*, siliceous scales and bristles. Among the included species are those for which physiological data regarding phosphorus uptake and growth kinetics are presented in Table 2-3. All drawings done to scale.

The range of cell and colony size among planktonic chrysophytes is equal to or greater than that of other taxonomic classes of freshwater phytoplankton. The diameter of unicells ranges from only a few micrometers (μm) for small species of *Chromulina, Ochromonas,* and *Spiniferomonas* to over 50 μm for large *Mallomonas* species. Effective cell diameter may be greatly exaggerated by a surrounding lorica or hallow of siliceous bristles (Fig. 2-2). The adoption of colonial growth forms also greatly increases the unit size of some chrysophytes. Linear or arborescent colonies of loricate *Dinobryon* species frequently measure 100–300 μm in maximum dimension; spherical or ovate colonies of *Synura* or *Chrysophaerella* cells are often 100–200 μm in diameter; spherical colonies of *Uroglena* may exceed 500 μm in diameter.

Although all the genera under discussion are flagellate, the effective swimming speed varies considerably. Actual rate measurements are lacking, but several observational generalizations are pertinent. In general, rapidity of movement varies inversely with cell size among unicellular chrysomonads, with the smallest species being as active as similarily sized green algal flagellates or cryptomonads, whereas the largest *Mallomonas* species (i.e., *M. caudata*) are ineffective swimmers compared to peridinoid dinoflagellates or large cryptomonads. Loricate and scale-covered chrysomonads swim more slowly than naked species of equivalent cell size. Swimming speeds of colonial chrysophytes do not appear to be either cell or colony-size dependent; spherical colonies of *Chrysosphaerella, Synura,* and *Uroglena* rotate actively and may swim as rapidly as small unicellular chrysomonads, whereas arborescent *Dinobryon* colonies achieve more modest speeds.

The ability to overcome ambient turbulent mixing and to move directionally in response to nutrient and light gradients has been proposed to be an important competitive advantage of flagellate algae in some habitats (Reynolds 1976, 1980, 1984a; Lewis 1977; Ilmavirta 1980, 1983; Salonen, Jones, & Arvola 1984). Freshwater dinoflagellates (Berman & Rodhe 1971; Tilzer 1973; Heaney & Talling 1980; Pollingher, Chapter 4, this volume) and, to a lesser degree, cryptomonads (Ilmavirta 1974; Happey-Wood 1976; Arvola 1984; Salonen et al. 1984) have been documented to be capable of regulating their vertical position in planktonic light fields and to undergo diurnal vertical migrations. Many chrysophyte flagellates possess the subcellular equipment for detecting directional light, and many exhibit phototaxis in culture. It is therefore worth considering whether these algae are also capable of actively regulating their vertical position in water columns.

Diurnal oscillations over vertical distances of 0.5 to 2 m have been reported for a number of epilimnetic and metalimnetic chrysophyte populations in highly colored forest ponds or sheltered small lakes (*Mallomonas caudata* and *Dinobryon cylindricum:* Weimann 1933; *Chrysococcus diaphanus:* Happey & Moss 1967; *Chrysococcus skujae:* Heynig 1967; *Mallomonas* sp.: Tilzer 1973; *Mallomonas tonsurata, Synura uvella,* and *Dinobryon* sp.: Ilmavirta 1974; *Mallomonas tonsurata* and *Ochromonas* sp.: Happey-Wood 1976; *Mallomonas caudata:* Nygaard 1977, Arvola 1984; *Chrysosphaerella longispina:* Pick & Lean 1984). The pattern of movement is somewhat dependent on ambient light intensity (Munch 1972), and not all coexisting chrysophytes participate to the same degree. Cells migrate generally upward in the morning and downward around noon on very bright days, or during late afternoon on days when sunlight is less intense. Although these studies do suggest that vertical migration is a possibility for some populations, the phenomenon appears restricted to specialized habitats characterized by extreme stability and a shallow epilimnion; vertical migration should thus not be considered a general characteristic of epilimnetic chrysophytes. Even within the same study, some chrysophyte species were observed to migrate, but others did not (Tilzer 1973; Ilmavirta 1974). Significantly, the only reports of surface accumulation of chrysophyte cells in large lakes during the ice-free season concern the "red tides" of *Uroglena* in Lake Biwa (Yoshida, Kawaguchi, & Kadota 1983); *Uroglena* is perhaps the strongest swimmer among the chrysophytes and perhaps the only form capable of overcoming epilimentic turbulence to any extent. For chrysophytes dominanting metalimnetic chlorophyll maxima in stratified lakes or for those forming stable stratified populations under ice (see Distribution section), regulation of constant light intensity or nutrient concentrations via phototactic behavior is likely to be very important for survival. It is perhaps the ability to *remain* within these narrowly defined strata rather than the capacity for extensive vertical migration that is the most ecologically important consequence of chrysophytes' ability to respond phototactically to gradients of light intensity.

NUTRITION

Three reviews of chrysophyte metabolism and nutrition have been published (Pringsheim 1952; Allen 1969; Aaronson 1980), and the subject is discussed generally in several algal physiology texts (Lewin 1962; Stewart 1974). General information regarding storage products and what is known about metabolic pathways is discussed with particular thoroughness by Aaronson (1980). Much of our current knowledge is, unfortunately, based upon studies of only a few species, particularly *Ochromonas danica* and *O. malhamensis*. If we can generalize from these limited data, however, it would appear that planktonic chrysophytes have a diverse and opportunistic strategy for satisfying their requirements for carbon and mineral nutrients.

The majority of chrysomonads are considered facultative photoauxotrophs that possess one or two golden-yellow chloroplasts per cell. The photosynthetic species that have been tested have an obligate requirement for several B vitamins, and can supplement (but generally not replace) their photosynthetic carbon production with the metabolism of simple organic molecules such as glucose, sucrose, fructose, pyruvate, and glycerol (Pringsheim 1952; Lehman 1976; Aaronson 1980). Macro- and micronutrient requirements typical for vascular plants also apply to chrysomonads; absolute requirements for iron and cobalt in chelated forms and a sensitivity to elevated potassium concentrations are particularly marked characteristics (Gusseva 1935; Lehman 1976; Van Donk 1983). Chrysophytes can use both ammonia and nitrate as nitrogen sources with the preferred source perhaps varying among species (Gusseva 1935; Lehman 1976). *Ochromonas* can utilize amino acids as a nitrogen source, and the use of urea and secretion of urease by the largely heterotrophic *O. malhamensis* have been described (Lui & Roels 1970). Orthophosphate and simple organophosphates such as glycerophosphate or glucose phosphate can support chrysophyte growth (Aaronson & Patni 1976; Lehman 1976; Aaronson 1980). Several species in culture and in natural limnetic populations produce an extracellular organophosphate degrading enzyme (acid phosphatase) in response to phosphorus limitation of growth or the presence of organophosphates under acidic conditions (Aaronson & Patni 1976; Healey & Hendzel 1979; Olsson 1983). However, a literature survey suggests that chrysophytes apparently do not secrete significant quantities of alkaline phosphatases in response to P-limitation under alkaline conditions as is typical of many other microalgae (i.e., Healey 1973). Perhaps the lack of this ecologically important enzyme contributes to chrysophyte's observed preference for neutral or slightly acid waters (see Distribution section).

Chrysophytes are generally quite "leaky" cells, with a variety of extracellular dissolved organics being released that sometimes amounts to 20–50% of the carbon fixed by photosynthesis (Nalewajko 1977; Aaronson 1980; Blaauboer, Van Keulen, & Cappenberg 1982). The utility of such excretions, if any, is unclear. A few are toxins, and it has been suggested that others may be extracellular enzymes employed in digesting food particles or macromolecules (Aaronson 1973, 1980). The possibility that some such excretions may act as chelators to make iron or other micronutrients more readily available deserves investigation.

Although photosynthetic chrysophytes are in the majority, many small colorless heterotrophs may also be found in lake plankton. Included among these are genera such as *Paraphysomonas* and *Spumella* (= *Heteroochromonas, Monas*) that are perhaps significant biomass contributors in lakes, but are poorly studied because of their very small size and because of a widely accepted view that they are unimportant. These colorless forms are generally outside the realm of discussion for this chapter, but are discussed by Fenchel (1986).

Since the publication of the aforementioned reviews of chrysophyte nutrition, a rather profound change in perspective has become necessary. Although ingestion of bacteria by chrysophytes has been known since the classic era of observation (see Aaronson 1973, 1980; Sanders & Porter 1987), and the presence of bacteria or putative food vesicles inside chrysophyte cells has been frequently described in electron microscopy studies (see Dürrschmidt 1987), the ecological importance of phagotrophy by photosynthetic chrysophytes has been considered small. Recent documentation of the great significance of algal phagotrophy in limnetic carbon cycling and the key role played by chrysomonads in these processes has, however, dramatically altered this view (Porter et al. 1985; Fenchel 1986; Sanders & Porter 1987; Boraas et al. in press; Porter in press). Almost every genus of planktonic chrysophyte tested, including both colorless and photosynthetic forms (but excepting *Mallomonas* and *Synura*), has been shown to be capable of actively ingesting treated beads, bacterial cells, or other algae. Many chrysophytes have been further shown to digest prey cells and incorporate the carbon and nutrients. Even elongate diatom cells have been demonstrated to be effectively ingested and digested by a much smaller colorless chrysomonad (Suttle et al. 1986). Bactivory was estimated to account for between 30–70% of the assimilated carbon for one metalimnetic *Dinobryon* population investigated (Bird & Kalff 1987). In another instance, phytoflagellate grazing was estimated to constitute up to 55% of total bactivory for all grazers combined (including cladocerans and rotifers) for an unstratified lake plankton assemblage dominated by chrysophytes (Sanders & Porter 1987). Rates of particle capture as high as 3 bacteria$\cdot$cell$^{-1}\cdot$h^{-1} were recorded for *Dinobryon* (Bird & Kalff 1986), and clearance rates of 5.8 nL$\cdot$cell$^{-1}\cdot$h^{-1} were estimated. Such rates far exceeded clearance by the ambient zooplankton and ciliates in the study; they indicate clearance of 1.5×10^6 times the cell volume and ingestion of 30 times the cell biomass each day. Even higher rates of particle capture and clearance have been estimated for unicellular chrysomonads (Porter in press). Indeed, Fenchel (1986) makes an argument for *obligate* clearing rates of at least 10^5 times the cell volume per hour for strictly phagotrophic microflagellates based upon conservative assumptions of a growth efficiency of 50%, a growth rate of one doubling each day, and steady-state planktonic bacterial "food" populations of 10^6 mL^{-1}. Clearing rates could be much lower for facultative phagotrophs, such as the photosynthetic chrysomonads, and in more eutrophic environments with higher bacterial standing crops. The relative importance of phagotrophic versus autotrophic nutrition in chrysomonads has been shown to be regulated by environmental factors such as temperature, light intensity, and concentrations of dissolved organic carbon (DOC) and food particles (Bird & Kalff 1987; Sanders & Porter 1987; Porter in press), and thus will vary through time and among lakes for the same chrysophyte.

It appears that many chrysophytes, and algal flagellates in general, should be considered opportunistic mixotrophs insofar as the ecological consequences of their nutrition is concerned. Facultative phagotrophy provides flagellates with the advantages (Boraas et al. in press; Porter in press) of 24-h feeding, a source of large amounts of nutrients quickly, and a source of micrometals such as iron in a readily utilizable form. Phagotrophy can also constitute a competitive advantage if the prey algal or bacterial cells are competing for the same limiting resources. It is likely that biologists' new awareness of algal mixotrophy will contribute to a better understanding of chrysomonad population dynamics, seasonal periodicity, and tolerances of low light, organic-rich habitats. It is an intriguing possibility that high rates of organic molecule leakage from chrysophyte cells in combination with chrysophyte bactivory may represent a "chrysophyte–bacterial loop" nutritional strategy that makes these algae particularly adapted for persistence in low-light habitats.

CHRYSOPHYTE DISTRIBUTION PATTERNS

Biogeographical Distribution

Classifying lakes on the basis of their most common or most characteristic phytoplankton species (i.e., "indicator" species) has been one of the most long-lived limnological activities (Pearsall 1932; Thunmark 1945; Nygaard 1949; Järnefelt 1952; Teiling 1955; Rawson 1956; Brook 1959; Hutchinson 1967; Hörnström 1981). In these studies, dominance of the phytoplankton by chrysophycean algae has consistently been proposed to be characteristic of lakes with a combination of cool summer water temperatures, low or moderate productivity, low nutrient availability, low alkalinity and conductivity, neutral or slightly acid pH, and often humic-colored water. Although not characteristic of the eutrophic temperate lakes that have been the subject of much limnological research, this combination of features actually describes a very great number of lakes. Chrysophytes are known to contribute a high relative proportion of the planktonic biomass in the softwater oligotrophic and mesotrophic lakes that cover north temper

Fig. 2-3. Relationship between the biogeographical distribution of some common planktonic chrysophytes and certain conservative lake characteristics of 1,250 lakes in Sweden. Plots are cumulative frequency distributions of species occurrence. The reference graph (solid line) represents the cumulative frequency distribution of each parameter for the 1,250 lakes sampled. From Rosén (1981); used with permission.

CHRYSOPHYCEAE

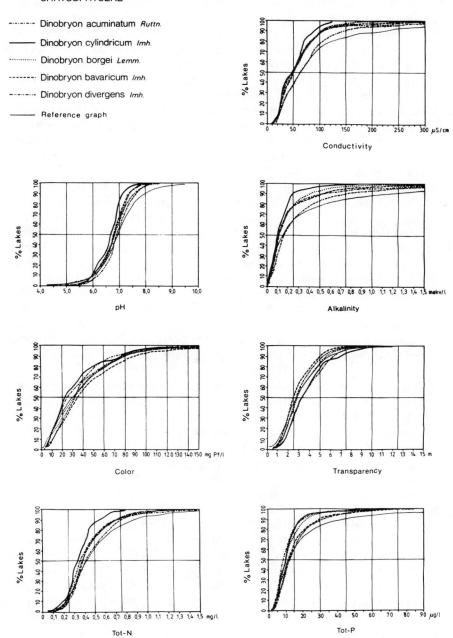

-·---·-- Dinobryon acuminatum *Ruttn.*

——— Dinobryon cylindricum *Imh.*

············ Dinobryon borgei *Lemm.*

------- Dinobryon bavaricum *Imh.*

-·---·-- Dinobryon divergens *Imh.*

——— Reference graph

ate North America and Scandinavia. They are also frequent codominants with other flagellates in the plankton of both alpine and arctic ponds or lakes (Moore 1979; Sheath 1986; Pavoni 1963; Tilzer 1973; Rott 1984), and they may be seasonally important in more productive temperate lakes. Indeed, if the relative proportion of oligotrophic to eutrophic lakes in the northern hemisphere is considered, chrysophytes are probably among the more important biomass contributors and primary producers in lake phytoplankton. Chrysophytes have also been shown recently to be important in the plankton of tropical oligotrophic lakes (Hecky & Kling 1981; Hecky 1984; Kalff & Watson 1986), although they are rare in highly productive lowland lakes (Lewis 1977, 1986).

One of the best examples of chrysophyte biogeographical distribution patterns is a very comprehensive survey of phytoplankton distribution among 1,250 small or moderate-sized Swedish lakes (Rosén 1981). These results are based on single quantitative samples taken during the late summer. The data pertinent to important chrysophycean species are reproduced here (Figs. 2-3 and 2-4), and it can be seen that the classic generalizations are well supported. With the exception of *Synura* and *Chrysosphaerella,* chrysophyte distributions are shifted to the left of the reference line for all parameters except water color and transparency. Other regional phytoplankton surveys with similar sampling designs that also support some or all of the same conclusions relative to the distribution of chrysophyte algae are those of Heinonen (1980), Hegewald, Hesse, & Jeeji-Bai (1981), Ilmavirta (1983, 1984), Eloronta (1986), and Earle, Duthie, & Scruton (1986). In addition, several regional studies that have concentrated solely on the distribution of chrysophytes also support the aforementioned generalizations, although their intent was primarily to emphasize differences among the species (Kristiansen 1975; Kies & Berndt 1984; Roijackers & Kessels 1986). Also, a survey of chrysophyte siliceous microfossils from the surficial sediments of lakes in northeastern North America support the proported chrysophyte preference for low pH, alkalinity, nutrients, and productivity (Smol et al. 1984).

Although such studies as these are strongly indicative of lakes in which chrysophyte species predominate, they may, to some extent, be misleading with regard to both biogeography and limiting factors. The reliance on a few summer samples or on fossilized remains of only scale-bearing species to characterize a lake may misrepresent the actual importance of algae, such as chrysophytes, that exhibit markedly periodic seasonal distributions. To more accurately test the validity of the stated generalizations, a new database was therefore constructed for the present study based upon seasonal or annual estimates of lake

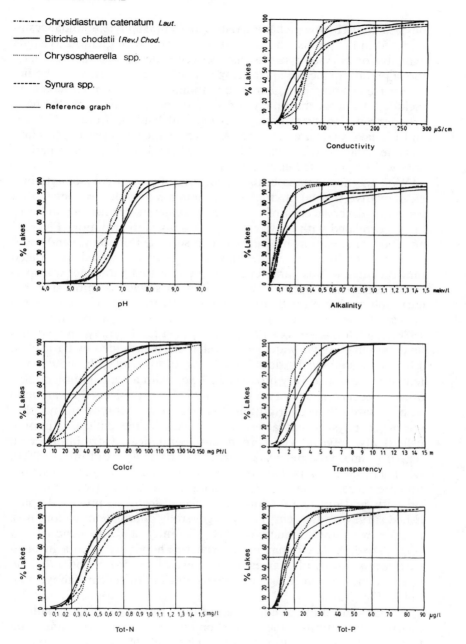

CHRYSOPHYCEAE

········· Chrysidiastrum catenatum Laut.

——— Bitrichia chodatii (Rev.) Chod.

············ Chrysosphaerella spp.

------ Synura spp.

——— Reference graph

Conductivity

pH

Alkalinity

Color

Transparency

Tot-N

Tot-P

Fig. 2-4. Continuation of Fig. 2-3. From Rosén (1981); used with permission.

physical and chemical characteristics and chrysophyte abundances as reported in the published literature. Only lakes with quantitative data available on chrysophyte biomass were considered, and there is therefore a predominance of north temperate lakes. However, water bodies as diverse as arctic ponds, prairie potholes, and tropical rift lakes are included. Data for multiple years are incorporated for lakes that have experienced significant changes in nutrient loading. In all, 142 lakes and 172 lake-years are included. A more detailed analysis of this chrysophyte database with regard to chrysophyte distribution and periodicity will be presented elsewhere.

With the new database, the mean importance of chrysophyte biomass relative to total phytoplankton biomass over the growing season was plotted versus each of the seven parameters implicated earlier as being correlated with chrysophyte geographical distribution (Figs. 2-5 and 2-6). The relationship between chrysophyte importance and water color derived from these lakes (Fig. 2-5a) does not suggest a restricted range relative to this parameter nor to the related parameter transparency (not shown). Rather, the distribution serves to emphasize that chrysophytes have a wide tolerance with regard to water color, which appears to be true for flagellate phytoplankton in general (Ilmavirta 1980, 1983, 1984; Arvola 1985; some potential diversity on a species level for chrysophyte water color preference is provided by Earle et al. 1986). Chrysophyte distribution with regard to mean epilimnetic water temperature is also very broad (Fig. 2-5b), with a contribution of 30% or more occurring over a range of 4°–26°C. This reflects their arctic to tropical latitudinal distribution. Likewise, consideration of morphometric or physical characteristics of lakes in the database (unpublished) also reveals very broad biogeographical distributions with regard to lake volume, surface area, mixing cycles, and mixed depth (Z_{epi}). The parameters that do seem to influence chrysophyte distribution are those associated with productivity, ionic chemistry, buffering capacity, and nutrient loading. As expected from the previously published survey studies, chrysophytes generally dominate or codominate in lakes with low productivity, low alkalinity, and low conductivity (Fig. 2-5c,d,e). Their maximum relative abundance occurs in lakes with mean annual pH values in the range of pH 5–7.5, and a distinctive decline in percentage biomass contribution occurs above that range (Fig. 2-5f).

Perhaps the most striking feature of chrysophyte biogeographical distribution patterns is the correlation between phosphorus concentration and the relative importance of chrysophyte algae in the phytoplankton. Using mean epilimnetic total phosphorus concentration as a readily available indicator of lake productivity or trophy in P-limited

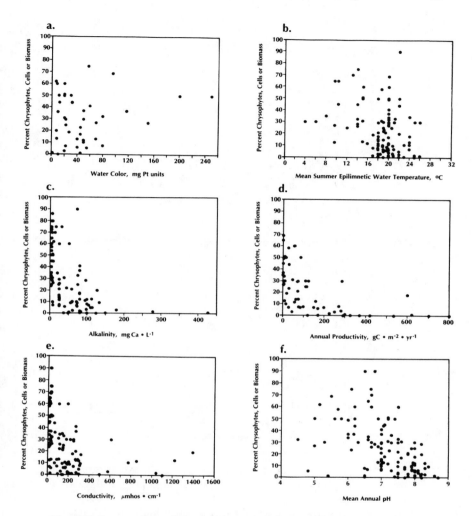

Fig. 2-5. Scatterplots of the relationships between some conservative lake characteristics and the relative importance of chrysophyte species. Values plotted represent mean seasonal estimates for individual lakes and years as derived from original literature and compiled in the chrysophyte database (see text for explanation).

lakes (Vollenweider 1969; Schindler 1977; Forsberg & Ryding 1980), Nicholls and co-workers demonstrated an inverse curvilinear relationship between chrysophyte percent abundance and total P for several systems of lakes in southern Ontario, Canada (Nicholls 1976; Nicholls,

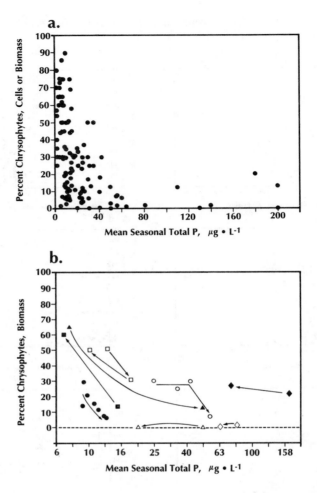

Fig. 2-6. Scatterplot of the relationship between mean seasonal epilimnetic total phosphorus and chrysophyte biogeographical distribution based on lakes included in the chrysophyte database. (a) Chrysophyte relative importance to phytoplankton. (b) Changes in chrysophyte importance in lakes that have experienced changes in phosphorus loading. Symbols: open diamond, L. Trummen, mean values for 1972–5 and 1976–8 (Cronberg 1982); closed diamond, L. Åsvalltjärn, mean values for 1972–4 and 1978–80 (Holmgren 1985); open circle, ELA L. 227, 1969–72 (Armstrong & Schindler 1971; Schindler et al. 1973; Findlay & Kling 1975); closed circle, Pilburger See 1975–81 (Rott 1983); open triangle, L. Mälaren region "C," mean values for 1964–70 and 1971–4 (Willén 1984); closed triangle, L. Hymenjaure, 1971 and 1974 (Holmgren 1984); open square, L. Langvatn, 1974, 1976, and 1978 (Reinertsen 1982); closed square, L. Vättern, mean values for 1966–70 and 1976–7 (Willén 1984).

I

Carney, & Robinson 1977; Nicholls et al. 1986). An analogous plot based upon the chrysophyte database (which includes only eight of Nicholls's data points) also supports this relationship (Fig. 2-6a), although there is considerably more vertical scatter when such a broad diversity of lakes is included. Similar findings were reported in a survey of Finnish lakes (Eloronta 1986, his Fig. 4), and chrysophyte microfossil studies would also tend to support an inverse relationship between chrysophyte importance in a lake and productivity or nutrient availability (Moss 1979; Munch 1980; Smol 1985).

Further support of a more experimental nature for the chrysophyte–phosphorus relationship can be found in studies of lakes in which phosphorus loading was purposely reduced or artificially elevated (Fig. 2-6b). Oligotrophic, chrysophyte-rich lakes that have been artificially eutrophied by phosphorous additions (i. e., ELA lake 227: Schindler et al. 1973; Lake Hymenjaure: Holmgren 1984; Lake Langvatn: Reinertsen 1982) experienced a decline in chrysophyte relative abundance. So did mesotrophic Pilburger See in response to increased cultural loading of phosphorus (Rott 1983). Conversely, decreased nutrient loading in naturally oligotrophic or mesotrophic lakes results in an increase in the relative abundance of chrysophyte algae. This is true for Lake Vättern (Olsén & Willén 1980; Willén 1984) and also for Lake Langvatn when the artificial eutrophication was discontinued (Reinertsen 1982). In contrast, lakes with a long history of eutrophic or hypereutrophic conditions, such as Trummen, Mälaren, and Norvikken, did not experience a persistent increase in chrysophyte importance in response to reduced nutrient loading (Cronberg 1982; Willén 1984; Ahlgren 1967, 1970, 1978; Ahlgren et al. 1979), perhaps because these lakes are still eutrophic even after sewage diversion, and the general biological and chemical constitution of such lakes continues to select against chrysophyte algae (see discussions in later sections). An apparent exception to these generalizations is the eutrophic, subarctic Lake Åsvalltjärn, which did experience a small increase in chrysophyte importance accompanying the initiation of sewage pretreatment (Fig. 2-6b; Holmgren 1985). However, the historical period of eutrophication has been much shorter for this lake, and the natural oligotrophic plankton may thus have retained greater resiliency. An excellent review of individual phytoplankton species changes accompanying changes in lake nutrient status is included in this paper by Holmgren.

The chrysophyte database analysis, the literature cited, and many other papers of more moderate scope that could be cited all emphasize the fact that chrysophytes, like several other taxonomic groups of phytoplankton, express certain generalities concerning their geographical distributions. The most notable generalization in this case is the predominance of chrysophytes in oligotrophic lakes with circumneutral pH, low nutrient loading, low total ions, and low conductivity. Such a generalization is suggestive of fundamental similarities, perhaps phylogenetically related, in the environmental requirements of this morphologically diverse assemblage of species. Such correlative studies therefore are extremely useful in generating hypotheses regarding conservative or broadly predictable characteristics that limit chrysophyte growth. They do not, however, provide *explanations* for the observed distribution patterns, and a number of cautionary notes in interpreting these data are important to recognize. First, correlations between chrysophyte relative abundance and environmental parameters do not, in themselves, indicate cause-and-effect relationships. These environmental parameters are actually complex characteristics that depend upon numerous specific attributes of lakes, and some of the parameters, for instance pH, conductivity, alkalinity, productivity, are highly intercorrelated (Oglesby 1977; Gorham, Dean, & Sanger 1983). Second, neither seasonal averages nor single samples adequately record the fine scale interaction of lake characteristics and chrysophyte distributions; it is thus likely that the actual distributional range with regard to the parameters is somewhat different. Third, even if we were to accept the horizontal ranges as being reasonably accurate tolerance limits for chrysophytes, there is still a great deal of vertical scatter within each of the plots. This strongly suggests that none of the parameters considered in this section is individually predominant in delimiting growth requirements of chrysophytes. Fourth, it must be remembered that individual chrysophyte species may demonstrate different trends within these relatively gross patterns. For instance, opposite trends with regard to conductivity and total phosphorus preferences have been noted for several chrysophyte species in subarctic lakes (Moore 1979).

As a final point of caution in interpreting the foregoing correlational results, it must be remembered that changes in the relative importance of chrysophytes along gradients of lake characteristics are not necessarily indicative of the *growth* response to the same gradient. Relative biomass and actual mean biomass trends may be quite different (i.e., Ilmavirta 1980). An important example here is the total phosphorus relationship (Fig. 2-6a). If mean seasonal chrysophyte biomass is considered rather than percent chrysophyte contribution to the phytoplankton, quite a different relationship between chrysophyte distribution and phosphorus results (Fig. 2-7). Chrysophyte biomass remains unchanged or decreases with increasing phosphorus loading among some lakes, but among others there is a direct relationship between increasing total phosphorus and chrysophyte biomass. This latter

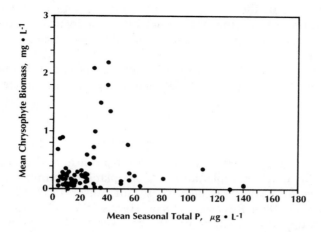

Fig. 2-7. Scatterplot of relationship between mean seasonal total phosphorus and mean chrysophyte biomass in the chrysophyte database lakes. The same lakes are involved as in Fig. 2-6a.

trend has also been implied for chrysophytes by some previous phytoplankton survey data (Eloronta 1986, his Fig. 2), and is the expected response of total phytoplankton biomass to eutrophication (Kalff & Knoechel 1978; Nicholls & Dillon 1978; Nicholls et al. 1986). The apparent dichotomy in chrysophyte biomass response to increasing phosphorus concentrations suggests that nutrient loading per se is not restricting chrysophyte biogeographic distribution. Gradients of phosphorus concentration and productivity among lakes are strongly correlated with many other aspects of lake chemistry, morphometry, and biological organization. It is all these factors in combination that dictate the balance between growth and loss rates for chrysophyte algae; total phosphorus or lake productivity are only indicators of the changing competitive and trophic arena of lakes along a gradient of increasing eutrophication.

On the basis of the information just presented, it is clear that comparisons of the biological and chemical characteristics of several types of lakes should provide insights into fundamental factors regulating chrysophyte distribution and importance. One critical comparison would be that of oligotrophic and eutrophic lakes, and the second would be that of eutrophic lakes with high chrysophyte biomass to eutrophic lakes that support very low chrysophyte standing stocks (à la Fig. 2-7). Such comparisons will be made in a later section. Critical insight can also be obtained from patterns of chrysophyte temporal periodicity in lakes, from chrysophyte responses to experimental manipulations of the planktonic environment, and from laboratory investigations of specific nutritional or physiological requirements. The next subsections provide a summary of chrysophyte seasonal periodicity patterns, and these subsections are followed by sections emphasizing specific studies of tolerance limits and growth-limiting factors as requisite elements for delimiting a chrysophyte growth strategy.

Seasonal Periodicity

General seasonal patterns of variation for physical and chemical factors and for total phytoplankton and zooplankton biomass are fairly well known for a variety of lake types (Hutchinson 1957, 1967; Round 1971; Wetzel 1975; Reynolds 1984a; Harris 1986). Among the taxonomic groups of phytoplankton, the periodicity of planktonic cyanobacteria algae and diatoms is the most well established, primarily because of the strong dependence of these groups on physical mixing cycles: Diatoms form large populations during periods of strong mixing, whereas cyanobacteria algae bloom during periods of high thermal stability (see Chapters 6 and 7). Seasonal periodicity patterns for flagellate algae in general, and chrysophyte flagellates in particular, are much more poorly understood because of the multiplicity of factors that regulate their growth.

Many papers discuss the seasonality of the group Chrysophyceae or specific chrysophytes in individual lakes. The diversity of descriptions suggests that there is a strong effect on chrysophyte seasonality of lake productivity, regional climate, and annual mixing cycle. However, no general synthesis of their seasonality incorporating these parameters has been published. Three authors (Reynolds 1980, 1984b; Rott 1984; Holmgren 1984, 1985) have attempted to graphically summarize overall phytoplankton seasonal patterns for a large number of lakes in such a manner that chrysophyte distributions can be easily extracted. At first inspection there appears to be little concordance among these summary statements with regard to the chrysophyte species, but it must be remembered that each author considered lakes of largely different morphometry in different regional climatic zones, and that there appear to be somewhat different operational definitions for oligo-, meso-, and eutrophic states. With these regional statements as a starting point, a scenario describing general seasonal distribution patterns for planktonic chrysophytes is proposed here (Table 2-1) and is supported with examples from representative lakes. Again, the previously mentioned database including 141 lakes is the basis for the synthesis. The operational definitions used here for trophic categories are established after the criteria of Wetzel (1975), Forsberg & Ryding (1980),

Table 2-1. *Planktonic chrysophyte seasonal periodicity.*[a]

PLANKTONIC CHRYSOPHYTE SEASONAL PERIODICITY

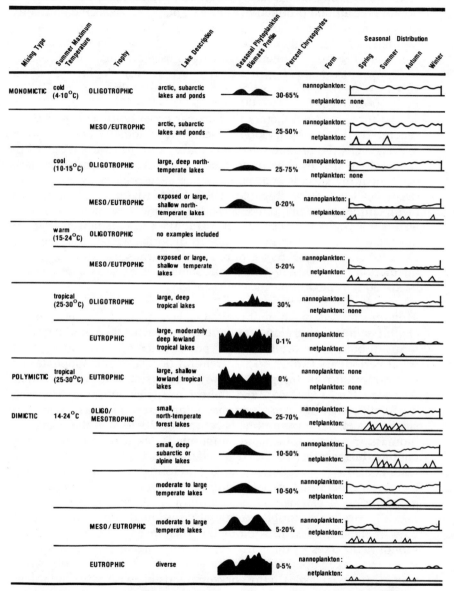

[a]Summary of the seasonal distribution of nanno- and netplanktonic chrysophyte algae in lakes categorized on the basis of mixing cycles and productivity. Basis for distinguishing productivity categories is explained in the text. Sources of information used in constructing this summary are cited in the text or have been otherwise compiled in the chrysophyte database (Sandgren unpubl.).

Heinonen (1980), and Rosén (1981), and are based upon a combination of lake annual productivity (P_r, g C·m^{-2}·yr^{-1}), specific productivity for the euphotic zone (P_z, g C·m^{-3}·yr^{-1}), and mean growth season estimates of chlorophyll a (*CHL*, mg^{-1}·m^{-3}) and phytoplankton wet biomass (B, g·m^{-3}). General distinctions used are oligotrophic – P_r = 1–20, P_z = 0.1–5, *CHL* = 0.2–5, B = 0.1–1; mesotrophic – P_r = 25–100, P_z = 3–10, *CHL* = 5–15, B = 0.5–3; eutrophic – P_r = 100–700, P_z = 10–200, *CHL* = 15–100, B = 2–30.

The three cardinal factors that appear consistently correlated with general attributes of chrysophyte seasonal periodicity are the annual mixing cycle, the productivity of the lake, and the morphology of the chrysophyte species. The importance of these as "system characteristics" dictating successional patterns has been previously emphasized (Margalef 1967, 1978; Reynolds 1980, 1984a,b). Other factors such as mean temperature, pH, or conservative chemical characteristics of the waters may influence *species* biogeography (e.g., Kristiansen 1986a,b), but seem to have little importance in dictating general seasonal patterns for chrysophytes as a group. For the purposes of this survey, the distinction between net- and nannoplanktonic algae is considered to be 20–30 μm maximum linear dimension.

Monomictic Lakes. In monomictic lakes, planktonic chrysophytes exhibit definite shifts along gradients of increasing lake productivity in terms of both the morphological types of species found and their seasonal periodicity (Table 2-1). A trend of decreasing chrysophyte relative importance in the plankton with increasing trophy is also demonstrated (i.e., Fig. 2-6a), and these tendencies appear to transcend broad difference in mean seasonal temperature and lake size. Very small, unicellular chrysomonads contribute 30–65% of the planktonic biomass in oligotrophic cold monomictic lakes or ponds of arctic regions during most of the year (Nauwerck 1968, 1980; Kalff 1967; Kalff et al. 1975; Alexander et al. 1980; Holmgren 1984; Sheath 1986). Common genera include *Chromulina, Ochromonas, Kephyrion, Pseudokephyrion, Dinobryon acuminatum*, and *D. crenulatum*. Morphologically similar nannoplanktonic chrysomonads also constitute 25–75% of the biomass throughout the year in cool, deep oligotrophic monomictic lakes of the north temperate zone such as large Scottish lochs and Lake Superior (Munawar et al. 1978; Maitland 1981; Munawar & Munawar 1986). Tiny chrysomonads likewise contribute significantly to the phytoplankton biomass throughout most of the year in large monomictic oligotrophic lakes from tropical regions such as Lake Tanganyika (Hecky & Kling 1981).

Cold monomictic lakes with elevated productivity are also codominated by nannoplanktonic chrysophytes (25–50%, Table 2-1). A distinctive additional feature of these more productive lakes is the presence of netplanktonic chrysophytes such as large-celled *Mallomonas* species and colonial *Dinobryon* and *Synura* species. These larger forms occur usually as brief monospecific peaks of either major or minor importance. For example, eutrophic, cold monomictic Lake Mývatn experiences brief spring or summer growths of *Dinobryon sociale* and *Uroglena americana,* and rather continual nannoplanktonic chrysophyte appearances (Jónasson & Adalsteinsson 1979). A similar pattern with peaks of *Synura uvella* and *Uroglena* occurred in eutrophic subarctic Lake Åsvalltjärn (Holmgren 1985) and in Meretta Lake, which had a peak of *Dinobryon sociale* (Kalff et al. 1975). Artificial nitrogen enrichment of subarctic Lake Magnusjaure, which is normally nannoplankton dominated, resulted in blooms of *Uroglena,* whereas phosphorus enrichment in neighboring Lake Hymenjaure promoted increased growth of chrysophyte nannoplankton instead (Holmgren 1984).

Warmer-water mesotrophic and eutrophic lakes of the monomictic type have reduced overall nannoplankton abundances and reduced mean chrysophyte relative abundances (0–20%, Table 2-1). The small unicellular chrysomonads are again present sporadically throughout most of the year at these reduced levels, with minimal importance during summer and maximal development during the winter season, at which time they may codominate the phytoplankton as in oligotrophic lakes. Brief chrysophyte netplankton blooms are again often present, occurring most frequently during the autumn or late spring or under the winter ice cover (Nebaeus 1984). Such patterns are found in shallow and productive north temperate lakes that freeze annually but do not stratify during the summer (i.e., cool monomictic lakes). Examples include Chautaugua Lake (Mayer et al. 1978), Lake Oneida (Mills 1975; Mills et al. 1978), Oude Waal pond D (Roijackers 1984, 1985), Lake Trummen (Cronberg 1982), Rice Lake and Scugog Lake (Nicholls 1976). Some extremely productive lakes of this cool monomictic type are apparently competely barren of chrysophytes (L. Tjeukemeer: Moed & Hoogveld 1982; L. Hjälmaren-Hemfjardel: Willén 1984), whereas small prairie potholes that are similarily hypereutrophic have significant chrysophyte representation (2–20%) by both net- and nannoplanktonic forms (Kling 1975; Barica 1975).

Moderately productive monomictic lakes that stratify but do not freeze (i.e., warm monomictic lakes) again share the dual patterns of fairly continual minor presence of nannoplanktonic forms and brief peaks of netplanktonic chrysophytes. Usually monospecific netplankton peaks can be found during either vernal and autumnal mixing or summer stratification, and these chrysophytes are found in association with a great diversity of other types of algae. Example lakes include Lake Biwa with its well-known *Uroglena* blooms (Yoshida et al. 1983; Yoshida, Matsumoto, & Kadota 1983; Tezuka 1984), Lake Seneca (Schaffner & Oglesby 1978), Lake Constance (Sommer 1986), and mesotrophic Laurentian Great Lakes (Munawar & Munawar 1981, 1986; Scavia & Robertsen 1984; Vande Castle 1985). Highly productive lakes in this warm monomictic category were not considered critically, but would include south temperate reservoirs and Lake Kinneret (Pollingher, Chapter 4 this volume; Pollingher 1986). Tropical polymictic lakes of high productivity generally have either no chrysophytes or a very sparse complement of net- and nannoplanktonic forms (L. Lanao: Lewis 1977, 1978; L. George: Burgis et al. 1973, Ganf 1974; L. Valencia: Lewis 1986; also see review by Kalff & Watson 1986).

Dimictic Lakes. Dimictic temperate lakes also show the trend of decreasing chrysophyte importance with increasing productivity. However, the differences between the biogeography of net- and nannoplanktonic forms is not marked (Table 2-1). Oligotrophic dimictic lakes are typically codominated by a combination of net- and nannoplanktonic chrysophytes that constitute 25–75% of the phytoplankton biomass. The small forms are present throughout the year, even under the ice (Nebaeus 1984). The large colonial species may be present during unstratified seasons, but their period of maximal importance in the plankton is largely restricted to the summer stratified season. This general pattern seems true for all types of oligotrophic and meso-oligotrophic dimictic lakes, including small north temperate forest lakes, alpine or subarctic lakes, and moderate to large temperate lakes. Typical chrysophyte genera include all those listed earlier for monomictic lakes as well as *Chrysococcus* and *Chrysosphaerella* species. Small forest lakes generally exhibit seasonal phytoplankton biomass profiles consisting of a series of small sequential maxima (ELA lakes: Schindler & Holmgren 1971, Kling & Holmgren 1972; Peter, Paul, and Tuesday Lakes: Elser, Elser, & Carpenter 1986, Carpenter, personal comunication; Botjärn and Vitalampa Lakes: Ramberg 1979; Randin, La Croix, and Metamek Lakes: Janus & Duthie 1979; Labrador lakes: Ostrofsky & Duthie 1975; Alinen Mustajärvi, Nimeton, and Horkkajärvi Lakes: Arvola 1983, 1984, Arvola & Rask 1984). These population maxima begin to develop immediately after ice melting or sometimes under clear late winter ice. Many such lakes are colored with humic material (dystrophic), which helps promote rapid heating of the water in the spring. This, in combination with their often-protected situation, contributes to the rapid onset of thermal stratification after ice melting, often within a few weeks. As a consequence, diatoms are of minor importance in the plankton, and the vernal phytoplankton pulse during the brief mixing period is usually dominated by green algal flagellates and small dinoflagellates (Ramberg 1979; Eloronta 1980; Ilmavirta 1984). Chrysophytes codominate the summer plankton. In these shallow and very stable water columns, netplanktonic chrysophyte species produce sequential brief and largely monospecific biomass maxima, and small chrysomonads and other flagellates constitute a continuous background (Ilmavirta 1982; references above).

Small dimictic oligotrophic lakes of alpine or subarctic regions usually have low surface-area-to-volume ratios and protected basins. Therefore they also develop rapid and very stable thermal stratification. Biomass profiles in such lakes tend to be unimodal with a maximum in early summer. Diatoms are again rare compared to lakes with more prolonged mixing, and nannoplanktonic chrysophytes are particularly important. The oligotrophic alpine lakes studied by Rott (1984) and Lake Langvatn (Reinertsen 1982) are of this type. More mesotrophic Pilburger See also has persistent populations of colonial chrysophytes that may occur during either early or late summer or under the winter ice (Rott 1983). In response to artificial nutrient additions, Lake Langvatn also developed summer netplanktonic chrysophyte populations (Reinertsen 1982).

Larger meso-oligotrophic lakes of temperate regions often have vernal mixing periods lasting much longer than the small lakes just discussed. Vernal mixing may last one or two months, and once thermal stratification is established the mixed layer is quite thick, in the range of 5–20 m or more. Such lakes frequently display summer unimodal phytoplankton biomass profiles, and it is usual to find a combination of nanno- and netplanktonic chrysophytes, other flagellates, and diatoms composing the plankton throughout the growing season (several Kawartha region lakes: Nicholls 1976; a number of New York Finger Lakes: Schaffner & Oglesby 1978; L. Vättern and Vänern: Willén 1984; L. Keurusselka zone V: Eloronta 1974; L. Konnevesi: Eloronta & Palomaki 1986; L. Pääjärvi: Ilmavirta & Kotimaa 1974). As in the richer alpine lakes, the large colonial chrysophyte species are able to establish populations that persist as codominant or common forms for two- or three-month periods during the summer after stratification becomes established (Table 2-1).

Mesotrophic and meso-eutrophic lakes of the dimictic type have characteristic bimodal seasonal phytoplankton biomass profiles, with a May/June maximum dominated by diatoms and an August/September maximum dominated by a succession of cyanobacteria, dinoflagellates, and an autumnal resurgence of diatoms. Chrysophytes compose 5–15% of the seasonal biomass in the lakes analyzed here (Conesus L.: Mills 1975; L. Erken: Nauwerck 1963, Rodhe, Vollenweider, & Nauwerck 1960; Hall L.: Munch 1972; Mondsee: Dokulil, & Skolaut 1986; L. Ormajärvi: Ilmavirta, Ilmavirta, & Kotimaa 1974). Nannoplankton forms have a ubiquitous seasonal distribution as in more oligotrophic lakes, but are always uncommon. Chrysophyte netplankters may be present at any season (e.g., Kristiansen 1975), but are usually of greatest importance during vernal mixing, with their importance relative to diatoms diminishing with increasing nutrient loading and increasing duration of the mixing period. Chrysophytes may dominate in spring in place of diatoms if the lake is small and the vernal mixing season short (e.g., Rott 1984). Small or moderate-sized lakes with approximately a month of vernal mixing usually have spring phytoplankton composed of a mixture of diatoms and chrysophytes, as in Hall Lake (Munch 1972) and Egg Lake, Washington (Fig. 2-8). Mesotrophic, dimictic lakes with still longer vernal mixing are completely dominated by diatoms or coccoid green algae during the spring, with only minor chrysophyte populations (i.e., Reynolds 1984b; Rott 1984). In many lakes with a month or more of vernal mixing, diverse chrysophyte species become dominant or codominant forms seemingly by default during the early summer "clear phase" that becomes established after diatom and vernal zooplankton populations collapse with the onset of thermal stratification (Hutchinson 1944; Nauwerck 1963; Lampert 1978; Sommer et al. 1986). Subsequent growth by coccoid green algae and cyanobacteria in such lakes displace the colonial chrysophytes for the rest of summer, although brief periods of storm-induced mixing may cause temporary "reversions" (Reynolds 1984b) to dominance by vernal chrysophytes and diatoms. *Dinobryon, Synura,* and *Uroglena* species may accompany or precede the autumnal return of diatoms with the late seasonal deepening of the thermocline. Productive dimictic lakes with prolonged autumnal and vernal mixing occasionally develop blooms of *Synura* or *Mallomonas* during these times (Munch 1972; Roijackers 1985), but clearly the most successful growth of large chrysophytes in dimictic lakes is during the earliest phase of summer stratification or under the winter ice (Wright 1964).

Eutrophic and hypereutrophic dimictic lakes have very reduced comparative importance of chrysophyte algae (0–5% of seasonal biomass, Table 2-1). Such lakes are characterized by greatly exaggerated bimodal biomass profiles or by continuously high biomass levels (Kalgaard L.: Sondergaard & Sand-Jensen 1979; L. Bysjön: Coveny et al. 1977; L. Esrom: Jónasson & Kristiansen 1967, Jónasson 1972; L. Norrviken: Ahlgren 1967, 1971; L. Mälaren region C: Willén 1984; L.

Vasikkalampi: Eloronta 1980; Irondequoit Bay: Bannister & Bubeck 1978; L. Lovojärvi: Ilmavirta et al. 1974, Keskitalo 1977; L. Mendota: Brock 1985). Both net- and nannoplanktonic chrysophytes may be present in these lakes, but only as minor components during winter or vernal mixing (Table 2-1).

Metalimnetic Chrysophytes. The discussion of chrysophyte seasonality thus far has concerned epilimnetic populations. A good deal of recent literature has documented the existence and importance to lake production of metalimnetic algal populations (Moll & Stoermer 1982). Chrysophytes are often dominant or codominant forms in these metalimnetic chlorophyll maxima, and it is possible to find large chrysophyte populations at depth in such lakes when the group is largely absent from the epilimnion. Only netplanktonic chrysophytes (*Chrysosphaerella, Synura,* large *Mallomonas* spp., and colonial *Dinobryon* spp.) form metalimnetic populations except in the most oligotrophic of lakes (i.e., Lake Superior), perhaps because of high grazing pressure from hypolimnetic zooplankton. Metalimnetic algae persist during the summer stratified season under conditions of reduced temperature and very low light, but are in close proximity to nutrient-rich hypolimnetic waters. They are often actively growing populations, not residual cells sinking out of the epilimnion. Lakes that develop metalimnetic maxima are characterized by low or moderate productivity, stable thermal stratification, and relatively clear water, so the depth of light penetration exceeds the depth of the epilimnion. Such conditions may be found in either dimictic or warm monomictic lakes. Examples include small forest lakes (Canadian Shield lakes: Schindler & Holmgren 1971, Fee 1976, 1978, Fee, Shearer, & DeClercq 1977, Pick, Nalewaijko, & Lean 1984, Bird & Kalff 1987; Finnish lakes: Ilmavirta 1983), alpine lakes (Tilzer 1973; Rott 1983), Lake Tahoe (Kiefer et al. 1972), Lake Vechten (Blaauboer 1982; Blaauboer et al. 1982), Mary Lake (Brook, Baker, & Klemer 1971), Store Gribsø (Nygaard 1977), Lake Superior (Fahnensteil & Glime 1983), and Lake Michigan (Brooks & Torke 1977; Moll, Brahce, & Peterson 1984).

The origin of metalimnetic chrysophyte populations is a matter of dispute (Brooks & Torke 1977; Moll & Stoermer 1982; Pick et al. 1984). In large lakes or alpine lakes with low surface-area-to-volume ratios they appear to result from sinking of vernal diatom and chrysophyte populations. But in small stratified lakes they more commonly seem to consist of new species that suddenly appear in the plankton and preferentially migrate to the metalimnion, perhaps originating as recruitment events from benthic resting cysts. Fee (1976) and Brooks and Torke (1977) have speculated that these summer deepwater populations may provide the source for autumnal chrysophyte and diatom epilimnetic populations that appear late in the growth season as mixing depth increases. The physiological characteristics of metalimnetic chrysophytes have been examined to some extent (Pick et al. 1984; Healey 1983). Bird and Kalff (1987) have suggested that metalimnetic chrysophytes are substantially supplementing photosynthesis with phagotrophic feeding because the measured rates of photosynthesis are inadequate to support the observed biomass and growth.

Seasonal Periodicity Summary. The foregoing discussion of chrysophyte seasonal distribution patterns can be summarized to generate concepts that can be incorporated into a statement of overall growth and life history strategies. It is clearly important to differentiate between large colonial chrysophytes and small chrysomonads because it is apparent that morphology or cell size effects the hierarchy of factors regulating chrysophyte seasonal periodicity. Nannoplanktonic chrysomonads have ubiquitous seasonal distributions, seemingly only restrained by the increased grazing pressure experienced in more productive lakes (Table 2-1; also see Grazing section). Colonial or large-celled chrysophytes also generally demonstrate decreased relative importance in the plankton along a gradient of increasing lake productivity, the major exception being their absence from oligotrophic monomictic lakes. In dimictic lakes, large chrysophytes may be present during well-mixed periods, but they seemingly only become dominant during early or late stages of the stratified season, regardless of lake productivity. However, in monomictic lakes of comparable productivity, the same large netplankton forms may produce brief blooms of either minor or major significance at any season, regardless of mixing conditions. The comparison of monomictic and dimictic lakes suggests that mixing conditions may be important for netplanktonic chrysophyte seasonality only in that it is highly correlated with other temporally varying factors. Grazing by herbivorous zooplankton or competition with codominating algae would seem good candidates, and these will be addressed in subsequent sections. Factors such as mean seasonal temperature, conservative water chemistry, water color, and mixing per se do not seem to be critical for determining general seasonal distribution patterns of planktonic chrysophytes (except for metalimnetic populations).

DEMOGRAPHIC PATTERNS

Important insight into factors regulating the growth of phytoplankton species can be deduced from population growth profiles (Kalff & Knoechel 1978). Demographic profiles characterized by a pattern of abrupt maxima and declines result from large imbalances between growth and loss processes. Broader population maxima with gradual periods of increase and decline are symptomatic of a close balance between the species and the environment over extended periods of time. On the basis of the seasonal periodicity discussion (Table 2-1), it is clear that the same genera of chrysophytes may demonstrate different demographic patterns in different types of lakes. For instance, netplankton forms in the summer plankton of large oligotrophic or mesotrophic dimictic lakes have fairly broad population maxima that persist for weeks or even months (Ruttner 1930; Mills 1975; Eaton and Kardos 1978; Willén 1984; Dokulil & Skolaut 1986; Munawar & Munawar 1986). The same species of *Uroglena, Mallomonas,* or *Dinobryon* exhibit brief and abrupt population maxima in small dimictic oligotrophic lakes or in more eutrophic lakes (Hutchinson 1944; Asmund 1955; Lehman 1976; Plinski & Magnin 1979; Cronberg 1982; Ito & Takahashi 1982; Van Donk 1983; Roijackers 1984; Sheath 1986). Nannoplanktonic forms seem to be particularly abrupt in their population fluctuations (Takahashi 1978; Reinertsen 1982; Holmgren 1984), suggesting brief release from rigorous control of growth by environmental parameters.

An example of mixed demographic patterns produced simultaneously by co-occurring chrysophyte species was derived from the Egg Lake data set (Figs. 2-8, 2-9). During this vernal period of moderate grazing activity by *Daphnia* and unstable thermal stratification (Lehman & Sandgren 1985), the dominant diatom *Asterionella* and several moderate- to large-sized chrysophytes developed generalist-type populations depicting gradual changes reflective of balanced growth (Fig. 2-9, upper panel). At the same time, a series of more specialized chrysophytes (both large and small forms) underwent a close-packed succession of abrupt, minor unialgal peaks suggestive of opportunistic growth (Fig. 2-9, lower panel). The extent to which overlap of specialist species' temporal distributions is minimized is striking. Clearly size alone does not dictate these patterns, and they must depend upon simultaneous interactions of the chrysophytes with several environmental factors, perhaps grazing and competitive interactions. A mixture of both generalist and opportunist demographic patterns is commonly found among co-occurring chrysophytes, although the most common pattern in chrysophyte-dominated lakes is a series of largely monospecific abrupt peaks suggestive of opportunistic growth. Clearly, a great deal of resolution of patterns such as those in Fig. 2-9 would be lost in limnological studies with the typical sampling interval of one week or more.

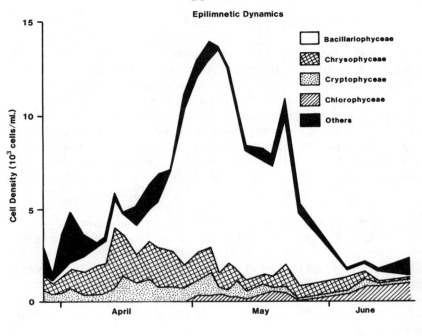

Egg Lake: 1976

Epilimnetic Dynamics

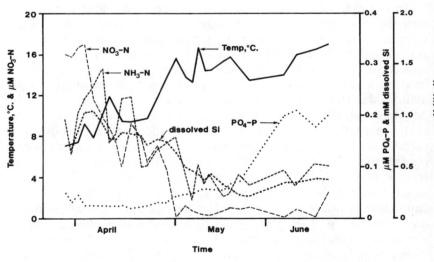

Fig. 2-8. Vernal epilimnetic water chemistry and phytoplankton population dynamics in Egg L., Washington in 1976. Values plotted are averages of two depths of each sampling day, and all data have been smooth with a three-point running mean. Data from Sandgren (1978) and Lehman & Sandgren (1985).

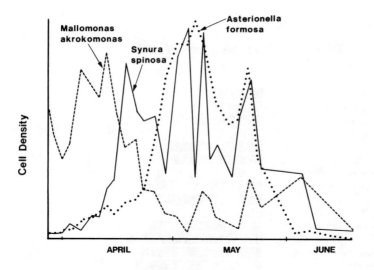

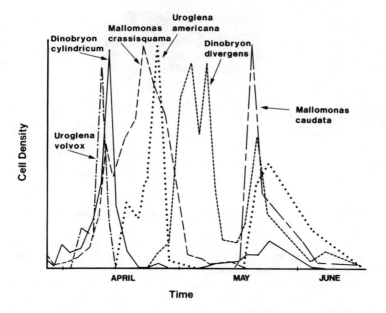

Fig. 2-9. Demographic patterns of some dominant and subdominant chrysophyte species in Egg L., Washington in 1976. Sampling frequency was at 2–3-day intervals, and density estimates were averaged from two or three epilimnetic sampling depths at one station. All data are scaled to the same vertical axes, so absolute cell densities are *not* comparable. (Upper panel) Tolerant or generalist species that exhibit demographic patterns suggestive of comparatively balanced growth, including the seasonal dominant phytoplankter *Asterionella formosa*. (Lower panel) Demography of some specialist species experiencing brief periods of opportunistic growth.

Geometric Niche Width

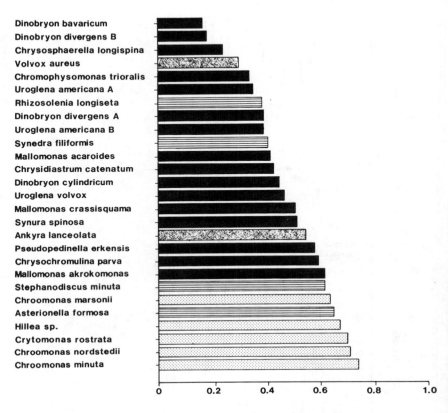

SPECIES   Relative Geometric Niche Width, 10 axes

Fig. 2-10. Calculated relative niche width for all species of phytoplankton in Egg and Sportsman Lakes with seasonal distributions confined to the periods of study (March–June 1976 and 1977). Estimates are based on a geometric mean calculated for each species from the ratio of the range of values for each environment over which positive growth occurred relative to the entire range of values existing during the periods of study. A value of 1.0 for a species would indicate observed net growth over the entire range of conditions occurring during the period of study. Method of calculation is detailed in Sandgren (1978). Environmental axes include zooplankton abundance, epilimnetic temperature, water column stability, abundance of *Asterionella,* total phytoplankton abundance, dissolved concentrations of

An estimate of the geometric niche width for positive growth was made for each species largely confined to the vernal period in Egg Lake (Fig. 2-10). As reflected in Fig. 2-9, some chrysophyte species are generalists with regard to this particular environment, and others are extreme specialists. In contrast, the cryptomonads, which are the other predominant flagellate members of this assemblage, are distinctly generalist in their growth requirements (Fig. 2-10) and demographic patterns.

GROWTH-LIMITING FACTORS

As has been repeatedly emphasized in recent years (Kalff & Knoechel 1978; Crumpton & Wetzel 1982; Reynolds et al. 1982; Lehman & Sandgren 1985), both loss and growth processes influence the seasonality and population dynamics of phytoplankton species. The preceding discussion of chrysophyte biogeography, periodicity, and demography has documented that at least netplankton species have relatively restricted seasonal distributions in most types of lakes, and that they frequently undergo abrupt population fluctuations. The same pattern is probably also correct for nannoplanktonic forms, but the dynamics are more difficult to discern because of problems in distinguishing species and accurately tracking their very rapid population pulses. If we accept as a premise that chrysophytes are very dynamic and temporally restricted components of lake plankton, then the obvious question in interpreting their growth strategy is *why* such patterns occur. What are the principle environmental variables that regulate their growth? Three frequently distinguished categories of factors (Lund 1965; Hutchinson 1967; Harrison & Turpin 1982; Paerl 1982) – tolerance limits and limiting nutrients, resource competition, grazing losses – will be discussed. Other factors of acknowledged potential importance for phytoplankton, such as sinking and respiratory losses, allelopathic interactions, and parasitism, will not be discussed because of the lack of information. Even for the factors discussed, specific experimental information concerning chrysophyte physiological tolerances and potentials is generally very limited, and it is therefore nec-

Caption to Fig. 2-10. (*cont.*)
silica, ammonia, nitrogen, and phosphorus. Black bars, chrysophytes; mottled bars, chlorophytes; stripped bars, diatoms; dotted bars, cryptomonads.

essary to rely on correlational evidence to an unfortunate extent or to generate new data, which has been done here in several instances.

Tolerance Limits and Limiting Nutrients

Light and Temperature. Classically, chrysophytes have been considered to be high-light, low-temperature forms (Findenegg 1943). The preceding distributional section of this paper would, however, more easily support a generalization of low-light, low-temperature preferences. Chrysophyte predominance year-round in arctic and north temperate lakes and their primary occurrence during the cool seasons in temperate lakes or in the cool summer metalimnion seems to support such a correlation. But the actual importance of light and temperature tolerances, either singly or in combination, in dictating such distribution patterns is not convincing. Confounding effects of temporally associated changes in grazing pressure and nutrient supply make cause-and-effect determinations difficult without experimentation (Hutchinson 1967; Kalff & Knoechel 1978).

The topic of temperature adaptation in phytoplankton has been admirably discussed by Li (1980). Chrysophyte temperature tolerances for growth and sexual reproduction have recently been reviewed in connection with a culture study of *Dinobryon* (Sandgren 1986). Those results are corroborated and extended to several other planktonic genera in a new study presented here (Fig. 2-11) and in the work of Munch (1972) and Healey (1983). Clonal variability and interspecific differences are notable attributes of chrysophyte thermal tolerances (also see Seaburg & Parker 1983). There is a strong linear temperature dependence of growth below 20°C in all cases (Fig. 2-11) in the data presented here. The mean Q_{10} value for growth between 10°–20°C was 3.5 (range 2–7); for the clones that grew at 5°C, the Q_{10} over 5°–20°C was 10 (range 5–18). This is reflective of the very high sensitivity of growth to temperature between 5° and 10°C as previously demonstrated for a different *Synura* species (Healey 1983). These values of Q_{10} are moderately high compared to cyanobacteria and marine diatoms (Healey 1983; Reynolds 1984a, his Fig. 68), more than double the comparable values for typical spring diatoms (*Stephanodiscus, Fragilaria, Asterionella;* Van Donk 1983), but are comparable to those of potentially co-occurring cryptomonads (Cloern 1977; Morgan & Kalff 1979). Many of these chrysophyte clones (Fig. 2-11) did not exhibit positive growth at 5°C, and this is in conflict with the frequent observations of similar species occurring under winter ice (Wright 1964; Cronberg 1982; Nebaeus 1984). Healey's (1983) observation that chrysophyte toler-

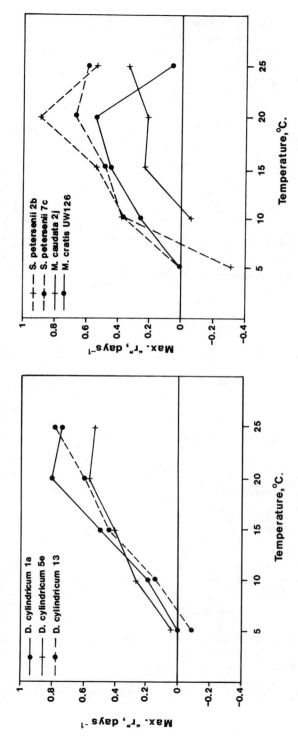

Fig. 2-11. Temperature tolerances for growth of some chrysophyte algae in culture. All clones were adapted to the experimental temperature for three weeks prior to estimation of maximum growth rate at very low cell density in triplicate batch cultures. All experiments were performed under continuous fluorescent illumination of 100 μmol·m^{-2}·s^{-1} and temperates were maintained ±1°C.

Veteran's Park Pond: 1985

Dinobryon cylindricum

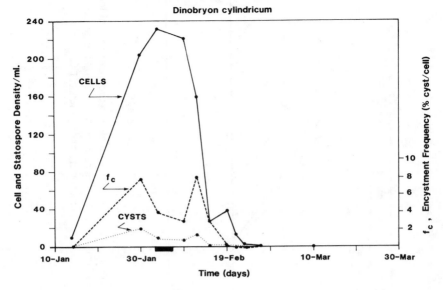

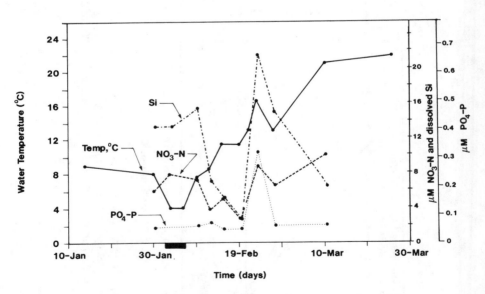

Fig. 2-12. Population dynamics and accompanying epilimnetic water chemistry for a natural population of *Dinobryon cylindricum* in Veteran's Park Pond, Arlington, Texas, in 1985. Pond area is approximately 1 ha; maximum depth 3 m. All samples were surface grab samples. Chemical analyses followed the methods outlined in Lehman

ance of low temperature increased with decreasing light intensity may help to explain this result, because these cultures were grown under relatively high light (100 μmol·m^{-2}·s^{-1} cool white fluorescent light), compared to winter conditions during which 5°C temperatures might be encountered (also see subsequent discussion of light tolerances).

It is clear that these species in culture have rather broad biokinetic tolerance ranges for temperature with maximum growth rates achieved around 20°C. The range of temperature supporting positive growth is similar to that of other freshwater planktonic algae (Goldman & Carpenter 1974; Cloern 1977; Rhee & Gotham 1981; Tilman, Mattson, & Langer 1981; Van Donk 1983; Mechling & Kilham 1982), with the exceptions of some high-temperature-tolerant green algae (Moss 1973a), cyanobacteria, and "weedy" pennate diatoms. Some chrysophyte species examined here exhibited possible signs of mild temperature stress on growth above 20°C, and similar stress has been noted previously for temperate chrysophytes (Knowles & Zingmark 1978; Seaburg & Parker 1983; Healey 1983). The biogeographic data in the chrysophyte database also shows the same trend of a broad maximum of chrysophyte importance in lakes between 10°–20°C mean epilimnetic temperature and a decline at higher temperatures (Fig. 2-5b). However, even if a modest reduction in growth rate does occur above 20°C for some chrysophytes, these data certainly do not support temperature tolerances alone as the factor responsible for the frequent lack of chrysophyte species in the plankton of temperate lakes during the summer.

A further test of chrysophyte temperature tolerances was performed on a natural population of *Dinobryon cylindricum* dominating the winter plankton of a small, south temperate pond (Fig. 2–12). There appears to be a strong temporal correlation between population dynamics and dramatic water temperature changes, positive growth occurring at low and decreasing temperatures, whereas population decline was associated with an increase in temperature of over 12°C in just two weeks. However, when subpopulations collected at two different times were incubated in the laboratory at constant light (50 μmol·m^{-2}·s^{-1}) over a range of temperatures, they exhibited maximum growth rates at 15°C (Fig. 2–19, *control* treatments), well above the

Caption to Fig. 2-12. (*cont.*)

and Sandgren (1985). Cell enumeration was accomplished using the Üntermohl method on replicate subsamples. Encystment frequency (f_c) denotes the calculated rate of statospore production on any sampling date à la Sandgren (1981). Black bar on time axis represents the period of ice cover.

temperature that might be proposed to be inhibitory based upon field observations alone. This optimal temperature and the observed growth rates are lower than those observed for *Dinobryon* in unialgal culture, perhaps because of either a lack of acclimation or clonal differences, but the general nature of the relationship is similar. As with the laboratory results given earlier, this study clearly suggests that physiological temperature tolerances for growth do not dictate chrysophyte periodicity or demography. If temperature has a regulating effect on chrysophyte seasonality or biogeography, then it must be in connection with light use, perennation success (see Reproduction section), regulation of zooplankton growth and feeding, or perhaps resource competition (Rhee & Gotham 1981; Tilman, Mattson, & Langer 1981; Van Donk 1983; Tilman & Kiesling 1984).

There is woefully little information available regarding the light requirements and photosynthetic characteristics of chrysophyte algae. I can find only a single publication (Healey 1983) that specifically focuses on the light requirements of a chrysophyte, and this species may not be typical in view of its limited biogeographical distribution. There are some publications concerning the photosynthetic characteristics of chrysophyte-rich lake plankton, but extrapolation of such data in formulating fundamental light characteristics of chrysophytes is always hazardous because of the mixture of species and the inability to explore the full potential range of the response. We are left with largely correlative information, and generalizations must therefore be viewed as hypotheses requiring testing, not as established fact.

Chrysophytes have a diverse complement of photosynthetic pigments very similar to those of diatoms. There is no basis for suspecting habitat preference based upon light quality. Certainly, the spectral composition shifts inherent in metalimnetic habitats or in humic-rich water (Fig. 2-5a) have no elective value. The results of Munch (1972) and Healey (1983) suggest that chrysophytes are tolerant of low light intensity for growth, but not unusually so in comparison to other freshwater algae, and therefore not particularly adapted for low-light environments. At least one common chrysophyte has the ability to substantially increase the chlorophyll content of cells at low light (Pick et al. 1984), a characteristic that is considered an important low-light adaptation but that is certainly not unusual among phytoplankton. Healey (1983) studied the light and temperature requirements of a metalimnetic clone of *Synura sphagnicola* in culture. He found this species to have an elevated light saturation level for growth ($100 \, \mu\mathrm{mol} \cdot \mathrm{m}^{-2} \cdot \mathrm{s}^{-1}$) as compared to previously studied phytoplankton, and that this saturation level was temperature insensitive over a broad temperature range ($5°–20°C$). The compensation light intensity for growth was similar to that of other phytoplankton (5–20 $\mu E \cdot m^{-2} \cdot s^{-1}$) and was directly related to temperature over the range tested. As previously discussed, growth was extremely sensitive to temperature below 10°C, but the actual growth rates under low temperature and light conditions were similar to or lower than rates published for nonchrysophyte species. Therefore, there appears to be no evidence for special chrysophyte adaptation to low-light growth conditions. Chrysophyte growth response to conditions of high light and temperature such as are found in the summer epilimnion of temperate lakes is unknown, as is their response to fluctuating light fields. Such data might help to explain their seasonal distributions. There is a critical need for research in this area and in establishing action spectra for growth and photosynthesis.

Ramberg (1979) has elegantly promoted the hypothesis that a combination of light and temperature optima, along with phosphorus utilization capacity, effectively segregates groups of phytoplankton species in Swedish lakes and can thus be used to explain the seasonal succession pattern. He proposed chrysophytes in his study lakes to be relatively temperature insensitive, but to dominate under low-light conditions while being outcompeted by coccoid green algae at high light. Ramberg further proposed that chrysophytes have high production efficiency (P_r/B ratios) at low light relative to green coccoid algae, but that the situation was reversed at higher light intensities. Holmgren (1984) distinguished between naked chrysophyte flagellates and loricate forms in his study of subarctic lakes, and proposed the former to be nutrient and light regulated while the latter were light and temperature regulated. Other modern studies that have also proposed light/temperature regulation of chrysophyte seasonality include Schindler et al. (1973), Ahlgren (1978), and Ilmavirta (1974, 1982, 1984). Although such ideas are enticing, they remain untested because they are based upon correlational studies. Again, there is a critical need for specific experimental studies of chrysophytes to determine their photosynthesis-versus-light relationships and production efficiency. Only then can the relevance of light as a potential distribution-regulating factor for chrysophytes be evaluated.

Alkalinity, pH, Carbon Sources. The importance of pH as a phytoplankton regulating factor is in its effects on pH-dependent nutrient uptake processes and on the equilibrium of nutrient or toxic metal ion species (Moss 1973b; Peterson, Healey, & Wagemann 1984). As a measure of the acid-buffering capacity of water, alkalinity reflects the ability of lake water to resist pH shifts and generally reflects the quantity of dissolved inorganic carbon (DIC) available for algal photosynthesis. Because the equilibrium between dissolved carbon dioxide, bicarbon-

ate, and carbonate ions is pH sensitive, a combination of alkalinity
and pH measurements provides estimates of both the amount and
molecular form of photosynthetically available carbon (Hutchinson
1957; Stumm & Morgan 1970).

Although chrysophytes as a group have conspicuously broad pH tolerances (Fig. 2-5f), the range is not inclusive of the entire pH range
found in natural lakes (Hutchinson 1957). Chrysophytes are generally
restricted to a pH range of 4.5–8.5, or a subset thereof. The best illustrations of this, other than Fig. 2-5f, include the work of Moss (1973b),
Bretthauer (1975), Kristiansen (1975), Takahashi (1978), Kies and
Berndt (1984), Smol et al. (1984), Roijackers and Kessels (1986), Reynolds (1986), and Steinberg and Hartmann (1986). The absence of
chrysophyte species of all types from lakes with mean pH greater than
8.5 is very conspicuous. In eutrophic lakes with a more neutral pH,
they are also consistently absent from the summer epilimnetic plankton during periods of elevated pH (>8.5) resulting from high photosynthetic activity. Sensitivity to elevated pH could be related to the
inability to secrete alkaline phosphatase as previously proposed (see
Nutrition section). But it may also be related to an inability to utilize
bicarbonate ions for photosynthesis and an inability to sequester an
adequate supply of dissolved free CO_2 under highly competitive conditions. When importance of chrysophytes is plotted as a function of
the mean quantity of calculated free CO_2 for lakes between pH 6 and
9 in the chrysophyte database, a minimum of at least $0.015 \text{ mmol} \cdot L^{-1}$
is indicated (Fig. 2-13a; calculated after the method of Mackereth,
Heron, & Talling 1978). Data presented by Hutchinson (1957, his Fig.
188) of actual (i.e., "analytical") free CO_2 estimates gathered by Birge
and Juday for Wisconsin lakes suggests that this value is only reached
below pH 8.

As previous authors have emphasized (Raven 1970; Lucas 1983;
Reynolds 1984a, 1986), all phytoplankton species are not equally effective at obtaining CO_2 for photosynthesis at high pH. Utilization of
bicarbonate ions, the overwhelmingly dominant form of dissolved carbon at high natural pH, as an alternative carbon source requires special
adaptations by cells. They must either have the ability to take up bicarbonate across the cell membrane and then to intracellularly convert it
into CO_2 for photosynthesis, or they must promote the extracellular
conversion of bicarbonate into carbon dioxide near the cell surface.
Some green algae and numerous species of cyanobacteria have been
demonstrated to use bicarbonate in photosynthesis; they possess either
intracellular or extracellular pools of carbonic anhydrase (an enzyme
capable of converting bicarbonate into dissolved CO_2), or they maintain a transmembrane pH gradient that performs the same function

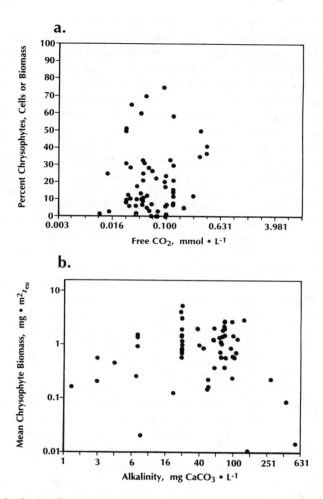

Fig. 2-13. (a) Scatterplot of the relationship between chrysophyte importance and potential free CO_2 among lakes in the chrysophyte database. Calculation of free CO_2 is based upon mean seasonal estimates of epilimnetic temperature, alkalinity, and pH after the method of Mackereth, Heron, and Talling (1978). (b) Scatterplot of relationship between mean chrysophyte biomass and alkalinity in the same lakes.

(Raven 1970; Shapiro 1973; Lehman 1978; Paerl 1982; Lucas 1983; Marcus, Volokita, & Kaplan 1984; Patel & Merrett 1986). However, neither the capacity to take up bicarbonate nor the presence of carbonic anhydrase has been demonstrated for any chrysophyte alga. Furthermore, these characteristics have been difficult to document for prymnesiophyte algae and diatoms (Sikes & Wheeler 1982; Patel & Merrett 1986), groups presumed to be phylogenetically related to chrysophytes in regard to their chloroplasts. The lack of ability to use bicarbonate as a source of carbon for photosynthesis may constitute a real physiological tolerance limit for chrysophytes; critical tests of their pH tolerances and ability to use bicarbonate are certainly needed.

Reasons for the avoidance of very acidic water (i.e., pH < 5) by chrysophytes have yet to be demonstrated specifically, but research on other algae clearly suggest that reduced nutrient uptake and increased metal toxicity are likely factors (e.g., Peterson et al. 1984). An excellent example of the decline in chrysophyte importance with decreasing pH is the artificial acidification of ELA Lake 223 (Findlay & Saesura 1980; Findlay 1984; Schindler et al. 1985). This lake, normally pH 7 and dominated by chrysophytes throughout the ice-free season, exhibited a marked decrease in the importance of chrysophytes relative to chlorophytes, cyanobacteria, and dinoflagellates as pH was decreased to 5. A number of studies with similar documentation of a decline in the relative importance of chrysophytes at pH values below 6 are cited in the reviews by Geelen and Leuven (1986) and Ravera (1986). It would seem important to concentrate future efforts toward detailing the physiological and toxicological effects of lake acidification on chrysophyte algae because of the dominant role played by chrysophytes in most softwater lakes potentially affected by acid deposition.

As previously described for the database lakes (Fig. 2-5c), alkalinity apparently restricts chrysophyte distribution, with a sharp decline in chrysophyte relative importance above an alkalinity of approximately 125 mg $CaCO_3 \cdot L^{-1}$. There appears to be a unimodal relationship between alkalinity and mean chrysophyte biomass integrated over the euphotic zone with maximal development between 10–125 mg $CaCO_3 \cdot L^{-1}$ (Fig. 2-13b). Although free CO_2 theoretically increases with increasing alkalinity (Stumm & Morgan 1970), there is a strong direct relationship in natural lakes between increasing pH and alkalinity (e.g., Moss 1973b; also chrysophyte database, Sandgren unpubl.). The result of this relationship is that free CO_2 actually decreases with increasing alkalinity in many lakes because pH also increases. High-pH, high-alkalinity lakes also tend to be very productive, so competition for free CO_2 also increases with increasing alkalinity. In light of the previous discussion, the apparent limiting nature of increasing alkalinity may well be in its direct relationship to pH and productivity and, therefore, its inverse relationship to available free CO_2 for chrysophyte photosynthesis.

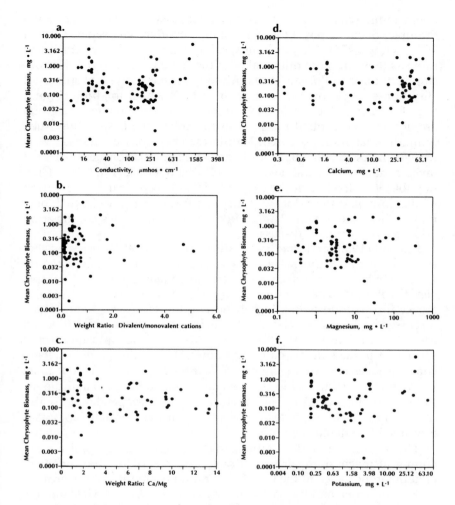

Fig. 2-14. Scatterplots of relationships between chrysophyte biomass and major ionic species of lakes in the chrysophyte database. All estimates are seasonal means.

Major Ions and Conductivity. There is a general decline in the relative importance of chrysophyte algae in lakes with increasing levels of conductivity (Fig. 2-5e) and with increasing total dissolved solids and total ionic content (Sandgren unpubl.). However, the trend in mean seasonal chrysophyte biomass is independent of these same factors (i.e., Fig. 2-14a). Several examples of chrysophyte-rich lakes or ponds with high conductivity or even brackish characteristics serve as additional examples of chrysophyte tolerance of high-conductivity waters (Willén 1962; Kling 1975). Chrysophyte species in these lakes are largely the same as those in more dilute lakes of the same region, so the relationship is not due to floristic shifts. Although some individual species of chrysophyte algae may indeed be inhibited by high total ion content (e.g., Provasoli, MacLaughlin, & Pinter 1954), chrysophytes as a group are not, and no explicit mechanism to explain the basis of any such limitation has been proposed. Within the chrysophyte database, as is true for lakes in general, there is high intercorrelation between total ions, conductivity, and alkalinity (Gorham et al. 1983; Sandgren unpubl.). It is likely that the trend of Fig. 2-5e is primarily a reflection of alkalinity and the associated affect on free CO_2 just discussed, or it may be associated with changes in zooplankton species composition (see later zooplankton discussion).

Pearsall (1932) and Pravosoli and co-workers (1954) suggested that the relative concentration of monovalent and divalent cations (M-to-D ratio) in lake water was important in influencing phytoplankton species distribution. The latter paper proposed that *Synura* showed a preference for medium with a high M-to-D ratio. However, Lund (1965) discussed problems associated with the interpretation of this ratio, and Moss (1972) detected no effect of the M-to-D ratio on a number of species examined in his cultures studies. The chrysophyte database contains lakes clustered primarily at low values of M to D, but there is no apparent relationship between chrysophyte biomass and the M-to-D ratio (Fig. 2-14b). The same lack of interdependence is also obtained when chrysophyte relative importance rather than biomass is examined (Sandgren unpubl.).

The two most abundant cations in most freshwater systems are calcium and magnesium. Provasoli and co-workers (1954) also considered the importance of the Ca:Mg ion ratio for phytoplankton growth, and documented an inverse relationship between *Synura* yield and the Ca:Mg ratio. Lehman, in his influential paper describing an artificial growth medium for *Dinobryon* (Lehman 1976), extended Provasoli's investigations. He manipulated both the relative and absolute concentrations of Ca and Mg in his DY III medium, and found growth of *Dinobryon sertularia* (his most sensitive species) to occur over the entire range of 4–140 mg·L^{-1} Ca and 2.5–30 mg·L^{-1} Mg, with greatest yields at approximately 5–12 mg·L^{-1} Mg and 12–40 mg·L^{-1} Ca. The optimal Ca:Mg ionic ratio was 3.2:1 by weight (2:1 molar). The chrysophyte database demonstrates no relationship between composite chrysophyte biomass and the Ca:Mg ratio of lakes over a broad range (Fig. 2-14c), although the relative importance of chrysophyte algae declines gradually above a weight ratio of 3–4 for the same lakes (Sand-gren unpubl.). These results with respect to both chrysophyte biomass and relative importance also hold for the relationships between cal-cium and magnesium concentrations individually (Fig. 2-14d,e; Sand-gren unpubl.). Lehman also proposed broad tolerances of *Dinobryon* for both sodium and chloride concentrations; this proposal is sup-ported by Munch (1972) and the chrysophyte database (Sandgren unpubl.).

The comparisons of both chrysophyte biomass and relative impor-tance relationships to all of these major ions support the general con-clusion raised earlier in connection with total phosphorus. That is, nei-ther the concentration of individual ions nor the total ionic strength appears to directly restrict chrysophyte growth. Rather, there appear to be other factors, probably biological, that are correlated with high dis-solved-ion content of lakes, and these other factors apparently favor other groups of phytoplankton over chrysophytes.

The role of potassium as a limiting factor deserves special attention because of its entanglement with a historic argument suggesting that phosphorus might be toxic for chrysophyte algae at elevated concen-trations. Rodhe (1948) proposed elevated phosphorus concentrations to be lethal to *Dinobryon* and *Uroglena* on the basis of the effects of additions of 5–10 $\mu g \cdot L^{-1}$ potassium phosphate to Lake Erken water. However, abundant evidence for luxuriant growth of *Dinobryon* and other chrysophytes at much higher concentrations of phosphorus in nature and in artificial media has since been published (Fig. 2-6; Leh-man 1976). And apparent toxicity responses of chrysophytes to phos-phorus added to mixed phytoplankton assemblages in artificial enclo-sures (e.g., Fig. 2-19; Van Donk 1983) can more easily be explained on the basis of competitive interactions and containment effects (see Competition section). Although the concentration of potassium added to Rodhe's experiment was very low, several investigators explored the possibility that potassium rather than phosphorus was responsible for the marked negative effect of elevated potassium phosphate on *Dino-bryon* growth. Munch (1972) observed growth inhibition of *D. diver-gens* at potassium concentrations of 3.3–4.0 $mg \cdot L^{-1}$; *D. cylindricum* exhibited clonal diversity in its response, one clone being inhibited at twice this concentration, whereas a second demonstrated no inhibition at 20 times this level. Lehman (1976) documented potassium toxicity beginning at 5–15 $mg \cdot L^{-1}$ for *D. cylindricum, D. sertularia,* and *D. sociale* var. *americanum,* with potassium tolerances increasing directly with the concentration of divalent cations (Ca, Mg) in the medium. Potassium concentrations in this range certainly occur in natural lakes, particularly in the autumn when potassium-rich deciduous leaf leachate enters the water, and both of these authors have suggested that potassium might influence the seasonal and biogeographic distributions of chrysophyte species. The chrysophyte database suggests no relationship between seasonal mean biomass and potassium over a wide range of concentrations (Fig. 2-14f), although chrysophyte percent importance exhibits a marked decrease among most lakes above a potassium concentration of 5 mg·L^{-1} (Sandgren unpubl.). The specific relationship between *Dinobryon* and potassium in the database has not yet been evaluated. Certainly, the distribution of individual chrysophyte species may be influenced by high potassium levels, and these relationships deserve future species-specific investigation, either along a gradient of lakes with increasing mean potassium concentrations or in individual lakes that exhibit seasonal oscillations in potassium levels.

Limiting Nutrients and Chelation. The diversity of potential subjects that could be discussed under this heading is broad. However, the paucity of information available about chrysophyte algae requires that the discussion be focused on growth limitation by four factors: nitrogen, phosphorus, iron, and the chelating ability of natural waters.

The long-term growth response of chrysophytes in lakes to artificial additions of nitrogen and phosphorus is usually negative (e.g., Fig. 2-6b). However, individual species in mixed assemblages demonstrate varied responses, suggesting diverse physiological types among chrysophytes. For instance, chrysophytes contributed approximately 50% of the epilimnetic biomass in L. Langvatn prior to enrichment with a common fertilizer containing nitrogen and phosphorus (Reinertsen 1982). In the year following fertilization, chrysophyte biomass decreased to 35%, but the relative importance of several large chrysophytes *(Chrysophaerella longispina, Mallomonas crassisquama)* increased, and there were many additional shifts in importance among the small chrysomonads. As a second example, fertilization of L. Hymenjaure by phosphorus alone for one growing season and both phosphorus and nitrogen in a second year also decreased the relative importance of chrysophytes but resulted in increased importance of an *Ochromonas* species apparently new to the lake (Holmgren 1984). Fertilization of the neighboring L. Magnusjaure with nitrogen alone resulted in a strong increase in chrysophyte biomass, primarily of *Uroglena,* which was relatively unimportant in the natural lake. Excellent summaries of individual species responses are included in both these papers.

Because of the complexity of interactions occurring in such seasonal studies, it is much better to investigate specific nutrient limitation on a species-specific basis by using short-term enrichments in which the actual growth response can be measured. In this way the physiological status of individual populations can be ascertained. A summary of such results for chrysophyte species from published studies is presented (Table 2-2). Species have been subdivided into morphological categories to explore morphology as a confounding factor.

It is clear from the totals column at the right of the table that nutrient limitation can play an important role in dictating growth rates and thus chrysophyte population dynamics. Only the two small chrysomonads failed to show evidence of nutrient limitation, perhaps because their high surface-area-to-volume ratio and small specific biomass might make them superior nutrient competitors (see Competition section). Of the total 123 experiments, 66% demonstrated nutrient limitation by at least one of the factors. In some instances the populations were nearly colimited, so the addition of two nutrients simultaneously (i.e., N + P or P + chelated Fe) gave greater positive response than the single most-limiting nutrient. Phosphorus was limiting in 31% of 121 trials, nitrogen in 7% of 88 trials, chelated iron or chelated micrometals mix containing iron in 27% of 86 trials, and addition of chelation alone (in the form of EDTA) resulted in enhanced growth in 70% of 20 trials. An important characteristic of chrysophyte nutrient limitation not evident from the table is that limitation is a periodic phenomenon, usually (but not always) occurring after the period of maximal population development. The limiting nutrient may change during the persistence of a chrysophyte species (e.g., *Uroglena* in L. Biwa), but in most instances chrysophyte populations are so short-lived that they experience limitation by only a single nutrient.

The importance of phosphorus as a limiting nutrient for chrysophyte growth and the relative unimportance of nitrogen is not surprising in view of the acknowledged general notion of phosphorus-limited phytoplankton growth in lakes (Schindler 1977). The adaptive importance of a finely tuned phosphorus uptake and utilization strategy for chrysophytes is obvious, and their phosphorus physiology will be considered relative to other phytoplankton in the following Competition section.

The considerable number of instances where near colimitation by phosphorus and nitrogen occurred in Egg and Sportsman Lakes (Table 2-2; Lehman & Sandgren 1985) attest to a close balance in the use of these nutrients by chrysophytes in some lakes. There is clear evidence that the N:P supply ratio for a lake, in combination with temperature and pH, can select for relative dominace by either cyanobacteria, diatoms, or green algae (Schindler 1975; Rhee & Gotham 1980; Smith

Table 2-2. *Nutrient limitation of natural chrysophyte populations.*[a]

Morphological category of species	Lake	Demonstrated limiting nutrient														No. exps. with limitation		Ref.
		P		N		N + P		Chelated micrometals		Chelation alone		Chelated Fe		N or P + chelated Fe				
		Yes	No	Yes	No	Yes	No	Yes	No	Yes	No	Yes	No	Yes	No	Yes	No	
A. Small unicells (<200 μm³)																		
1. *Pseudopedinella erkensis*	Egg	0	4	0	4	0	4	0	3							0	4	Lehman & Sandgren 1985
	Sportsman	0	2	0	2	0	2	0	2							0	2	Lehman & Sandgren 1985
2. *Chromophysomonas trioralis*	Egg	0	3	0	3	0	3	0	3							0	3	Lehman & Sandgren 1985
B. Moderate to large unicells (>200 μm³)																		
1. *Mallomonas akrokomonas*	Egg	1	3	2	2	2	2	0	3							3	1	Lehman & Sandgren 1985
	Sportsman	0	2	0	2	0	2	0	2							0	2	Lehman & Sandgren 1985
2. *Mallomonas crassisquama*	Egg	0	4	1	3	1	3	0	3							1	3	Lehman & Sandgren 1985
	Sportsman	0	2	0	2	0	2	0	2							0	2	Lehman & Sandgren 1985

															Reference
3. *Mallomonas caudata*	Egg	0	4	0	4	2	2	0	3					2	2	Lehman & Sandgren 1985
	Sportsman	0	2	0	2	0	2	0	2					0	2	Lehman & Sandgren 1985
C. Colonial forms																
1. *Uroglena americana*	Egg	1	2	1	2	1	2	0	3					1	2	Lehman & Sandgren 1985
	Sportsman	0	2	0	2	0	2	0	2					0	2	Lehman & Sandgren 1985
	L. Biwa[b]	11	24	0	35	0		20	15	6	4	18	17	33		Ishida et al. 1982
2. *Uroglena volvox*	Egg	0	1	0	1	0	1	0	1					0	1	Lehman & Sandgren 1985
3. *Dinobryon cylindricum*	Egg	0	2	0	2	0	2	0	2					0	2	Lehman & Sandgren 1985
	Sportsman	0	2	0	2	0	2	0	2					0	2	Lehman & Sandgren 1985
	Hall	7	5	1	2	2	4				1	1	1	9	3	Munch 1972
	Peter					1	0					4		1	0	Bergquist & Carpenter 1986
4. *Dinobryon divergens*	Egg	0	2	1	1	1	1	0	1					1	1	Lehman & Sandgren 1985
	Hall	9	3	0	2	0	3				0	6	3	10	2	Munch 1972

Table 2-2. *cont.*

Morphological category of species	Lake	Demonstrated limiting nutrient															No. exps. with limitation		Ref.
		P		N		N + P		Chelated micrometals		Chelation alone		Chelated Fe		N or P + chelated Fe					
		Yes	No	Yes	No	Yes	No	Yes	No	Yes	No	Yes	No	Yes	No			Yes	No	
5. *Dinobryon sertularia*	L. Maarsseveen	2	7							7	2							9	0	Van Donk 1983
	Peter					1	0											1	0	Bergquist & Carpenter 1986
6. *Synura petersenii*	Hall	0	2							1	0	2		1	0			3	0	Munch 1972
7. *Synura spinosa*	Egg	0	4	0	4	2	2	0	3									2	2	Lehman & Sandgren 1985
8. *Chrysosphaerella longispina*	Egg	0	1	0	1	0	1	0	1									0	1	Lehman & Sandgren 1985
	Jacks[c]	6	1	0	4													6	0	Pick et al. 1984
	Totals	37	84	6	82	13	39	20	53	14	6	3	10	23	18			82	41	

[a] Short-term experimental additions of nutrients to either mixed species or unialgal populations used to estimate nutrient limitation by species growth response. Specifics of experimental designs are reported in the individual references. Many of the species included are illustrated in Fig. 2-2.

[b] Number of experiments and results approximated from graphs in publication.

[c] Nutritional status determined from estimates of P and N debt rather than growth response.

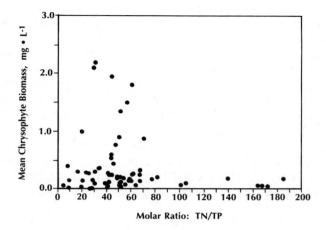

Fig. 2-15. Scatterplot of relationship between mean seasonal chryso-phyte biomass and mean seasonal total N:P molar ratio for lakes of the chrysophyte database.

1983; Kilham & Kilham 1984). It is therefore important to investigate the possibility that chrysophytes also may exhibit growth preferences with regard to N:P supply. On the basis of the lakes included in the chrysophyte database, there is no clear preference for lakes along an N:P supply gradient in terms of percent chrysophyte importance (Sandgren unpubl.) such as Smith (1983) has published for cyanobac-teria. But mean chrysophyte biomass does exhibit an interesting rela-tionship to the N:P ratio (Fig. 2-15), with higher biomass observed in lakes where the summer mean total N:P ratio is between 30 and 60. This range is comparable to those demonstrated to be optimal for growth of other eukaryotic algae (Kilham & Kilham 1984), and this finding supports the supposition made by Reynolds (1984b) that chry-sophytes characteristically require high N:P ratios for growth. Almost all of the high-chrysophyte biomass data points in Fig. 2-15 are derived from north temperate or subarctic lakes artificially enriched with phos-phorus, and the points are the same as those that appear as high-chry-sophyte biomass points in the plot of biomass versus total phosphorus (Fig. 2-7). Although the long-term response of chrysophyte algae in most lakes to phosphorus addition is negative, as demonstrated by Fig. 2-6a, it is clear from Figs. 2-7 and 2-15 that chrysophyte biomass can respond positively during the first one or two seasons of enrichment so long as the N:P supply ratio apparently remains within the range cited earlier and other growth-limiting factors (i.e., grazing pressure) do not simultaneously increase.

A number of published reports have supported the importance of iron, particularly chelated iron, as a potential limiting nutrient for freshwater phytoplankton (Rodhe 1948; Schelske 1962; Schelske, Hooper, & Haertl 1962; Sakamoto 1971; Goldman 1972; Jackson & Hecky 1980; DeHaan, Veldhuis, & Moed 1985). Although Schindler (1971) has argued against iron availability as a biomass-limiting factor, it may still be important in influencing the species mix of phytoplankton. The chelating ability of natural waters or the production of specific chelating agents for iron (siderophores) by some species have also been proposed to be generally important in determining phytoplankton species composition (e.g., Murphy, Lean, & Nalewajko 1976; DeHaan, DeBoer, & Hoogveld 1981; DeHaan et al. 1984). This is because chelation alters both the availability and toxicity of micrometals.

The results (Table 2-2) of Munch (1972), Ishida and co-workers (1982), and Van Donk (1983) suggest that iron may be a particularly important limiting micrometal for chrysophytes, and that the chelating ability of lake water may have a real impact on chrysophyte seasonality through its effect on iron availability for chrysophyte growth. Munch (1972) demonstrated that the growth of *Dinobryon cylindricum* and *Synura petersenii* was repeatedly stimulated by the addition of the chelator EDTA (in addition to iron or phosphorus) to Hall Lake water over separate periods of more than two months. She proposed that the chelated iron requirement of four species she studied was species-specific, with a possible ranking of increasing need being *Asterionella formosa, Dinobryon divergens, Dinobryon cylindricum, Synura petersenii.* The Lake Biwa study of *Uroglena americana* (Ishida et al. 1982) demonstrated that growth of this species could be stimulated by the addition of chelated iron for periods of many weeks during the growth season. A similar stimulation could be invoked by the addition of chelator alone (EDTA) in all attempted trials, suggesting that ambient iron concentration was adequate for growth if it were available in a utilizable form. Van Donk (1983) supported the importance of chelation capacity of lakes in an instance of apparent iron-limited growth of *Dinobryon divergens.* Growth of this species during the summer months in Lake Maarseeveen was stimulated by the addition of chelated iron for two consecutive years. She also suggested that diatoms may play a key role in chrysophyte seasonality by leaking organic chelators necessary for growth of chrysophytes. The common co-occurrence of chrysophytes and diatoms in the plankton as well as the frequent instances of chrysophyte blooms following vernal diatom maxima tend to support this notion, and a similar sort of indirect chemical interaction, but of a negative kind, is known for the iron metabolism of cyanobacteria and green algae (Murphy et al. 1976). A second suggestion Van Donk made was that the near absence of chrysophytes from hardwater, high-alkalinity lakes might be explained by the low levels of dissolved organic carbon and thus low levels of chelated iron in such lakes. Conversely, the great importance of chrysophytes in highly colored lakes rich in humic and fulvic acids may also be related to iron or other micrometals being more readily available in chelated forms to chrysophytes in such lakes.

All these comments can be summarized as suggesting that chrysophytes may be inferior to some other algae (cyanobacteria, diatoms, green algae) in sequestering iron in clear-water lakes. There is clearly a need for a well-constructed experimental study of chrysophyte seasonality and growth that incorporates measurement and manipulation of chelation capacity, iron concentration, and molecular forms of iron.

Resource Competition

Physiological models of phytoplankton growth and the theoretical basis of resource competition as it applies to microalgae have been given an excellent review by Tilman (1982) and are elegantly placed in the broader context of phytoplankton physiology and ecology by Turpin (Chapter 8, this volume). An understanding of competitive interactions between algal species requires consideration of all aspects of population growth, the entire physiological "packaging plan," and not just a single characteristic. At least three cardinal physiological attributes are necessary to adequately describe the competitive ability of phytoplankton: their ability to *store* resources in excess of immediate needs, the ability to *sequester* resources over the often broad range of concentrations available in patchy natural habitats, and, ultimately, the ability to use the resources for the production of new cells *(growth)*. Nutrient storage capacity is defined in terms of the cellular quota necessary for maintenance and growth. Q_0 is the mean cellular internal concentration of a nutrient at which the population growth rate is zero, and Q_{max} is the maximum potential internal store of the nutrient. The ratio of these has been termed the luxury storage coefficient (Droop 1974; Elrifi & Turpin 1985), and it is a convenient measure for assessing a species' ability to continue increasing its population size in the absence of an immediately available external nutrient supply. Sequestering of nutrients by algal cells has been observed empirically to follow saturation kinetics and can thus be described usefully with two parameters: V_{max}, the saturated specific uptake rate, and K_m, the external concentration of nutrient at which the rate is one half V_{max}. Although K_m has been used alone to compare the "affinity" of phytoplankton species for a nutrient, such usage does not well describe their sequestering ability over the natural range of nutrient concentrations. Instead, Healey (1980) has suggested that the ratio V_{max}/K_m is an appropriate competitive scaler of nutrient uptake capacity for phytoplankton because such a ratio adequately expresses the uptake rate at low, ambient concentrations. Apparent growth kinetics of phytoplankton also empirically follow a saturation curve. Under steady-state conditions, growth can be adequately described by the Monod formulation with the key parameters being μ_{max}, the maximum physiologically attainable growth rate, and K_μ, the external nutrient concentration at which the growth rate is half of the maximal rate. Again, a Healey ratio of μ_{max}/K_μ can be used to describe a species' growth ability at low nutrient concentration. Under nonsteady-state conditions, both the storage coefficients and uptake rates vary temporally, and growth can only be adequately described with a combination of all three sets of parameters.

Because of the limited data available on chrysophyte nutrient physiology, only consideration of their characteristics under phosphorus-limited conditions will be discussed here. The previous discussion of nutrient limitation suggested that nitrogen seldom limits chrysophyte growth, but future consideration of micrometals, perhaps silicon, and certainly light utilization is clearly warranted.

A literature search for physiological characteristics describing phosphorus utilization by chrysophytes revealed only a few pertinent data. These data were therefore supplemented with new laboratory determinations made for several clones of *Dinobryon, Synura,* and *Mallomonas;* they have been summarized here (Table 2-3). All six of the physiological parameters have been previously demonstrated to be cell-size dependent among phytoplankton in general (Kennedy 1984; Sandgren & Kennedy unpubl.), but there are insufficient data to meaningfully consider any size dependency among these few chrysophyte species. However, it is possible to compare the physiological characteristics of these chrysophytes to other phytoplankton by calculating the expected values of similar-sized cells from published size-dependent regressions. On this basis, the chrysophytes can be shown to be physiologically different from other freshwater microalgae in a number of important ways.

The minimum phosphorus quotas for growth (Q_0) are somewhat lower for chrysophytes than what would be expected for phytoplankton of the same size based upon the analyses of Shuter (1978) and Kennedy (1984). Considering primarily the diatoms, green algae, and cryptomonads that constitute the planktonic algal groups most frequently co-occurring with chrysophytes (Reynolds 1980; Cronberg 1982; Sand-

Table 2-3. *Characteristics of growth and phosphorus uptake kinetics for P-limited populations of chrysophyte flagellates.*[a]

Species	Clone	Mean cell volume (μm^3)	μ_{max} (day^{-1})	K_μ (μm)	V_{max} (10^{-9} $\mu mol \cdot cell^{-1} \cdot min^{-1}$)	k_m (μm)	Q_{max} (10^{-9} $\mu mol \cdot cell^{-1}$)	Q_0 (10^{-9} $\mu mol \cdot cell^{-1}$)
Dinobryon cylindricum	1	—	0.90	—	—	—	—	2.40
	5	272	0.51	0.014[b]	—	—	18.5	1.77
	7	—	0.58	—	—	—	—	2.15
	13	290	0.75	0.021[b]	—	—	21.0	1.87
Dinobryon bavaricum		80	—	—	0.34[c]	0.72[c]	—	—
					0.10[d]	0.11[d]	—	—
					0.22[d]	0.01[d]	—	—
Dinobryon sociale		—	—	—	2.32[c]	0.39[c]	—	—
Dinobryon divergens		—	—	—	—	0.10–0.27[c]	—	—
Synura petersenii	2b	374	0.51	0.003	5.1	1.19	90.0	3.04
	7c	431	0.76	0.001	21.8	1.35	55.2	1.96
Mallomonas cratis	UW-126	1,516	0.55	0.001	14.2	0.36	152	7.90
Mallomonas caudata	2j	10,625	0.30	—	—	—	—	—

[a]All clones used by the author for determination of physiological characteristics are maintained in his personal culture collection. Characteristics were determined under standard conditions of 17°C and 150 $\mu mol \cdot m^{-2} \cdot s^{-1}$ continuous cool white fluorescent illumination. Cell volumes were estimated from exponentially growing cells. Experimental methodology followed the procedures of Kilham (1978). Most of the species are illustrated in Fig. 2-2.

[b]Calculated using uptake kinetic data of Lehman 1976.

[c]Lehman 1976.

[d]Smith and Kalff 1982.

gren unpubl.), it is possible to make the stronger statement that these chrysophytes have markedly lower minimal phosphorus requirements for growth per cell. One exception to this generalization is the ubiquitous diatom *Asterionella formosa,* which has a size-specific requirement similar to these chrysophytes. The maximum storage capacity (Q_{max}) and the luxury storage coefficients for phosphorus (Q_{max}/Q_0) of chrysophytes are high compared to those of other eukaryotic algae, again with the exception of *Asterionella,* which has an extremely high coefficient (see later discussion). These chrysophytes also have, on average, low half saturation constants for phosphorus uptake as compared to 17 other freshwater phytoplankton species of similar cell size (as reported in Lehman 1976; Gotham & Rhee 1981; Smith & Kalff 1982; Kennedy 1984). Chrysophyte values of K_m are comparable to those of *Asterionella* and several small *Chlamydomonas* species, all of which are considered to be very efficient phosphorus scavengers. Chrysophyte values for the maximum phosphorus uptake rate (V_{max}) are comparable or slightly higher than those estimated for this same group of 17 species, and therefore the Healey ratio calculated for phosphorus sequestering capacity is as high or higher than competing eukaryotic algae. The proposal of Lehman (1976) and Reinertsen (1982), among others, that chrysophytes are extremely effective competitors for low concentrations of phosphorus would thus seem well supported by these data. Although these chrysophytes have the apparent advantages of a superior phosphorus uptake capacity under conditions of low ambient supply rates and very high phosphorus storage capacity relative to immediate cellular requirements, the estimated maximum growth rates of these chrysophyte clones are similar to those of many competing algae. μ_{max} values are of the same magnitude as the freshwater planktonic diatoms, cyanobacteria, and cryptomonads that have been studied (Tilman & Kilham 1976; Cloern 1977; Gotham & Rhee 1981; Kilham 1984), and they are much lower than rates estimated for weedy coccoid green algae and green flagellates (Gotham & Rhee 1981; Kennedy 1984). In highly eutrophic situations where phosphorus is always in excess, chrysophytes would certainly be outcompeted by such green algae. However, the Healey ratio for growth (μ_{max}/K_μ) of chrysophytes is high, suggesting that they would be effective competitors under phosphorus-limited conditions.

For comparing the competitive ability for phosphorus of chrysophytes to those of other freshwater phytoplankton, I have adopted the three ratios Q_{max}/Q_0, V_{max}/K_m, and μ_{max}/K_μ) as descriptors of their ability to store and sequester phosphorus and to grow under phosphorus-limited conditions. (For other schemes used to compare resource com-

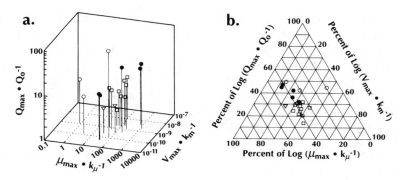

Fig. 2-16. Comparison of phosphorus physiology of planktonic chrysophytes to some other freshwater phytoplankton species. Sources of data and methodology are explained in the text and Table 2-3. (a) Three-dimensional plot of species characteristics against axes representing capacity for phosphorus storage, uptake, and P-limited growth. (b) Triangular representation of the same data for comparison of the allocation among the three axes characteristics. Plot (*b*) assumes that the length of each log axis equals 100% allocation. Symbols: black dots, chrysophytes; open circles, diatoms; open triangles, cyanobacteria; open squares, volvocene green algal flagellates.

petency of phytoplankton see Kilham & Kilham 1980; Sommer 1984; Reynolds, Chapter 10 this volume). Very few species have been sufficiently studied to have published estimates of all six parameters necessary to generate these three ratios. However, data for three diatoms and two cyanobacteria were extracted from Tilman and Kilham (1976) and Gotham and Rhee (1981), and comparable estimates for 10 species of unicellular and colonial green flagellates were used from Kennedy (1984). These data are plotted here in three-dimensional space (Fig. 2-16a), along with similar estimates for five chrysophyte clones from Table 2-3. The comparatively high phosphorus storage capacity of chrysophytes is apparent from the relatively longer length of the vertical lines connecting each data point to the horizontal plane. Only *Asterionella* has a higher value, whereas several large-celled *Chlamydomonas* species have similar capacities. There is wide scatter among the chrysophytes along the resource sequestering axis (V_{max}/K_m), suggesting no phylogenetically based clustering with regard to this ratio for chrysophytes, in contrast to the green algal flagellates included.

There is a general cell-size dependency of the phosphorus sequestering ratio among both chrysophytes and green flagellates, with larger cells having larger ratios (Table 2-3; Kennedy 1984). But the data presented in this manner provide no support for the preceding hypothesis that chrysophytes have an advantage in sequestering phosphorus. Chrysophytes occupy the right-hand edge of the data array regardless of their value of V_{max}/K_m, suggesting that they are growth (μ_{max}/K_μ) selected regardless of their nutrient sequestering capacity (or cell size). Consideration of the entire nutrient utilization package of cells in this way thus puts a different interpretation on potential resource competition among these species than does consideration of only individual physiological characteristics. Although chrysophytes are very effective at sequestering phosphorus under phosphorus-limited conditions, they are comparatively much more effective in a competitive sense at storing phosphorus and using it for production of new cells.

This same conclusion can be deduced from a two-dimensional "allocation" argument if the length of each logarithmic axis in Fig. 2-16a is presumed to represent the physiological range of variation for the corresponding ratio. Then the relative competence of each species along each axis can be expressed as a percentage, and a triangular diagram illustrating the comparative importance of each attribute to the physiological strategy of the species can be derived (Fig. 2-16b). Clustering of chrysophytes along with diatoms in the upper part of the diagram suggests that these organisms have a survival strategy that emphasizes growth and storage rather than nutrient uptake. Cyanobacteria, on the other hand, fall in the center with equally weighted allocation (in the universe as defined here), whereas green flagellates cluster more toward the lower center, suggesting an emphasis on sequestering phosphorus and growth as their strategy. Such an allocation diagram thus indicates that species may have emphasized certain characteristics over others in their evolution, either as a consequence of their phylogenetic heritage or cell size constraints. However, the actual physiological values (i.e., Fig. 2-16a) must be considered to predict competitive interactions.

The preceding analysis supports the hypothesis that chrysophytes with the physiological characteristics in Table 2-3 should be competitively dominant under conditions of phosphorus limitation, regardless of the supply schedule (i.e., continuous or stochastic). Only diatoms such as *Asterionella* and small, rapidly growing green algae such as *Scenedesmus* and *Chlamydomonas reinhardtii* could be predicted to outcompete chrysophytes. This physiological analysis perhaps helps to explain the consistently great importance of chrysophytes in oligotrophic, generally phosphorus-limited lakes (see Distribution section).

Clearly, a comparison of physiological characteristics for cryptomonads and chrysophytes in the manner described here would be very interesting when adequate data become available because these two groups of similar algal flagellates often co-occur in such lakes. The failure of chrysophytes to establish more than low-density populations or brief periodic blooms in more productive lakes points to the greater importance of other limiting factors, such as grazing losses and micrometal or light stress, for regulating chrysophyte abundance in these lakes.

Effects of Herbivorous Zooplankton

Herbivorous zooplankton and their potential impact on phytoplankton populations are discussed by Lehman (Chapter 9, this volume) and the review articles cited therein. For the purposes of analyzing the effect of zooplankton on chrysophyte temporal and spatial distribution, only a few general comments need be made here. Grazing by filter-feeding animals is generally size selective, with the upper threshold of potentially edible algal cells increasing as the animal body size increases. Large cladocerans such as *Daphnia* spp. can process larger particles than can small cladocerans such as *Bosmina* and *Chydorus*, which in turn can filter larger phytoplankton than planktonic rotifers. Behavioral and morphological specializations in feeding can complicate this pattern (e.g., DeMott & Kerfoot 1982). The amount of water cleared by an individual animal also increases directly with body size, so a few *Daphnia* can have proportionally greater impact on overall phytoplankton density than a much larger population of rotifers or small cladocerans. Zooplankton grazing can therefore drastically alter the size distribution and species composition of phytoplankton populations, with the result being highly dependent on the composition of the zooplankton assemblage itself (Langeland & Reinertsen 1982; Lehman & Sandgren 1985; Bergquist, Carpenter, & Latino 1985). Empirical evidence exists for a direct relationship between phytoplankton and zooplankton biomass in lakes (McCauley & Kalff 1981), and theoretical predictions suggest that zooplankton animal size should increase with increasing overall phytoplankton abundance (Carpenter & Kitchell 1984). Lakes are often dominated by either small- or large-bodied zooplankton species, depending on the abundance of phytoplankton and planktivorous fish (Neill 1984; Vanni 1986). Furthermore, competitive dominance of highly efficient *Daphnia* species over the small and less efficient *Bosmina* species has been demonstrated at the elevated levels of phytoplankton food availability characteristic of productive lakes (DeMott & Kerfoot 1982; Neill 1984; Kerfoot,

DeMott, & DeAngelis 1985; Vanni 1986). *Daphnia* has also been demonstrated to kill rotifers in the course of their normal feeding activities and even to be actively carnivorous on rotifers (Burns & Gilbert 1986). Generally, then, the zooplankton assemblage of oligo- or mesotrophic lakes low in phytoplankton biomass but rich in chrysophytes is often dominated by less efficient rotifers and small cladocerans that specialize on small flagellates. More eutrophic lakes with elevated total phytoplankton biomass, but greatly reduced importance of chrysophytes, are more typically dominated by efficient *Daphnia* species, particularly during the summer.

Empirical evidence for zooplankton grazing on chrysophyte algae is considerable, particularly for very small unicells. Small chrysomonads (<10 μm) are a common component of the nannoplankton-size particle fraction that is the preferred food resource of all herbivorous filter feeders (Pechlander 1971; Langeland & Reinertsen 1982; Reynolds et al. 1982; Reynolds, Chapter 10, this volume). However, because the efficiency with which such small cells are handled is dependent on the composition of the zooplankton assemblage, the overall impact of grazing on the small-size fraction of phytoplankton may be markedly different. *Daphnia*-dominated lakes reduce the abundance of the nannoplankton fraction, whereas rotifer- and small-cladoceran-dominated systems may actually experience a proportional enrichment of the nannoplankton-size fraction by a mechanism that is unclear but probably involves differential utilization of recycled nutrients generated by grazing activities (Langeland & Reinertsen 1982; Bergquist et al. 1985). Likewise, the response of large unicells or colonial chrysophytes to grazing clearly depends on the size of animals and cells or colonies. An example is provided in Table 2-4, based upon two enclosure studies in which the species-specific responses of chrysophytes to zooplankton abundance manipulations were estimated. Larger, siliceous-scale-covered monads were occasionally grazed by *Daphnia,* but more frequently exhibited no response. The very large *D. pulex* present during part of the Egg Lake study (Lehman & Sandgren 1985) were able to handle *Uroglena* and *Dinobryon* colonies. Smaller animals of the same species present later in the same study as well as the smaller zooplankton species present in the Tuesday Lake study of Bergquist et al. (1985) were not. As a result, the large colonial chrysophytes occasionally actually benefited from the presence of zooplankton by using remineralized limiting nutrients released by the grazing of animals on competing small algae in the assemblage (Porter 1976; Lehman 1980; Carpenter & Kitchell 1984; Sterner 1986).

Although increased size clearly represents a potential refuge for chrysophytes from grazing, the existing data suggest that even the largest colonial chrysophytes are subjected to grazing losses in the presence of large zooplankton. The efficacy of coloniality or the siliceous spines of mallomonadacean chrysophytes as grazer-avoidance mechanisms is therefore highly dependent on the composition and size spectrum of the zooplankton assemblage. It must also be remembered that size-dependent filtering of phytoplankton is not the only mechanism of zooplankton feeding. Rotifers and ciliates, for instance, may severely impact the growth of even large colonial chrysophyte populations by selectively and repeatedly removing individual cells from colonies (Munch 1972; Sandgren unpubl.). Feeding on large freshwater chrysophytes by predatory copepods has not been investigated to my knowledge, but the mechanism would be similar.

With these comments and results in mind, it is instructive to examine the relationship between chrysophyte distribution and zooplankton abundance in the lakes included in the chrysophyte database (Fig. 2-17). There is a direct relationship between total phytoplankton biomass and zooplankton biomass for these lakes (Fig. 2-17a), as has been previously demonstrated for other lakes by McCauley & Kalff (1981). Unlike their study, however, there is a positive intercept of the relationship with the phytoplankton axis ($b = 0.35$ mg$\cdot$L^{-1}; $P = 0.0004$ for regression). The choice of the zooplankton density parameter for this comparison does not change the result greatly. Similar relationships exist for these database lakes between phytoplankton biomass and either the biomass or animal density of total zooplankton, of total cladocerans, of total rotifers, or of *Daphnia* alone. The importance of chrysophytes in the plankton declines among the lakes with increasing animal biomass or abundance (Fig. 2-17b), again regardless of whether total zooplankton or one of the foregoing subcategories is considered. Because of the strong correlations between phosphorus supply, phytoplankton abundance, and zooplankton abundance in this data set, the decreasing relative importance of chrysophytes is, in effect, the same one represented in Fig. 2-6a for increasing phosphorus concentrations. There is considerable variability in chrysophyte importance at low animal density or biomass levels (Fig. 2-17b), and it is clear that chrysophyte-dominated lakes are those which lack *Daphnia*.

The net result of increasing phytoplankton biomass (Fig. 2-17a) but decreasing chrysophyte importance (Fig. 2-17b) as a function of zooplankton abundance among these lakes is a flat relationship between actual chrysophyte biomass and zooplankton biomass (Fig. 2-17c). Again, the shape of this relationship remains unchanged regardless of which expression of zooplankton abundance (biomass or animal density) or which category of zooplankton (total zooplankton, total cladocerans, *Daphnia,* or total rotifers) is selected. Chrysophyte biomass

Table 2-4. *Response of natural chrysophyte populations to grazing by herbivorous zooplankton.*

Morphological category of species	Lake	Dominant grazer	Grazing response			No. of exps.	Ref.
			"+"	"−"	None		
A. Small unicells (<200 μm³)							
1. *Pseudopedinella erkensis*	Egg & Sportsman	*Daphnia pulex*	0	4	2	6	Lehman & Sandgren 1985
B. Moderate to large unicells (>200 μm³)							
1. *Mallomonas akrokomonas*	Egg & Sportsman	*Daphnia pulex*	0	1	5	6	Lehman & Sandgren 1985
2. *Mallomonas crassisquama*	Egg & Sportsman	*Daphnia pulex*	0	0	3	3	Lehman & Sandgren 1985
3. *Mallomonas caudata*	Egg & Sportsman	*Daphnia pulex*	0	0	2	2	Lehman & Sandgren 1985
4. *Mallomonas* sp.	Peter	*Daphnia pulex, Diaptomus oregonensis*	0	1	0	1	Bergquist et al. 1985

C. Colonial forms

1. *Uroglena americana*	Egg & Sportsman	*Daphnia pulex*	0	1	5	6	Lehman & Sandgren 1985
2. *Dinobryon cylindricum*	Egg & Sportsman	*Daphnia pulex*	0	0	1	1	Lehman & Sandgren 1985
3. *Dinobryon divergens*	Egg & Sportsman	*Daphnia pulex*	1	2	3	6	Lehman & Sandgren 1985
4. *Dinobryon* spp.	Peter	*Daphnia pulex, Diaptomus oregonensis*	3	0	1	4	Bergquist et al. 1985
	Tuesday	*Bosmina longirostris,* rotifers	0	0	4	4	Bergquist et al. 1985
5. *Synura spinosa*	Egg & Sportsman	*Daphnia pulex*	1	1	3	5	Lehman & Sandgren 1985
6. *Chrysosphaerella longispina*	Egg & Sportsman	*Daphnia pulex*	0	0	1	1	Lehman & Sandgren 1985
	Peter	*Daphnia pulex, Diaptomus oregonensis*	1	0	0	1	Bergquist et al. 1985
	Tuesday	*Bosmina longirostris,* rotifers	0	0	1	1	Bergquist et al. 1985

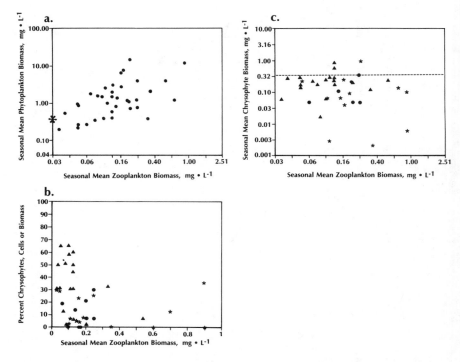

Fig. 2-17. Scatterplots depicting relationships between phytoplankton and zooplankton abundance among the lakes in the chrysophyte database. (a) Mean seasonal total phytoplankton biomass versus mean total zooplankton biomass. Y-intercept of a least squares regression is indicated by the star. (b) Chrysophyte mean relative importance versus mean total zooplankton biomass. (c) Chrysophyte mean seasonal biomass versus mean total zooplankton biomass. Dashed line in (c) represents regression intercept of panel (a). Symbols for panels b and c: star, lakes with *Daphnia;* triangles, lakes dominated by copepods or small cladocerans; dots, lakes with no information regarding zooplankton composition.

clearly does not increase directly with zooplankton abundance, as does overall phytoplankton biomass (i.e., Fig. 2-17a). The dashed horizontal line in Fig. 2-17c represents the intercept biomass level of the phytoplankton/zooplankton biomass regression (Fig. 2-17a). It is clear that chrysophyte biomass among the study lakes rarely exceeds this "threshold." This relationship, in combination with the increasingly restricted seasonal periodicity of chrysophytes observed in lakes of increasing productivity and zooplankton abundance (Table 2-1), is suggestive evidence for the strong general regulatory effect of zooplankton on chrysophyte algae of all size classes. In light of the previous arguments made here concerning chrysophyte phosphorus physiology and general physiological tolerance limits, it seems likely that the most important factor regulating chrysophyte importance along lake trophic gradients is grazing by herbivorous zooplankton, particularly species of *Daphnia*. Chrysophyte dominance in oligotrophic lakes may very well be dependent on the lack of large efficient cladoceran grazers such as *Daphnia* in such lakes.

REPRODUCTIVE STRATEGIES

Life Cycle

The information reviewed in previous sections regarding chrysophyte seasonal distribution patterns suggests that they characteristically grow during only a few months of the year in productive lakes. Even in oligotrophic lakes dominated by chrysophytes, the individual species have short-lived populations, and a succession of both dominant and secondary species occurs during the growing season. For species such as these that have seasonally restricted development of vegetative cell populations, there must be a crucial reliance on resting cysts for successful perennation. Chrysophytes produce a characteristic siliceous resting cyst, termed a *statospore* or stomatocyst, which is frequently observed in association with vegetative cell populations.

The genetic and strategic significance of the statospore in the chrysophyte life history has long been a matter of dispute. However, several recent examinations of statospore production in culture (Sandgren 1981, 1983a, 1986; Sandgren & Flanagin 1986) have clarified the life cycle sufficiently that a general pattern for chrysophytes can be proposed (Fig. 2-18). Vegetative chrysophyte cells are haploid. They contribute to vegetative population growth through binary fission, but under some conditions individual cells can be induced to form a statospore. Either uni- or binucleate statospores may be produced from single vegetative cells, suggesting that such cysts must be asexual or autogamic (self-sexual). Sexual reproduction is heterothallic in the species that have been examined in culture, requiring the presence of two compatible clones. Hologametic fusion of vegetative cells results in the formation of a short-lived binucleate, quadriflagellate planozygote that subsequently undergoes encystment to become a binucleate hypnozygote or sexual statospore. Duration of the encystment process varies, lasting from a few minutes to half a day. Statospore germination results in the formation of one, two, or four initial vegetative cells,

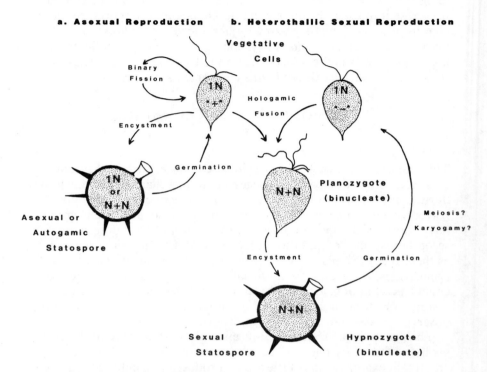

Fig. 2-18. Generalized life cycle diagram for planktonic chrysophytes. Ploidy level of morphological stages is indicated. See text for further explanation.

depending upon the species. The actual processes of karyogamy and meiosis that must accompany germination have not yet been examined. Sexual and asexual statospores are morphologically indistinguishable for a species, but the morphology of statospores varies greatly among chrysophyte species. On the basis of this life cycle, the statospore must be interpreted as a very opportunistic type of resting cyst, apparently capable of being initiated by several different stimuli. This flexibility perhaps underscores the importance of resting cysts for the survival of planktonic chrysophyte species.

To understand the role of statospores in the chrysophyte perennation strategy, we must consider encystment cues, size of cyst refuge populations, survivorship potential, and germination requirements. Each of these aspects will be discussed. Most of the available information is based on investigations of only a few species and largely concerns sexual statospores; there is a clear need for future research to test the applicability of the generalizations proposed here for other planktonic chrysophytes.

Encystment Cues and Dynamics

Environmental stress upon vegetative cells, such as that caused by nutrient limitation or temperature changes, has often been proposed to be the critical trigger initiating sexual or asexual cyst production in planktonic microalgae, such as cyanobacteria, dinoflagellates, and green algae (reviewed in Sandgren 1983a, 1986; Sandgren & Flanagin 1986). The typical dynamic pattern of cell and cyst production observed in such cases is for the vegetative growth rate to drop as the encystment rate increases, the result being a peak of cyst production occurring as the vegetative cell population declines. This pattern has been referred to as *extrinsic* encystment (Sandgren & Flanagin 1986). Cyst production in such instances is primarily under *exogenous* control, being dependent on physical or chemical conditions that influence vegetative growth of the cells. Encystment contributes to the decline of such algal populations, but it is not an important loss factor for the vegetative population during the period of vegetative growth and persistence.

In comparison, planktonic chrysophytes have a very different mechanism for controlling encystment and a resulting different dynamic pattern of cyst production. It was not possible to induce statospore formation or to alter the rate of encystment for *Dinobryon cylindricum* in culture by employing a variety of nutrient stresses: phosphorus, nitrogen, vitamins, and micrometals (Sandgren 1981). When the effects of temperature on encystment were studied, encystment rate was shown to be *directly* related to the population growth rate during periods of positive growth, and statospore yield was determined to be dependent on the density of the least dense mating clone (Sandgren 1986). Subsequent analysis of sexual cyst production in *Synura petersenii* (Sandgren & Flanagin 1986) demonstrated that statospore formation was *density dependent,* requiring a minimum threshold number of cells. Above this threshold density, cyst yield was directly dependent on vegetative cell density, which is dependent on the net growth rate of the slowest-growing mating clone. The rate of cyst formation was constant above the threshold, and both encystment frequency and the threshold cell density at which sexual encystment was initiated were proposed to

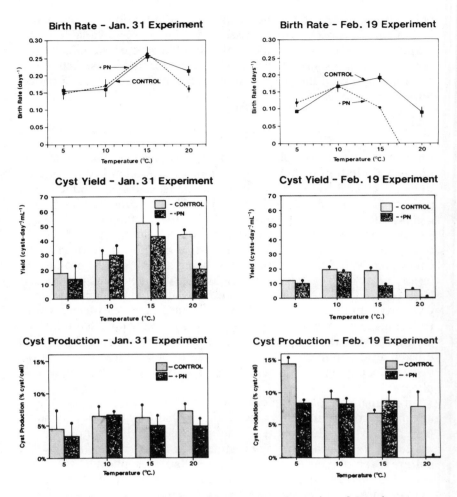

Fig. 2-19. In vitro growth and encystment dynamics of *Dinobryon cylindricum* from Veteran's Park Pond (Fig. 2-12) on two separate dates. Populations were collected with a plankton net and incubated in filtered pond water over a gradient of temperatures with or without added phosphorus and nitrogen. (Upper panels) Birth rates corrected for sexual encystment losses. (Middle panels) Average daily cyst yields over the eight-day incubations. (Lower panels) Cyst production rates (new cysts/total new cells × 100) over the same eight days.

be species- or perhaps clone-specific characteristics. The possible existence of an extracellular mating hormone controlling clonal compatibility and gametogenesis was proposed (Sandgren 1981; Sandgren & Flanagin 1986). Sexual statospore production in chrysophytes therefore appears to be under *endogenous* control, being directly dependent on the cell density of compatible mating clones and only indirectly dependent on external factors that regulate population growth rate.

A further test of the role of cell density and temperature on *Dinobryon* statospore formation was conducted on populations from the Texas pond discussed previously with regard to temperature effects on vegetative growth (Figs. 2-12, 2-19). In situ population dynamics (Fig. 2-12) demonstrated that this population was actively forming statospores during the first experiment (Jan. 31), but had essentially stopped encysting by the time of the second experiment (Feb. 19), at which time the vegetative cell population density had dropped by approximately one order of magnitude. Populations on both dates were concentrated by net hauls and incubated in the laboratory over a range of temperatures, with and without added nutrients (Fig. 2-19.) It is obvious from the resulting growth that the population was not nutrient limited on either date (upper panels, Fig. 2-19), and unpublished information suggests that rotifer grazing may have been primarily responsible for the *Dinobryon* population decline. It is also obvious that both experiments resulted in cyst production at approximately the same rates of 5–7% of vegetative cell numbers per day (lower panels, Fig. 2-19), and that encystment was relatively insensitive to temperature even without prior acclimation. The simple act of artificially increasing the cell density was apparently sufficient to reinitiate encystment during the mid-February experiment, a time when the natural population density dropped below the suggested encystment threshold for this population. Total cyst yields in these two trials vary among temperature treatments (Fig. 2-19, central panels) in a manner that is directly related to population growth rate and, therefore, to the production rate of new cells. To test the validity of interpreting these results as indicative of a density-dependent encystment pattern with a required minimum threshold density level for encystment initiation, a third manipulation was performed on Feb. 20 that consisted of a serial dilution series of *Dinobryon* cell density. It demonstrated that it was possible to halt statospore production by maintaining the cell density below approximately 200 cells·mL^{-1} over the course of the trial (Sandgren unpubl.). If cyst production for all three of these experiments is plotted as a function of either population growth rate or actual cell numbers produced in three separate experiments, a direct relationship results, with the best correlation being with actual cell production (Fig. 2-20). The data points again suggest that a positive cell density of several hundred cells per milliliter is required to initiate encystment, although the regression line has a negative X-intercept because of the large variance among the results at high cyst production levels for the three separate experiments. Thus, density-

Encystment Dynamics:
Veterans Park Pond, 3 Experiments

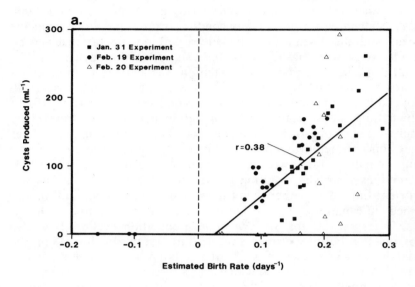

a.

■ Jan. 31 Experiment
● Feb. 19 Experiment
△ Feb. 20 Experiment

r=0.38

Cysts Produced (ml.⁻¹)

Estimated Birth Rate (days⁻¹)

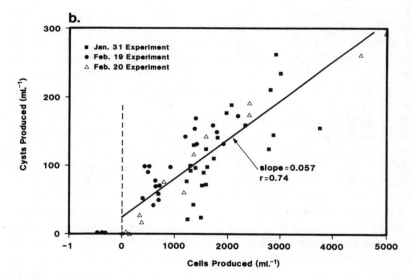

b.

■ Jan. 31 Experiment
● Feb. 19 Experiment
△ Feb. 20 Experiment

slope=0.057
r=0.74

Cysts Produced (ml.⁻¹)

Cells Produced (ml.⁻¹)

Fig. 2-20. Analysis of encystment dynamics for three in vitro experiments performed with the Veteran's Park Pond *Dinobryon* population (Figs. 2-12, 2-19). (a) Regression of cysts produced during the experimental incubations versus birth rates corrected for encystment

dependent statospore production can be demonstrated for natural chrysophyte populations as it has been for clones in cultures.

As opposed to the extrinsic pattern of encystment observed for phytoplankton reproduction under exogenous control, the density-dependent production of sexual statospores by chrysophytes would be expected to result in a direct relationship between vegetative cell density and cyst density. This pattern has been described as *intrinsic* encystment (Sandgren & Flanagin 1986). The meager published data on the encystment dynamics of natural chrysophyte populations appears to support a hypothesis suggesting that the intrinsic encystment pattern is commonly observed (Ruttner 1930; Nauwerck 1968; Munch 1972; Sheath, Hellebust, & Sawa 1975). The encystment dynamics of six chrysophytes present during the spring of 1976 in Egg Lake all generally conform to the intrinsic pattern as well (Fig. 2-21; see also Figs. 2-8, 2-9). The examples of *Synura spinosa* and *Mallomonas akrokomonas* (Fig. 2-21, upper panels) are particularly striking in this regard. Encystment dynamics of natural chrysophyte populations may not follow a simple density-dependent model if the growth rates or relative cell densities of compatible clones are very dissimilar (Sandgren 1986). The second population maxima of the two Egg Lake *Dinobryon* species that produced no statospores (Fig. 2-21, lower panels) could perhaps be examples of such a situation. Also, asexual statospore production may not follow the density-dependent model (Sandgren 1981) and may indeed conform to the extrinsic pattern, as suggested by the dynamics of a *Mallomonas eoa* population (Cronberg 1973). There is certainly a need for a comparative examination of sexual and asexual encystment dynamics in culture.

Unlike the extrinsic encystment events of many other phytoplankton species, the intrinsic statospore production of chrysophytes does constitute a potentially important loss factor for planktonic populations because it occurs during the period of active growth and persistence (Sandgren & Flanagin 1986). Loss rates of 0.15 day^{-1} for *Synura* and 0.45 day^{-1} for *Dinobryon* in culture or 0.06 day^{-1} for the Texas pond *Dinobryon* population (Fig. 2-12) can have a significant impact on the population dynamics of these species whose maximal potential growth rates are about 0.6–0.9 day^{-1}. For chrysophytes, it is therefore important to include encystment losses along with resource limitation and herbivorous grazing as significant negative factors in population

Caption to Fig. 2-20. (*cont.*)
losses. (b) Regression of cysts produced versus total cells produced (i.e., change in cell density plus cells estimated to have been lost to sexual encystment). The nature of the experiments is explained in the text.

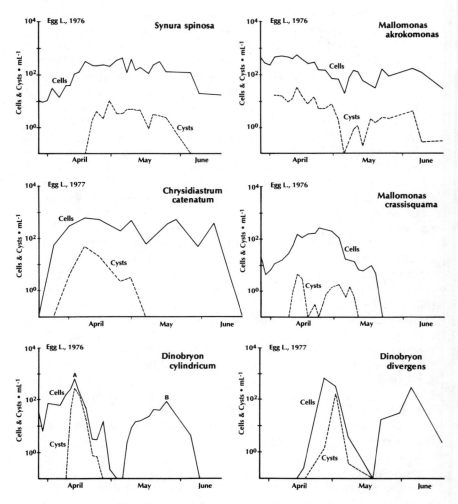

Fig. 2-21. Demographic patterns of some planktonic chrysophyte populations from Egg L., Washington during spring 1976. Solid lines indicate cell density and dashed lines indicate statospore density. Note log *Y*-scale.

regulation (Reynolds et al. 1982; Crumpton & Wetzel 1982; Lehman & Sandgren 1985).

Although loss of cells via encystment during population growth has a negative impact on vegetative persistence, density-dependent, endogenously controlled encystment may be the only viable reproduc-

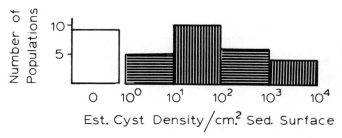

Fig. 2-22. Frequency distribution of the estimated sizes of chryso-phyte resting cyst populations in Egg L., 1976 and 1977. Population size was estimated by integrating epilimnetic cyst density estimates over the observed period of encystment and then correcting for the curvature of the basin sediment surface. From Sandgren (1983a); used with permission.

tive option for planktonic chrysophytes. Most of these species are actually present in the plankton of lakes for only a few weeks annually, and they exhibit opportunistic demographic patterns characterized by rapid population increases and declines. For species such as these, reliance on exogenous factors such as nutrient stress or fluctuating physical variables for encystment triggers might be too risky to ensure successful perennation. Instead, chrysophytes have adopted a "bet-hedging" type of reproductive strategy (Stearns 1976) that is more likely to ensure production of some resting cysts before the population disappears. This proposed integration of demographic and reproductive strategies is perhaps more marked for chrysophytes than for other planktonic microalgae because of their distinctly limited seasonal distribution.

Size of Refuge Statospore Populations

Sandgren (1983a) estimated the probable size of refuge or "seed" populations of statospores deposited on the sediments of Egg Lake for 34 chrysophyte populations present during the spring of 1976 and 1977 (Fig. 2-22). Although not all populations produced statospores, almost all species included in this analysis produced cysts at some time over the period of study. The estimated density of cysts settling onto the sediment surface varied widely among species, being dependent on both the intensity and duration of the encystment event. Many of these refuge populations were extremely large when extrapolated over the entire sedimentary surface area, and it would seem likely that they should be adequate to repopulate the plankton. However, there is no

STATOSPORE SURVIVORSHIP:

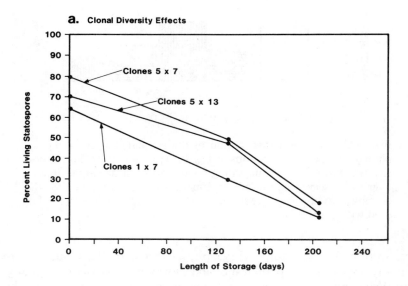

a. Clonal Diversity Effects

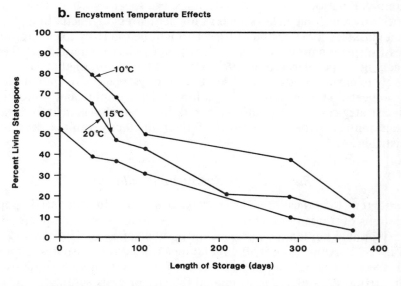

b. Encystment Temperature Effects

Fig. 2-23. Some factors influencing the formation and survivorship of *Dinobryon cylindricum* sexual statospore populations produced in vitro. (a) Effects of genetic differences as reflected in various clonal mating crosses. (b) Effects of ambient temperature during cyst formation. Information of the effects of these factors on statospore morphology is included in Sandgren (1983b); temperature effects on encystment are more thoroughly discussed in Sandgren (1986).

real basis for deciding what "adequate" for potential recruitment means from the benthic cyst populations. Many factors related to viability and survivorship remain to be ascertained. For instance, it is questionable whether statospores deposited onto different regions of the lake sedimentary surface are equally viable through time. Shallow littoral sediments provide a lighted, oxic environment, but also one subject to great bioturbation and potential burial or ingestion of surficial cyst populations. Deepwater anoxic and highly reducing sedimentary environments, on the other hand, may greatly reduce survivorship or eliminate required excystment cues (but see later discussion). For chrysophytes, as for essentially all other microalgae with benthic cyst stages, there is clearly a great deal we do not know about the requirements for successful alternation between benthic and planktonic habitats.

Statospore Survivorship and Germination Requirements

The only available experimental information on statospore survivorship is derived from studies of sexual *Dinobryon* statospores produced in laboratory cultures and then stored for varying lengths of time under oxic conditions in the dark at 4°C. Both the effect of genetic or clonal composition of the statospores and the effect of environmental temperature during encystment on survivorship were evaluated (Fig. 2-23). Both of these variables had an impact on the initial percentage of potential viable, mature statospores (t_0, Fig. 2-23), but neither had an effect on survivorship because the lines are all roughly parallel. These results demonstrate that viable statospores do remain after more than one year storage in such storage conditions, and they suggest that small percentages of these stored cysts can germinate. Observations by Cronberg (1982) on blooms of chrysophytes generated from benthic cyst populations in Lake Trummen after the superficial sediments were removed by dredging suggest that statospores can remain viable for decades in sediments that are periodically anerobic. Statospores thus appear to have more than sufficient survivorship potential to serve as a perennation mechanism for annually recurring, seasonally restricted species.

A preliminary report of germination requirements for *Dinobryon* statospores produced in culture was previously published (Sandgren 1986), but results from a more extensive matrix of germination environments are reported here. The importance of light, exogenous nutrients, and the germination temperature regime were tested on sexual statospores from two different clonal crosses produced at 15°C and stored for 200 days at 4°C (Table 2-5). The percentage of statospores germinating was relatively low for both populations. There was clearly

Table 2-5. *Germination requirements of Dinobryon cylindricum zygotic statospores.*

Germination temperature	Clones 5 × 7 Light/Dark			Clones 5 × 13 Light/Dark		
	Control	− Nutrients	+ Nutrients	Control	− Nutrients	+ Nutrients
5	5/0	6/2	7/0	13/7	0/11	15/17
10	8/0	0/0	28/0	17/8	53/14	61/25
15	0/0	1/0	7/0	33/0	54/0	42/0
20	5/0	3/0	0/0	21/0	20/0	24/0

Statospores stored 204 days at 4 C° in the dark. Data expressed as percentage germinating of 300 cysts counted after 7 days in germination conditions.

a clonal effect on ability to germinate, with statospores of clones 5 × 13 being capable of much higher success rates. Neither nutrient levels nor germination temperature regime had a consistent effect on ability to germinate. Surprisingly, light promoted germination, but it was not necessary. Statospores of both populations germinated in the dark at low temperatures.

If these results are applicable to natural populations, then chrysophytes may not use a distinctive set of environmental cues to initiate germination. Instead, some small number of statospores may be excysting at any time, and establishment of vegetative cell populations in the plankton would therefore depend upon the germinating cell encountering conditions favorable for vegetative growth. This explanation is in harmony with the observed slow decline in viable statospores observed in the survivorship studies. Such a recruitment strategy is extremely opportunistic in that it provides a mechanism by which chrysophytes can persist by capitalizing on brief windows of time in the plankton during which competitive or grazing stress is temporarily reduced.

If this mechanism of recruitment from benthic cyst populations is accurate, it suggests that the seasonal periodicity patterns of chrysophytes are largely dependent on the growth and loss rates of the vegetative populations as dictated by ambient rate-limiting processes; benthic recruitment only provides the *potential* for vegetative growth but does not dictate the seasonal pattern. Therefore, there should be no strict seasonal distribution pattern for chrysophytes that holds true across a broad spectrum of lakes, a prediction that is certainly supported by the observed distribution (see Distribution section). Survivorship estimates and germination dynamics for statospores in natural sedimentary habitats are critically needed to confirm or dispute this proposed repopulation strategy.

CONCLUSION: THE CHRYSOPHYTE STRATEGY

Despite their broad diversity of cell size, colonial morphology, and cell coverings, planktonic chrysophytes are predominantly flagellate algae with a shared phylogenetic background and, thus, similar nutritional and reproductive potential. So long as proper consideration is given to size-dependent characteristics, it is therefore reasonable to compare chrysophytes to other groups of freshwater planktonic algae with regard to distribution patterns, tolerance limits, limiting factors, and reproductive dynamics. On the basis of the information presented here, it is possible to formulate some generalizations concerning the ecology of planktonic chrysophyte algae that help to delineate their "survival strategy."

Chrysophytes are nutritionally opportunistic, possibly being able to switch between autotrophic, heterotrophic, and phagotrophic modes in response to cell requirements and environmental conditions. In contrast to some classically held beliefs, their temperature, light, and mineral nutrient requirements are generally similar to those of other planktonic algae. Exceptional characteristics that have been discussed are sensitivity to elevated potassium concentrations, possible lack of ability to secrete alkaline phosphatase, comparatively inefficient mechanism for sequestering dissolved iron in clear-water lakes, and perhaps an obligate requirement for dissolved free CO_2 as the only form of dissolved carbon used in photosynthesis. All these features require additional confirmation. An analysis of their ability to compete for phosphorus, the most frequent growth-limiting nutrient in freshwaters, suggests that they are superior to most other phytoplankton species considered (common species of diatoms, cyanobacteria, and green flagellates) with regard to both their storage and growth potential under conditions of low phosphorus availability. They can thus be considered both storage- and growth-selected species. In addition, they have abilities to sequester phosphorus that are comparable to those of these same competing species. With these characteristics, chrysophytes should be proficient at growth under conditions of both homogeneous and patchy phosphorus supply. Furthermore, they appear to have a broad optimal range of N:P supply ratios that promote growth. All these nutritional characteristics suggest that chrysophytes should be very effective competitors with other phytoplankton species, except perhaps in extremely eutrophic or hardwater systems where absolute maximum growth rate and potential CO_2 limitation may come into play. On the basis of their opportunistic growth-promoting processes, they should therefore be very widely distributed.

The observed biogeographical distribution of planktonic chrysophytes, however, is much more restricted than these nutritional characteristics might suggest. Lake characteristics such as water temperature, water color, major ionic composition, and annual physical mixing cycles do not appear to directly influence chrysophyte distribution. But other characteristics such as lake trophic state, pH, and the bicarbonate buffering system seem to be very important. As a group, chrysophytes dominate only in lakes characterized by low productivity, low alkalinity and conductivity, and neutral to slightly acidic pH. The importance of chrysophytes in the plankton, but not always their actual biomass, tends to decrease dramatically as these highly intercorrelated factors increase in lakes along trophic gradients. Such a trend, together with the above nutritional comments, suggests that it is not the increased nutrient loading or intensified competition typical of productive lakes that limits chrysophyte populations, but something correlated with them. Elevated pH and subsequent decline in free CO_2 may be one such factor. Grazing by filter-feeding zooplankton seems to be another, and grazing losses may well be the single most important factor regulating chrysophyte biomass in lakes. Nannoplanktonic chrysophytes suffer severe grazing losses from all classes of filter feeders. Larger netplanktonic forms are able to avoid grazing losses and may actually profit nutritionally from grazing activities in situations dominated by small- or moderate-sized animals. But the increase in cladoceran animal size and feeding efficiency that accompanies increasing lake productivity appears to effectively eliminate chrysophytes of all sizes from productive lakes. Chemostat studies of mixed phytoplankton assemblages with and without grazers could test these hypotheses.

Seasonal periodicity of chrysophytes is influenced by the same gradient of productivity and associated factors. The periods of maximum chrysophyte development shift progressively away from the summer period of maximum overall phytoplankton development toward cooler seasons as productivity increases, and may shift to include summer metalimnetic populations in some instances. The typical demographic pattern of chrysophytes can be described as brief episodes of rapid opportunistic growth followed by subsequent rapid decline. This pattern seems to be characteristic of these algae even in chrysophyte-dominated lakes in which a rapid succession of dominant chrysophyte species is typically observed. Such dynamics are suggestive of the close interplay between their opportunistic potential for rapid, competitive growth and the cyclical onset of severe losses by grazing.

Possibly in response to these predictably brief episodes of vegetative growth, chrysophytes have evolved an equally opportunistic reproductive or perennation strategy. Apparently identical resistant resting cysts may be formed by either sexual or asexual processes. Sexual reproduction is heterothallic and is regulated by an endogenous, density-dependent mechanism rather than by exogenous environmental conditions, as is the case for most other groups of phytoplankton. This pattern of encystment, which is initiated by low-threshold cell densities, apparently ensures the production of resting cysts during even brief appearances of vegetative cell populations. Resting cyst survivorship characteristics are adequate to provide a seed source for populations that reappear once annually. Available evidence suggests that cyst germination requires no external cueing. Repopulation of the plankton depends upon the successful vegetative growth of the few new cells excysted at any given time and is thus primarily dependent on ambient conditions in the epilimnion rather than those near the sediment surface.

Craig D. Sandgren

The chrysophyte survival strategy as outlined here emphasizes the integrated importance of species morphology, vegetative growth and loss processes, and perennation mechanisms as determined by sexual reproduction. Although individual characteristics discussed occur in other groups of phytoplankton, particularly other flagellate groups, it seems clear that the strategic "package" is unique to the chrysophytes. The various attributes of chrysophyte biology discussed integrate into a comprehensive pattern that tells us more about the ecology of these algae than do the individual pieces considered separately. There are thus clear advantages to integrating phycological and limnological as well as proximate and ultimate perspectives into our view of phytoplankton ecology.

REFERENCES

Aaronson, S. (1973). Particle aggregation and phagotrophy by *Ochromonas. Arch. Mikrobiol.,* 92, 39–44.
  (1980). Descriptive biochemistry and physiology of the Chrysophyceae (with some comparisons to Prymnesiophyceae). In: *Biochemistry and Physiology of the Protozoa,* 2d ed., vol. 3, eds. M. Levandowski and S. H. Hutner, pp. 117–69. New York: Academic Press.
Aaronson, S. and Patni, N. J. (1976). The role of surface and extracellular phosphatases in the phosphorus requirement of *Ochromonas. Limnol. Oceanogr.,* 21, 838–45.
Adam, D. P. and Mahood, A. D. (1981). Chrysophyte cysts as potential environmental indicators. *Geol. Soc. Amer. Bull.,* 92, 839–44.
Ahlgren, G. (1970). Limnological studies of Lake Norrviken, a eutrophicated Swedish lake. II. Phytoplankton and its production. *Schweiz. Z. Hydrol.,* 32, 354–96.
  (1978). Response of phytoplankton and primary production to reduced nutrient loading in Lake Norrviken. *Verh. Internat. Verein. Limnol.,* 20, 840–5.
Ahlgren, I. (1967). Limnological studies of Lake Norrviken, a eutrophicated Swedish lake. I. Water chemistry and nutrient budget. *Schweiz. Z. Hydrol.,* 29, 53–90.
Ahlgren, I., Ramberg, L., Hansson, M., Lundgren, A., Lindstrom, K., Persson, G., and Pettersson, K. (1979). Lake metabolism studies and results at the Institute of Limnology in Uppsala. *Arch. Hydrobiol. Beih.,* 13, 10–30.
Alexander, V., Stanley, D. W., Daley, R. J., and McRoy, C. (1980). Primary producers. In: *Limnology of Tundra Ponds. Barrow, Alaska,* US/IBP Synthesis Series, vol. 13, ed. J. E. Hobbie, Stroudsburg, PA: Dowden, Hutchinson & Ross.
Allen, M. B. (1969). Structure, physiology and biochemistry of the Chrysophyceae. *Ann. Rev. Microbiol.,* 23, 29–46.

Andersen, R. A. (1987). Synurophyceae classis nov., a new class of algae. *Amer. J. Bot.,* 74, 337–53.

Armstrong, F. A. J. and Schindler, D. W. (1971). Primary chemical characterization of waters in the Experimental Lakes Area, northwestern Ontario. *Can. J. Fish. Aq. Sci.,* 28, 171–87.

Arvola, L. (1983). Primary production and phytoplankton in two small, polyhumic forest lakes in southern Finland. *Hydrobiologia,* 101, 105–10.

(1984). Diel variation in primary production and the vertical distribution of phytoplankton in a polyhumic lake. *Arch. Hydrobiol.,* 101, 503–19.

(1985). On the factors affecting phytoplankton species composition in highly coloured small forest lakes. *Lammi Notes,* 12, 1–4.

Arvola, L. and Rask, M. (1984). Relations between phytoplankton and environmental factors in a small, spring-meromictic lake in southern Finland. *Aqua Fenn.,* 14, 129–38.

Asmund, B. (1955). Electron microscope observations of *Mallomonas caudata* and some remarks on its occurrence in four Danish ponds. *Bot. Tidsskr.,* 52, 163–8.

Bannister, T. T. and Bubeck, R. C. (1978). Limnology of Irondequoit Bay, Monroe County, New York. In: *Lakes of New York State,* vol. II, ed. J. A. Bloomfield, pp. 106–223. New York: Academic Press.

Barica, J. (1975). Geochemistry and nutrient regime of saline eutrophic lakes in the Erickson-Elphinstone District of southwestern Manitoba. *Can. Fish. Mar. Serv. Tech. Rept.* 511.

Bergquist, A. M. and Carpenter, S. R. (1986). Limnetic herbivory: effects on phytoplankton populations and primary production. *Ecology,* 67, 1351–60.

Bergquist, A. M., Carpenter, S. R., and Latino, J. C. (1985). Shifts in phytoplankton size structure and community composition during grazing by contrasting zooplankton assemblages. *Limnol. Oceanogr.,* 30, 1037–45.

Berman, B. and Rodhe, W. (1971). Distribution and migration of *Peridinium* in Lake Kinneret. *Mitt. Internat. Verein. Limnol.,* 19, 266–76.

Bird, D. F. and Kalff, J. (1986). Bacterial grazing by planktonic lake algae. *Science,* 231, 493–5.

(1987). Algal phagotrophy: regulating factors and importance relative to photosynthesis in *Dinobryon* (Chrysophyceae). *Limnol. Oceanogr.,* 32, 277–84.

Blaauboer, M. C. I. (1982). The phytoplankton species composition and the seasonal periodicity in Lake Vechten. *Hydrobiologia,* 95, 25–36.

Blaauboer, M. C. I., Van Keulen, R., and Cappenberg, Th. E. (1982). Extracellular release of photosynthetic products by freshwater phytoplankton populations, with special reference to the algal species involved. *Freshwater Biol.,* 12, 559–72.

Bold, H. C. and Wynne, M. J. (1985). *Introduction to the Algae: Structure and Reproduction,* 2d ed. Englewood Cliffs, NJ: Prentice-Hall.

Boraas, M., Estep, K. W., Johnson, P. W., and Sieburth, J. McN. (In press). Phagotrophic phototrophs: the ecological significance of mixotrophy. *J. Protozool.*

Bretthauer, R. (1975). Laboratory experiments to determine the limits of tolerance of flagellates to some abiotic factors. *Verh. Internat. Verein. Limnol.,* 19, 2043–50.

Brock, T. D. (1985). *A Eutrophic Lake: Lake Mendota, Wisconsin.* New York: Springer-Verlag.

Brook, A. J. (1959). The status of desmids in the plankton and the determination of phytoplankton quotients. *J. Ecol.,* 47, 429–45.

Brook, A. J., Baker, A. L., and Klemer, A. R. (1971). The use of turbidimetry in studies of the population dynamics of phytoplankton populations with special reference to *Oscillatoria agardhii* var. *isothrix. Mitt. Internat. Verein. Limnol.,* 19, 244–52.

Brooks, A. S. and Torke, B. G. (1977). Vertical and seasonal distribution of chlorophyll *a* in Lake Michigan. *J. Fish. Res. Bd. Can.,* 34, 2280–7.

Burgis, M. J., Darlington, J. P. E. C., Dunn, I. G., Ganff, G. G., Gwahaba, J. J., and McGowan, L. M. (1973). The biomass and distribution of organisms in Lake George, Uganda. *Proc. R. Soc. Lond. B.,* 184, 271–98.

Burns, C. W. and Gilbert, J. J. (1986). Direct observations of the mechanisms of interference between *Daphnia* and *Keratella cochlearis. Limnol. Oceanogr.,* 31, 869–66.

Carney, H. J. and Sandgren, C. D. (1983). Chrysophycean cysts: indicators of eutrophication in the recent sediments of Frains Lake, Michigan, U.S.A. *Hydrobiologia,* 101, 195–202.

Carpenter, S. R. and Kitchell, J. F. (1984). Plankton community structure and limnetic primary production. *Am. Nat.,* 124, 159–72.

Cloern, J. E. (1977). Effects of light and temperature on *Cryptomonas ovata* (Cryptophyceae) growth and nutrient uptake rates. *J. Phycol.,* 13, 389–95.

Coveney, M. F., Cronberg, G., Enell, M., Larsson, K. and Olofsson, L. (1977). Phytoplankton, zooplankton and bacteria – standing crop and production relationships in a eutrophic lake. *Oikos,* 29, 5–21.

Cronberg, G. (1973). Development of cysts in *Mallomonas eoa* examined by scanning electron microscopy. *Hydrobiologia,* 43, 29–38.

    (1982). Phytoplankton changes in Lake Trummen induced by restoration. *Folia Limnol. Scand.,* 18, 1–119 + appendixes.

    (1986). Chrysophyceaen cysts and scales in lake sediments: a review. In: *Chrysophytes: Aspects and Problems,* eds. J. Kristiansen and R. A. Anderson, pp. 281–316. New York: Cambridge University Press.

Crumpton, W. G. and Wetzel, R. G. (1982). Effects of differential growth and mortality in the seasonal succession of phytoplankton populations in Lawrence Lake, Michigan. *Ecology,* 63, 1729–39.

DeHaan, H., DeBoer, T., and Hoogveld, H. L. (1981). Metal binding capacity in relation to hydrology and algal periodicity in Tjeukemeer, The Netherlands. *Arch. Hydrobiol.,* 92, 11–23.

DeHaan, H., DeBoer, T., Voerman, J., Kramer, H. A., and Moed, J. R. (1984). Size classes of "dissolved" nutrients in shallow, alkaline, humic, and eutrophic Tjeukemeer, The Netherlands, as fractionated by ultrafiltration. *Verh. Internat. Verein. Limnol.,* 22, 876–81.

DeHaan, H., Veldhuis, M. J. W., and Moed, J. R. (1985). Availability of dissolved iron from Tjeukemeer, The Netherlands, for iron-limited growing *Scenedesmus quadricuada. Water Res.,* 19, 235–9.

DeMott, W. R. and Kerfoot, W. C. (1982). Competition among cladocerans: nature of the interaction between *Bosmina* and *Daphnia. Ecology,* 63, 1949–66.

Dokulil, M. and Skolaut, C. (1986). Succession of phytoplankton in a deep stratifying lake: Mondsee, Austria. *Hydrobiologia,* 138, 9–24.

Droop, M. R. (1974). Nutrient status of algal cells in continuous cultures. *J. Mar. Biol. Assoc. U.K.,* 54, 541–55.

Dürrschmidt, M. (1987). Fine structure, taxonomy and distribution of silica-scaled Chrysophyceae. In: *Phykotalk,* vol. II, ed. E. Kumar. India: University of Varawasi Press.

Earle, J. C., Duthie, H. C., and Scruton, D. A. (1986). Analysis of the phytoplankton composition of 95 Labrador lakes, with special reference to natural and anthropogenic acidification. *Can. J. Fish. Aquat. Sci.,* 43, 1804–11.

Eaton, S. W. and Kardos, L. P. (1978). The limnology of Canadaigua Lake. In: *Lakes of New York,* vol. I, ed. J. A. Bloomfield, pp. 226–312. New York: Academic Press.

Eloranta, P. (1974). Studies on the phytoplankton in Lake Keurusselka, Finnish Lake District. *Ann Bot. Fennici.,* 11, 13–24.

(1980). Winter phytoplankton in a pond warmed by a thermal power station. *Ann. Bot. Fennici.,* 17, 264–75.

(1986). Phytoplankton structure in different lake types in central Finland. *Holarct. Ecol.,* 9, 214–24.

Eloranta, P. and Palomaki, A. (1986). Phytoplankton in Lake Konnevesi with special reference to eutrophication of the lake by fish farming. *Aqua Fennica,* 16, 37–45.

Elrifi, I. R. and Turpin, D. H. (1985). Steady-state luxury consumption and the concept of optimum nutrient ratios: a study with phosphate and nitrate limited *Selenastrum minutum* (Chlorophyta). *J. Phycol.,* 21, 592–602.

Elser, J. J., Elser, M. M., and Carpenter, S. R. (1986). Size fractionation of algal chlorophyll, carbon fixation and phosphatase activity: relationships with species-specific size distributions and zooplankton community structure. *J. Plank. Res.,* 8, 365–83.

Fahnensteil, G. L. and Glime, J. M. (1983). Subsurface chlorophyll maximum and associated *Cyclotella* pulse in Lake Superior. *Int. Revue ges. Hydrobiol.,* 68, 605–16.

Fee, E. J. (1976). The vertical and seasonal distribution of chlorophyll in lakes of the Experimental Lakes Area, northwestern Ontario: implications for primary production estimates. *Limnol. Oceanogr.,* 21, 767–83.

(1978). Studies of hypolimnion chlorophyll peaks in the Experimental Lakes Area, northwestern Ontario. Can. Fish. Mar. Serv. Tech. Rept. 754.

Fee, E. J., Shearer, J. A., and DeClercq, D. R. (1977). In vivo chlorophyll profiles from lakes in the Experimental Lakes Area, northwestern Ontario. Can. Fish. Mar. Ser. Tech. Rpt. 703.

Fenchel, T. (1986). The ecology of heterotrophic microflagellates. In: *Advances in Microbial Ecology*, vol. 9, ed. K. C. Marshall, pp. 57–97. New York: Plenum Press.

Findenegg, I. (1943). Untersuchungen über die Ökologie und die Produktions verhältnisse des Planktons im Kärntner Seengebiet. *Int. Revue ges. Hydrobiol.*, 43, 368–429.

Findlay, D. L. (1984). Effects on phytoplankton biomass, succession and composition in Lake 223 as a result of lowering pH levels from 5.6 to 5.2. Data from 1980 to 1982. Can. Manu. Rept. Fish. Aquat. Sci. 1761.

Findlay, D. L. and Kling, H. J. (1975). Seasonal successions of phytoplankton in Seven Lake Basins in the Experimental Lakes Area Northwestern Ontario following artificial eutrophication. Can. Fish. Mar. Serv. Tech. Rept. 513.

Findlay, D. L. and Saesura, G. (1980). Effects on phytoplankton biomass, succession and composition in Lake 223 as a result of lowering pH levels from 7.0 to 5.6. Data from 1974 to 1979. Can. Manu. Rept. Fish. Aquat. Sci. 1585.

Forsberg C. and Ryding, S-O. (1980). Eutrophication parameters and trophic state indices in 30 Swedish waste-receiving lakes. *Arch. Hydrobiol.*, 89, 189–207.

Ganf, G. G. (1974). Phytoplankton biomass and distribution in a shallow eutrophic lake (Lake George, Uganda). *Oecologia*, 16, 9–29.

Geelen, J. F. M. and Leuven, R. S. E. W. (1986). Impact of acidification on phytoplankton and zooplankton communities. *Experimentia*, 42, 486–94.

Goldman, C. R. (1972). The role of minor nutrients in limiting the productivity of aquatic ecosystems. In: *Nutrient and Eutrophication,* ed. G. E. Likens, pp. 21–44. Lawrence, Kansas: American Society of Limnology and Oceanography Special Symposium No. 1.

Goldman, J. C. and Carpenter, E. J. (1974). A kinetic approach to the effect of temperature on algal growth. *Limnol. Oceanogr.*, 19, 756–66.

Gorham, E., Dean, W. E., and Sanger, J. E. (1983). The chemical composition of lakes in the north-central United States. *Limnol. Oceanogr.*, 28, 287–301.

Gotham, I. J. and Rhee, G-Y. (1981). Comparative kinetic studies of phosphate-limited growth and phosphate uptake in phytoplankton in continuous culture. *J. Phycol.*, 17, 257–65.

Gusseva, K. A. (1935). The conditions of mass development and physiology of nutrition of *Synura. Mikrobiol.*, 4, 44–55. (English summary)

Happey, C. and Moss, B. (1967). Some aspects of the biology of *Chrysococcus diaphanus* in Abbott's Pond, Somerset. *Br. Phycol. Bull.*, 3, 269–79.

Happey-Wood, C. M. (1976). Vertical migration patterns in phytoplankton of mixed species composition. *Br. Phycol. J.*, 11, 355–69.

Harris, G. P. (1986). *Phytoplankton Ecology: Structure, Function and Fluctuation*. New York: Chapman and Hall.

Harrison, P. J. and Turpin, D. H. (1982). The manipulation of physical, chemical and biological factors to select from natural phytoplankton communities. In: *Marine Mesocosms: Biological and Chemical Research in*

Experimental Ecosystems, eds. G. D. Grice and M. R. Reeve, pp. 275–89. New York: Springer-Verlag.

Healey, F. P. (1973). Inorganic nutrient uptake and deficiency in algae. *CRC Crit. Rev. Microbiol.,* 3, 69–113.

(1980). Slope of the Monod equation as an indicator of advantage in nutrient competition. *Microb. Ecol.,* 5, 281–6.

(1983). Effect of temperature and light intensity on the growth of *Synura sphagnicola. J. Plank. Res.,* 5, 767–73.

Healey, F. P. and Hendzel, L. L. (1979). Fluorometric measurement of alkaline phosphatase activity in algae. *Freshwat. Biol.,* 9, 429–39.

Heaney, S. I. and Talling, J. F. (1980). *Ceratium hirundinella* – ecology of a complex, mobile and successful plant. Freshwat. Biol. Ass. Ann. Rept., 48, 27–40.

Hecky, R. E. (1984). African lakes and their trophic efficiencies: a temporal perspective. In: *Trophic Interactions within Aquatic Ecosystems,* eds. D. G. Meyers and J. R. Strickler, pp. 405–48. AAAS Selected Symposium 85, American Association for the Advancement of Science. Boulder, CO: Westview Press.

Hecky, R. E. and Kling, H. J. (1981). The phytoplankton and protozooplankton of the euphotic zone of Lake Tanganyika: species composition, biomass, chlorophyll content and spatio-temporal distribution. *Limnol. Oceanogr.,* 26, 548–64.

Hegewald, E., Hesse, M., and Jeeji-Bai, N. (1981). Ecological and physiological studies on plankton algae from certain Hungarian waters. *Arch. Hydrobiol. Suppl.,* 60, 172–201.

Heinonen, P. (1980). Quantity and composition of phytoplankton in Finnish inland waters. *Publ. Water. Res. Inst. Finland,* 37, 1–91.

Heynig, H. (1967). Beitrage zur Taxonomie und Ökologie der Gattung *Chrysococcus* Klebs (Chrysophyceae). *Arch. Protistenk.,* 110, 259–79.

Holmgren, S. K. (1984). Experimental lake fertilization in the Kuokkel Area, northern Sweden. Phytoplankton biomass and algal composition in natural and fertilized subarctic lakes. *Int. Revue ges. Hydrobiol.,* 69, 781–817.

(1985). Phytoplankton in a polluted subarctic lake before and after nutrient reduction. *Water. Res,* 19, 63–71.

Hörnström, E. (1981). Trophic characterization of lakes by means of qualitative phytoplankton analysis. *Limnologica,* 13, 249–61.

Hutchinson, G. E. (1944). Limnological studies in Connecticut. 7. A critical examination of the supposed relationship between phytoplankton periodicity and chemical changes in lake waters. *Ecology,* 25, 3–25.

(1957). *A Treatise on Limnology. Vol. 1. Geography, Physics and Chemistry.* New York: Wiley.

(1967). *A Treatise on Limnology. Vol. 2. Introduction to Lake Biology and the Limnoplankton.* New York: Wiley.

Ilmavirta, V. (1974). Diel periodicity in the phytoplankton community of the oligotrophic Lake Pääjärvi, southern Finland. I. Phytoplankton primary production and related factors. *Ann. Bot. Fennici,* 11, 136–77.

(1980). Phytoplankton in 35 Finnish brown water lakes of different trophic status. In: *Developments in Hydrobiology*, vol. 3, eds. M. Dokulil, H. Metz, and D. Jewson, pp. 121–30. The Hague: Dr. W. Junk.

(1982). Dynamics of phytoplankton in Finnish lakes. *Hydrobiologia*, 86, 11–20.

(1983). The role of flagellated phytoplankton in chains of small brownwater lakes in southern Finland. *Ann. Bot. Fennici*, 20, 187–95.

(1984). The ecology of flagellated phytoplankton in brown-water lakes. *Verh. Internat. Verein. Limnol.*, 22, 817–21.

Ilmavirta, V., Ilmavirta, K., and Kotimaa, A. L. (1974). Phytoplankton primary production during the summer stagnation in the eutrophicated lakes Lovojärvi and Ormajärvi, southern Finland. *Ann. Bot. Fennici*, 11, 121–32.

Ilmavirta, V. and Kotimaa, A. L. (1974). Spatial and seasonal variations in the phytoplanktonic primary production and biomass in the oligotrophic Lake Pääjärvi, southern Finland. *Ann. Bot. Fennici*, 11, 112–20.

Ishida, Y., Kimura, B., Nakahara, H., and Kadota, H. (1982). Analysis of major nutrient effecting *Uroglena americana* bloom in the northern Lake Biwa, by use of algal bioassay. *Bull. Jap. Soc. Sci. Fish.*, 48, 1281–7.

Ito, H. and Takahashi, E. (1982). Seasonal fluctuation of *Spiniferomonas* (Chrysophyceae, Synuraceae) in two ponds on Mt. Rokko, Japan. *Jap. J. Phycol.*, 30, 272–8.

Jackson, T. A. and Hecky, R. E. (1980). Depression of primary productivity by humic matter in lake and reservoir waters of the boreal forest zone. *Can. J. Fish. Aquat. Sci.*, 37, 2300–17.

Janus, L. L. and Duthie, H. C. (1979). Phytoplankton and primary production of lakes in the Metamek Watershed, Quebec. *Int. Revue ges. Hydrobiol.*, 64, 89–98.

Järnefelt, H. (1952). Plankton als Indikator der Trophiegruppen der Seen. *Ann. Acad. Scient. Fenn. Ser. A. IV, Biol.*, 18, 1–29.

Jónasson, P. M. (1972). Ecology and production of the profundal benthos in relation to phytoplankton in Lake Esrom. *Oikos Suppl.* 14.

Jónasson, P. M. and Adalsteinsson, H. (1979). Phytoplankton production in shallow eutrophic Lake Mývatn, Iceland. *Oikos*, 32, 113–38.

Jónasson, P. M. and Kristiansen, J. (1967). Primary and secondary production in Lake Esrom. Growth of *Chironomus anthracinus* in relation to seasonal cycles of phytoplankton and dissolved oxygen. *Int. Revue ges. Hydrobiol.*, 52, 163–217.

Kalff, J. (1967). Phytoplankton abundance and primary production rates in two arctic ponds. *Ecology*, 48, 558–65.

Kalff, J. Kling, H. J., Holmgren, S. H., and Welch, H. E. (1975). Phytoplankton, phytoplankton growth and biomass cycles in an unpolluted and in a polluted polar lake. *Verh. Internat. Verein. Limnol.*, 19, 487–95.

Kalff, J. and Knoechel, R. (1978). Phytoplankton dynamics in oligotrophic and eutrophic lakes. *Ann. Rev. Ecol. Syst.*, 9, 475–95.

Kalff, J. and Watson, S. (1986). Phytoplankton and its dynamics in two tropical lakes: a tropical and temperate zone comparison. *Hydrobiologia,* 138, 161–76.

Kennedy, K. C. (1984). The influence of cell size and the colonial habit on phytoplankton growth and nutrient uptake kinetics: a critical evaluation using a phylogenetically related series of volvocene green algae. M.S. Thesis, University of Texas-Arlington.

Kerfoot, W. C., DeMott, W. R., and DeAngelis, D. L. (1985). Interactions among cladocerans: food limitation and exploitive competition. *Arch. Hydrobiol. Beih.,* 21, 431–52.

Keskitalo, J. (1977). The species composition and biomass of phytoplankton in the eutrophic Lake Lovojärvi, southern Finland. *Ann. Bot. Fennici,* 14, 71–81.

Kiefer, D. A., Holm-Hansen, O., Goldman, C. R., Richards, R., and Berman, T. (1972). Phytoplankton in Lake Tahoe: deep-living populations. *Limnol. Oceanogr.,* 17, 418–22.

Kies, L. and Berndt, M. (1984). Die *Synura*-Arten (Chrysophyceae) Hamburgs und seiner nordöstlichen Umgebund. *Mitt. Inst. Allg. Bot. Hamburg,* 19, 99–122.

Kilham, S. S. (1978). Nutrient kinetics of freshwater planktonic algae using batch and semicontinuous methods. *Mitt. Internat. Verein. Limnol.,* 21, 147–57.

(1984). Silicon and phosphorus growth kinetics and competitive interactions between *Stephanodiscus minutus* and *Synedra* sp. *Verh. Internat. Verein. Limnol.,* 22, 435–9.

Kilham, P. and Kilham, S. S. (1980). The evolutionary ecology of phytoplankton. In: *The Physiological Ecology of Phytoplankton,* Studies in Ecology, vol. 7, ed. I. Morris, pp. 571–97. London: Blackwell Scientific.

(1984). The importance of resource supply rates in determining phytoplankton community structure. In: *Trophic Interactions within Aquatic Ecosystems,* eds. D. G. Meyers and J. R. Strickler, pp. 7–27. AAAS Selected Symposium 85, American Association for the Advancement of Science. Boulder, CO: Westview Press.

Kling, H. (1975). Phytoplankton successions and species distribution in prairie ponds of the Erickson-Elphinstone District, southwestern Manitoba. Can. Fish. Mar. Serv. Tech. Rept. 512.

Kling, H. and Holmgren, S. K. (1972). Species composition and seasonal distribution in the experimental lakes area, northwestern Ontario. Can. Fish. Mar. Serv. Tech. Rept. 337.

Knowles, S. C. and Zingmark, R. G. (1978). Mercury and temperature interactions on the growth rates of three species of freshwater phytoplankton. *J. Phycol.,* 14, 104–9.

Kristiansen, J. (1975). On the occurrence of the species of *Synura* (Chrysophyceae). *Verh. Internat. Verein. Limnol.,* 19, 2709–15.

Kristiansen, J. (1986a). Identification, ecology and distribution of silica-scale-bearing Chrysophyceae, a critical approach. In: *Chrysophytes – Aspects and Problems,* eds. J. Kristiansen and R. A. Andersen, pp. 229–39. New York: Cambridge University Press.

Kristiansen, J. (1986b). Silica-scale bearing chrysophytes as environmental indicators. *Br. Phycol. J.,* 21, 425–36.

Lampert, W. (1978). Climatic conditions and planktonic interactions as factors controlling the regular succession of spring algal bloom and extremely clear water in Lake Constance. *Verh. Internat. Verein. Limnol.,* 20, 969–74.

Langeland, A. and Reinertsen, H. (1982). Interactions between phytoplankton and zooplankton in a fertilized lake. *Holarct. Ecol.,* 5, 253–72.

Lehman, J. T. (1976). Ecological and nutritional studies on *Dinobryon* Ehrenb.: seasonal periodicity and the phosphate toxicity problem. *Limnol. Oceanogr.,* 21, 646–58.

 (1978). Enhanced transport of inorganic carbon into algal cells and its implication for the biological fixation of carbon. *J. Phycol.,* 14, 33–42.

 (1980). Release and cycling of nutrients between planktonic algae and herbivores. *Limnol. Oceanogr.,* 25, 620–32.

Lehman, J. T. and Sandgren, C. D. (1985). Species-specific rates of growth and grazing loss among freshwater algae. *Limnol. Oceanogr.,* 30, 34–46.

Lewin, R. A. (1962). *Physiology and Biochemistry of Algae.* New York: Academic Press.

Lewis, W. M. (1977). Net growth rate through time as an indicator of ecological similarity among phytoplankton species. *Ecology,* 58, 149–57.

 (1978). Dynamics and succession of the phytoplankton in a tropical lake: Lake Lanao, Philippines. *J. Ecol.,* 66, 849–80.

 (1986). Phytoplankton succession in Lake Valencia, Venezuela. *Hydrobiologia,* 138, 189–203.

Li, W. K. W. (1980). Temperature adaptation in phytoplankton: cellular and photosynthetic characteristics. In: *Primary Productivity in the Sea,* vol. 19, Environmental Science Research Series, ed. P. G. Falkowski, pp. 259–79. New York: Plenum Press.

Lucas, W. J. (1983). Photosynthetic assimilation of exogenous HCO_3^- by aquatic plants. *Ann. Rev. Plant Physiol.,* 34, 71–104.

Lui, N. S. T. and Roels, O. A. (1970). Nitrogen metabolism of aquatic organisms. I. The assimilation and formation of urea by *Ochromonas danica. Arch. Biochem. Biophys.,* 139, 269–77.

Lund, J. W. G. (1965). The ecology of freshwater phytoplankton. *Biol. Rev.,* 40, 231–93.

Mackereth, F. J. H., Heron, J., and Talling, J. F. (1978). *Water Analysis: Some Revised Methods for Limnologists.* Scientific Publication No. 36. Windermere Laboratory, Ambleside, Cumbria: Freshwater Biological Association.

Maitland, P. S. (1981). *The Ecology of Scotland's Largest Lochs: Lomond, Awe, Ness, Morar and Shiel.* The Hague: Dr. W. Junk.

Marcus, Y., Volokita, M., and Kaplan, A. (1984). The location of the transporting system for inorganic carbon and the nature of the form translocated in *Chlamydomonas reinhardtii. J. Exper. Bot.,* 35, 1136–44.

Margalef, R. (1967). Some concepts relative to the organization of plankton. *Oceanogr. Mar. Biol. Ann. Rev.,* 5, 257–89.

(1978). Life-forms of phytoplankton as survival alternatives in an unstable environment. *Oceanol. Acta,* 1, 493–509.

Mayer, J. R., Barnard, W. M., Metzger, W. J., Storch, T. A., Erlandson, T. A., Luensman, J. P., Nicholson, S. A., and Smith, R. I. (1978). Chautauqua Lake – watershed and lake basins. In: *Lakes of New York State,* vol. II., ed. J. A. Bloomfield, pp. 2–105. New York: Academic Press.

McCauley, E. and Kalff, J. (1981). Empirical relationships between phytoplankton and zooplankton biomass in lakes. *Can. J. Fish. Aquat. Sci.,* 38, 458–63.

Mechling, J. A. and Kilham, S. S. (1982). Temperature effects on silicon limited growth of the Lake Michigan diatom *Stephanodiscus minutus* (Bacillariophyceae). *J. Phycol.,* 18, 199–205.

Mills, E. L. (1975). Phytoplankton composition and comparative limnology of four Finger Lakes, with emphasis on lake topology. Ph.D. Thesis. Ithaca, New York: Cornell University.

Mills, E. L., Forney, J. L., Clady, M. D., and Schaffner, W. R. (1978). Oneida Lake. In: *Lakes of New York State,* vol. II., ed. J. A. Bloomfield, pp. 367–455. New York: Academic Press.

Moed, J. R. and Hoogveld, H. L. (1982). The algal periodicity in Tjeukemeer during 1968–1978. *Hydrobiologia,* 95, 223–34.

Moll, R. A., Brahce, M. Z., and Peterson, T. P. (1984). Phytoplankton dynamics within the subsurface chlorophyll maximum of Lake Michigan. *J. Plank. Res.,* 6, 751–66.

Moll, R. A. and Stoermer, E. F. (1982). A hypothesis relating trophic status and subsurface chlorophyll maxima in lakes. *Arch. Hydrobiol.,* 94, 425–40.

Moore, J. W. (1979). Factors influencing the diversity, species composition and abundance of phytoplankton in twenty one arctic and subarctic lakes. *Int. Revue. ges. Hydrobiol.,* 64, 485–99.

Morgan, K. C. and Kalff, J. (1979). Effect of light and temperature interactions on growth of *Cryptomonas erosa* (Cryptophyceae). *J. Phycol.,* 15, 127–34.

Moss, B. (1972). The influence of environmental factors on the distribution of freshwater algae: an experimental study. I. Introduction and the influence of calcium concentration. *J. Ecol.,* 60, 917–32.

(1973a). The influence of environmental factors on the distribution of freshwater algae: an experimental study. III. Effects of temperature, vitamin requirements and inorganic nitrogen compounds on growth. *J. Ecol.,* 61, 179–92.

(1973b). The influence of environmental factors on the distribution of freshwater algae: an experimental study. II. The role of pH and the carbon dioxide-bicarbonate system. *J. Ecol.,* 61, 157–77.

(1979). Algal and other fossil evidence for major changes in the Strumpshaw Broad, Norfolk, England in the last two centuries. *Br. Phycol. J.,* 14, 263–83.

Munawar, M. and Munawar, I. F. (1981). A general comparison of the taxonomic composition and size analysis of the phytoplankton of the North American Great Lakes. *Verh. Internat. Verein. Limnol.,* 21, 1695–716.

(1986). The seasonality of phytoplankton in the North American Great Lakes, a comparative synthesis. *Hydrobiologia,* 138, 83–115.

Munawar, M., Munawar, I. F., Culp, L. R., and Dupuis, G. (1978). Relative importance of nannoplankton in Lake Superior biomass and community metabolism. *J. Great Lakes Res.* 4: 462–480.

Munch, C. S. (1972). An ecological study of the planktonic chrysophytes of Hall Lake, Washington. Ph.D. Thesis. Seattle: University of Washington.

(1980). Fossil diatoms and scales of Chrysophyceae in the recent history of Hall Lake, Washington. *Freshwat. Biol.,* 10, 61–6.

Murphy, T. P., Lean, D. R., and Nalewajko, C. (1976). Blue-green algae: their excretion of iron-selective chelators enables them to dominate other algae. *Science,* 192, 900–2.

Nalewajko, C. (1977). Extracellular release in freshwater algae and bacteria: extracellular products of algae as a source of carbon for heterotrophs. In: *Aquatic Microbial Communities,* ed. J. Cairns, Jr., pp. 589–625. New York: Garland.

Nauwerck, A. (1963). Die Beziehungen zwischen Zooplankton und Phytoplankton in See Erken. *Symb. Bot. Upsal.,* 17(5).

(1968). Das Phytoplankton des Latnjajaure 1954–1965. *Schweiz. Z. Hydrol.,* 30, 188–216.

(1980). Die pelagische Primarproduktion im Latnjajaure, Schwedisch Lappland. *Arch. Hydrobiol. Suppl.,* 57, 291–323.

Nebaeus, M. (1984). Algal water-blooms under ice-cover. *Verh. Internat. Verein. Limnol.,* 22, 719–24.

Neill, W. E. (1984). Regulation of rotifer densities by crustacean zooplankton in an oligotrophic montane lake in British Columbia. *Oecologia,* 48, 164–77.

Nicholls, K. H. (1976). The phytoplankton of the Kawartha Lakes (1972 and 1976). Chapter 2. In: *Kawartha Lakes Management Study,* pp. 29–45. Toronto: Ontario Ministry of the Environment.

Nicholls, K. H., Carney, E. C., and Robinson, G. W. (1977). Phytoplankton of an inshore area of Georgian Bay, Lake Huron, prior to reductions in phosphorus loading. *J. Great Lakes Res.,* 3, 79–92.

Nicholls, K. H. and Dillon, P. J. (1978). An evaluation of phosphorus-chlorophyll-phytoplankton relationships for lakes. *Int. Revue ges. Hydrobiol.,* 63, 171–84.

Nicholls, K. H., Heintsch L., Carney, E., Beaver J., and Middleton, D. (1986). Some effects of phosphorus loading reductions on the phytoplankton in the Bay of Quinte, Lake Ontario. *Can. Spec. Publ. Fish. Aquat. Sci.* 86.

Nygaard, G. (1949). Hydrobiological studies on some Danish ponds and lakes. II. The quotient hypothesis and some new or little known phytoplankton organisms. *K. danske Vidensk. Selsk. Biol. Skr.,* 7, 1–293.

(1977). Vertical and seasonal distribution of some motile freshwater plankton algae in relation to some environmental factors. *Arch. Hydrobiol. Suppl.,* 51, 67–76.

Oglesby, R. T. (1977). Phytoplankton summer standing crop and annual production as functions of phosphorus loading and various physical factors. *J. Fish. Res. Bd. Can.,* 34, 2255–70.

Olsén, P. and Willén, E. (1980). Phytoplankton response to sewage reduction in Vättern, a large oligotrophic lake in central Sweden. *Arch. Hydrobiol.,* 89, 171–88.

Olsson, H. (1983). Origin and production of phosphatases in the acid Lake Gårdsjön. *Hydrobiologia,* 101, 49–58.

Ostrofsky, M. L. and Duthie, H. C. (1975). Primary productivity and phytoplankton of lakes on the Eastern Canadian Shield. *Verh. Internat. Verein. Limnol.,* 19, 732–8.

Paerl, H. W. (1982). Factors limiting productivity of freshwater ecosystems. In: *Advances in Microbial Ecology,* vol. 6, ed. K. C. Marshall, pp. 75–110. London: Plenum Press.

Patel, B. N. and Merrett, M. J. (1986). Inorganic carbon uptake by the marine diatom *Phaeodactylum tricornutum. Planta,* 169, 222–7.

Pavoni, M. (1963). Die Bedeutung des Nannoplanktons in Vergleich zum Netplankton. *Schweiz. Z. Hydrol.,* 25, 220–341.

Pearsall, W. H. (1932). Phytoplankton in the English lakes. The composition of the phytoplankton in relation to dissolved substances. *J. Ecol.,* 20, 241–61.

Pechlaner, R. (1971). Factors that control the production rate and biomass of phytoplankton in high-mountain lakes. *Mitt. Internat. Verein. Limnol.,* 19, 125–45.

Peterson, H. G., Healey, F. P., and Wagemann, R. (1984). Metal toxicity to algae: a highly pH dependent phenomenon. *Can. J. Fish. Aquat. Sci.,* 41, 974–9.

Pick, F. R. and Lean, D. R. S. (1984). Diurnal movements of metalimnetic phytoplankton. *J. Phycol.,* 20, 430–6.

Pick, F. R., Nalewajko, C., and Lean, D. R. S. (1984). The origin of a metalimnetic chrysophyte peak. *Limnol. Oceanogr.,* 29, 125–34.

Plinski, M. and Magnin, E. (1979). Analyse écologique du phytoplancton de trois lacs des Laurentides (Quebec, Canada). *Can. J. Bot.,* 57, 2791–9.

Pollingher, U. (1986). Phytoplankton periodicity in a subtropical lake (Lake Kinneret, Israel). *Hydrobiologia,* 138, 127–38.

Porter, K. G. (1976). Enhancement of algal growth and productivity by grazing zooplankton. *Science,* 192, 1332–4.

(In press). Phagotrophic phytoflagellates in microbial food webs. *Hydrobiologia.*

Porter, K. G., Sherr, E. B., Sherr, B. F., Pace, M., and Sanders, R. W. (1985). Protozoa in planktonic food webs. *J. Protozool.,* 32, 409–15.

Pringsheim, E. G. (1952). On the nutrition of *Ochromonas. Quart. J. Micros. Sci.,* 93, 71–96.

Provasoli, L., McLaughlin, J. J. A., and Pinter, I. J. (1954). Relative and limiting concentrations of major mineral constituents for the growth of algal flagellates. *Trans. N.Y. Acad. Sci.,* 16, 412–7.

Ramberg, L. (1979). Relations between phytoplankton and light climate in two Swedish forest lakes. *Int. Revue ges. Hydrobiol.,* 64, 749–82.

Raven, J. A. (1970). Exogenous inorganic carbon sources in plant photosynthesis. *Biol. Rev.,* 45, 167–221.

Ravera, O. (1986). Effects of experimental acidification on freshwater environments. *Experimentia,* 42, 507–16.

Rawson, D. S. (1956). Algal indicators of trophic lake types. *Limnol. Oceanogr.,* 1, 18–25.

Reinertsen, H. (1982). The effect of nutrient addition on the phytoplankton community of an oligotrophic lake. *Holarct. Ecol.,* 5, 225–52.

Reynolds, C. S. (1976). Succession and vertical distribution of phytoplankton in response to thermal stratification in a lowland lake, with special reference to nutrient availability. *J. Ecol.,* 64, 529–51.

(1980). Phytoplankton assemblages and their periodicity in stratifying lake systems. *Holarct. Ecol.,* 3, 141–59.

(1984a). *The Ecology of Freshwater Phytoplankton.* Cambridge: Cambridge University Press.

(1984b). Phytoplankton periodicity: the interactions of form, function and environmental variability. *Freshwat. Biol.,* 14, 111–42.

(1986). Experimental manipulations of the phytoplankton periodicity in large limnetic enclosures in Blelham Tarn, English Lake District. *Hydrobiologia,* 138, 43–64.

Reynolds, C. S., Thompson, J. M., Ferguson, A. J. D., and Wiseman, S. W. (1982). Loss processes in the population dynamics of phytoplankton maintained in closed systems. *J. Plank. Res.* 4: 561–600.

Rhee, G-Y. and Gotham, I. J. (1980). Optimum N:P ratios and coexistence of planktonic algae. *J. Phycol.,* 16, 486–9.

(1981). The effect of environmental factors on phytoplankton growth: temperature and the interaction of temperature and nutrient limitation. *Limnol. Oceanogr.,* 26, 635–48.

Rodhe, W. (1948). Environmental requirements of freshwater plankton algae. Experimental studies in the ecology of phytoplankton. *Symb. Bot. Upsal.,* 10, 1–149.

Rodhe, W., Vollenweider, R. A., and Nauwerck, A. (1960). The primary production and standing crop of phytoplankton. In: *Perspectives in Marine Biology,* ed. A. A. Buzzati-Traverso, pp. 299–322. Berkeley: University of California Press.

Roijackers, R. M. M. (1984). Some structural characteristics of the phytoplankton in the Oude Waal near Nijmegan, The Netherlands. *Verh. Internat. Verein. Limnol.,* 22, 1687–94.

(1985). Phytoplankton studies on a nymphaeid-dominated system, Ph.D. Thesis. Wageningen, The Netherlands: Agricultural University.

Roijackers, R. M. M. and Kessels, H. (1986). Ecological characteristics of scale-bearing Chrysophyceae from The Netherlands. *Nord. J. Bot.,* 6, 373–85.

Rosén, G. (1981). Phytoplankton indicators and their relations to certain chemical and physical factors. *Limnologica,* 13, 263–90.

Rott, E. (1983). Sind die Veränderungen im Phytoplanktonbild des Pilburger See Auswirkungen der Tiefenwasserableitung? *Arch. Hydrobiol. Suppl.,* 67, 29–80.

(1984). Phytoplankton as biological parameters for the trophic characterization of lakes. *Verh. Internat. Verein. Limnol.,* 22, 1078–85.

Round, F. E. (1971). The growth and succession of algal populations in freshwaters. *Mitt. Internat. Verein. Limnol.,* 19, 70–99.

Ruttner, F. (1930). Das Plankton des Lunzer Untersees: seine Verteilung in Raum und Zeit während der Jahre 1908–1913. *Int. Revue ges. Hydrobiol.,* 23, 1–304.

Sakamoto, M. (1971). Chemical factors involved in the control of phytoplankton production in the Experimental Lakes Area, northwest Ontario. *J. Fish. Res. Bd. Can.,* 28, 203–13.

Salonen, K., Jones, R. I., and Arvola, L. (1984). Hypolimnetic phosphorus retrieval by diel vertical migrations of lake phytoplankton. *Freshwat. Biol.,* 14, 431–8.

Sanders, R. W. and Porter, K. G. (1987). Phagotrophic phytoflagellates. *Adv. Microbiol. Ecol.,* 10.

Sandgren, C. D. (1978). Resting cysts of the Chrysophyceae: their induction development and strategic significance in the life history of planktonic species. Ph.D. Thesis. Seattle: University of Washington.

(1981). Characteristics of sexual and asexual resting cyst (statospore) formation in *Dinobryon cylindricum* Imhof. *J. Phycol.,* 17, 199–210.

(1983a). Survival strategies of chrysophycean flagellates: reproduction and formation of resistant resting cysts. In: *Survival Strategies of the Algae,* ed. G. Fryxell, pp. 23–48. Cambridge: Cambridge University Press.

(1983b). Morphological variability in populations of chrysophycean resting cysts. I. Genetic (interclonal) and encystment temperature effects on morphology. *J. Phycol.,* 19, 64–70.

(1986). Effects of environmental temperature on the vegetative growth and sexual life history of *Dinobryon cylindricum* Imhof. In: *Chrysophytes: Aspects and Problems,* eds. J. Kristiansen and R. A. Andersen, pp. 207–25. Cambridge: Cambridge University Press.

Sandgren, C. D. and Flanagin, J. (1986). Heterothallic sexuality and density dependent encystment in the chrysophycean alga *Synura petersenii* Korsh. *J. Phycol.,* 22, 206–16.

Scavia, D. and Robertsen, A. (1984). North American Great Lakes. In: *Lakes and Reservoirs,* vol. 23, *Ecosystems of the World,* ed. F. B. Taub, pp. 135–76. New York: Elsevier.

Schaffner, W. R. and Oglesby, R. T. (1978). Limnology of eight Finger Lakes: Hemlock, Canadice, Honeoye, Keuka, Seneca, Owasco, Skaneateles and Otisco. In: *Lakes of New York,* Vol. I, ed. J. A. Bloomfield, pp. 313–470. New York: Academic Press.

Schelske, C. L. (1962). Iron, organic matter and other factors limiting primary productivity in a marl lake. *Science,* 136, 45–6.

Schelske, C. L., Hooper, F. F., and Haertl, E. J. (1962). Responses of a marl lake to chelated iron and fertilizer. *Ecology,* 43, 646–53.

Schindler, D. W. (1971). Carbon, nitrogen and phosphorus and the eutrophication of freshwater lakes. *J. Phycol.,* 7, 321–9.

(1975). Whole-lake eutrophication experiments with phosphorus, nitrogen
     and carbon. *Verh. Internat. Verein. Limnol.,* 19, 3221–31.

(1977). Evolution of phosphorus limitation in lakes. *Science,* 195, 260–2.

Schindler, D. W. and Holmgren, S. K. (1971). Primary production and phy-
     toplankton in the Experimental Lakes Area, northwestern Ontario, and
     other low-carbonate waters, and a liquid scintillation method for deter-
     mining ^{14}C activity in photosynthesis. *J. Fish. Res. Bd. Can.,* 28, 189–
     201.

Schindler, D. W., Kling, H., Schmidt, R. V., Prokopowich, J., Frost, V. E.,
     Reid, R. A., and Capel, M. (1973). Eutrophication of Lake 227 by addi-
     tion of phosphate and nitrate: the second, third and fourth years of
     enrichment, 1970, 1971 and 1972. *J. Fish. Res. Bd. Can.,* 30, 1415–40.

Schindler, D. W., Mills, K. H., Malley, D. F., Findlay, D. L., Shearer, J. A.,
     Davies, I. J., Turner, M. A., Linsey, G. A., and Cruikshank, D. R. (1985).
     Long-term ecosystem stress: the effects of years of experimental acidifi-
     cation on a small lake. *Science,* 228, 1395–401.

Seaburg, K. G. and Parker, B. C. (1983). Seasonal differences in the tempera-
     ture ranges of growth of Virginia algae. *J. Phycol.,* 19, 380–6.

Shapiro, J. (1973). Blue-green algae: why they become dominant. *Science,* 179,
     382–4.

Sheath, R. G. (1986). Seasonality of phytoplankton in northern tundra ponds.
     Hydrobiologia, 138, 75–83.

Sheath, R. G., Hellebust, J. A., and Sawa, T. (1975). The statospore of *Dino-
     bryon divergens* Imhof: formation and germination in a subarctic lake.
     J. Phycol., 11, 131–8.

Shuter, B. (1978). Size dependence of phosphorus and nitrogen subsistence
     quotas in unicellular microorganisms. *Limnol. Oceanogr.,* 23, 1248–55.

Sikes, C. S. and Wheeler, A. P. (1982). Carbonic anhydrase and carbon fixation
     in coccolithophorids. *J. Phycol.,* 18, 423–6.

Smith, E. H. and Kalff, J. (1982). Size-dependent phosphorus uptake kinetics
     and cell quotas in phytoplankton. *J. Phycol.,* 18, 275–84.

Smith, V. H. (1983). Low nitrogen to phosphorus ratios favor dominance by
     bluegreen algae in lake phytoplankton. *Science,* 221, 669–71.

Smol, J. P. (1980). Fossil synuracean (Chrysophyceae) scales in lake sediments:
     a new group of paleoindicators. *Can. J. Bot.,* 58, 458–65.

(1985). The ratio of diatom frustules to chrysophycean statospores: a useful
     paleolimnological index. *Hydrobiologia,* 123, 199–208.

Smol, J. P., Charles, D. F., and Whitehead, D. R. (1984). Mallomonadacean
     microfossils provide evidence of recent lake acidification. *Nature,* 307,
     628–30.

Sommer, U. (1984). The paradox of the plankton: fluctuations of phosphorus
     availability maintain diversity of phytoplankton in flow-through cul-
     tures. *Limnol. Oceanogr.,* 29, 633–6.

(1986). The periodicity of phytoplankton in Lake Constance (Bodensee) in
     comaparison to other deep lakes of central Europe. *Hydrobiologia,* 138,
     1–7.

Sommer, U., Gliwicz, Z. M., Lampert, W., and Duncan, A. (1986). The PEG-model of seasonal succession of planktonic events in fresh waters. *Arch. Hydrobiol.,* 106, 433–71.

Sondergaard, M. and Sand-Jensen, K. (1979). Physico-chemical environment, phytoplankton biomass and production in oligotrophic, softwater, Lake Kalgaard, Denmark. *Hydrobiologia,* 63, 241–53.

Stearns, S. C. (1976). Life history tactics: a review of the ideas. *Quart. Rev. Biol.,* 51, 3–47.

Steinberg, C. and Hartmann, H. (1986). A biological paleoindicator for early lake acidification: Mallomonadacean (Chrysophyceae) scale abundance in sediments. *Naturwissenschaften,* 73, 37–9.

Sterner, R. W. (1986). Herbivore's direct and indirect effects on algal populations. *Science,* 231, 605–7.

Stewart, W. D. P. (1974). *Algal Physiology and Biochemistry.* Botanical Monographs, vol. 10. Berkeley: University of California Press.

Stumm, W. and Morgan, J. J. (1970). *Aquatic Chemistry.* New York: Wiley.

Suttle, C. A., Chan, A. M., Taylor, W. D., and Harrison, P. J. (1986). Grazing of planktonic diatoms by microflagellates. *J. Plank. Res.,* 8, 393–8.

Takahashi, E. (1978). *Electron Microscopical Studies of the Synuraceae (Chrysophyceae) in Japan: Taxonomy and Ecology.* Tokyo: Tokai University Press.

Teiling, E. (1955). Some mesotrophic phytoplankton indicators. *Verh. Internat. Verein. Limnol.,* 12, 212–15.

Tezuka, Y. (1982). Seasonal variations of dominant phytoplankton, chlorophyll *a* and nutrient levels in the pelagic regions of Lake Biwa. *Jap. J. Limnol.,* 45, 26–37.

Thunmark, S. (1945). Zur Soziologie des Süsswasserplanktons. *Fol. Limnol. Scand.,* 3, 1–66.

Tilman, D. (1982). *Resource Competition and Community Structure.* Princeton: Princeton University Press.

Tilman, D. and Kiesling, R. L. (1984). Freshwater algal ecology: taxonomic tradeoffs in the temperature dependence of nutrient competitive abilities. In: *Current Perspectives in Microbial Ecology,* eds. M. J. Klug and C. A. Reddy, pp. 314–19. Washington, D. C.: American Society of Microbiology.

Tilman, D. and Kilham, S. S. (1976). Phosphate and silicate growth and uptake kinetics of the diatoms *Asterionella formosa* and *Cyclotella meneghiniana* in batch and semicontinuous culture. *J. Phycol.,* 12, 375–83.

Tilman, D., Mattson, M., and Langer, S. (1981). Competition and nutrient kinetics along a temperature gradient: an experimental test of a mechanistic approach to niche theory. *Limnol. Oceanogr.,* 26, 1020–33.

Tilzer, M. M. (1973). Diurnal periodicity in the phytoplankton assemblage of a high mountain lake. *Limnol. Oceanogr.,* 18, 15–30.

Vande Castle, J. R. (1985). A study of long-term changes in the phytoplankton community and the seasonal change of alkaline phosphatase activity in Lake Michigan. Ph.D. Thesis, University of Wisconsin-Milwaukee.

Van Donk, E. (1983). Factors influencing phytoplankton growth and succession in Lake Maarsseveen (I). Ph.D. Thesis. Amsterdam: University of Amsterdam.

Vanni, M. J. (1986). Competition in zooplankton communities: suppression of small species by *Daphnia pulex*. *Limnol. Oceanogr.*, 31, 1039–56.

Vollenweider, R. A. (1969). Möglichkeiten und Grenzen elementarer Modelle der Stoffbilanz von Seen. *Arch. Hydrobiol.*, 66, 1–36.

Weimann, R. (1933). Hydrobiologische und Hydrobiographische Untersuchungen an zwei teichartigen Gewässern. *Beih. Bot. Cantralbl.*, 51, 397–476.

Wetzel, R. G. (1975). *Limnology*. Philadelphia: Saunders.

Willén, E. (1984). The large lakes of Sweden: Vänern, Vättern, Mälaren and Hjälmaren. In: *Lakes and Reservoirs,* vol. 23, *Ecosystems of the World,* ed. F. B. Taub, pp. 107–34. New York: Elsevier.

Willén, T. (1962). Studies on the phytoplankton of some lakes connected with or recently isolated from the Baltic. *Oikos,* 13, 169–99.

Wright, R. T. (1964). Dynamics of a phytoplankton community in an ice-covered lake. *Limnol. Oceanogr.*, 9, 163–78.

Yoshida, Y., Kawaguchi, K., and Kadota, H. (1983). Studies on a freshwater red tide in Lake Biwa. III. Patterns of horizontal distribution of *Uroglena americana*. *Jap. J. Limnol.*, 44, 293–7.

Yoshida, Y., Matsumoto, T., and Kadota, H. (1983). Studies on a freshwater red tide in Lake Biwa. II. Relation between occurrence of red tide and environmental factors. *Jap. J. Limnol.*, 44, 28–35.

Yoshida, Y., Mitamura, O., Tanaka, N., and Kadota, H. (1983). Studies on a freshwater red tide in Lake Biwa. I. Changes in the distribution of phytoplankton and nutrients. *Jap. J. Limnol.*, 44, 21–7.

Chapter 3

ECOLOGY OF THE CRYPTOMONADIDA: A FIRST REVIEW

DAG KLAVENESS

INTRODUCTION

A large amount of information concerning the ecology of the Crypto-monadida is scattered throughout the literature, often as only a few lines in a larger context. Only a fraction is cited here. Some ideas about the preferences of phototrophic cryptomonads existed before 1950, but the total knowledge of the biology of cryptomonad flagellates to that time was very strongly biased toward the colorless, heterotrophic *Chilomonas paramecium* Ehr. from which a tremendous amount of basic cell biology and good experimental techniques (cf. Cosgrove 1950) have been learned (also see Danforth 1967; Kudo 1977). The emphasis of this review will therefore be on the more modern literature.

This essay briefly reviews the ecology of cryptomonad flagellates as extracted from the literature and from personal experience. There is no previous review on the ecology of the Cryptomonadida, but the related review papers of Gantt (1980) and Antia (1980) have important information. For further notes on ecology and additional references, the second edition of Huber-Pestalozzi (1968), the works of Skuja (1956) and Pringsheim (1963, 1968), and also the books of Round (1981) and Reynolds (1984a) should be consulted. Although emphasis is on freshwater forms, the homogeneity of the class and the very few ecological investigations undertaken justify discussing results from both the marine and limnic environments. Based upon knowledge of present distributions and the ecological range of cryptomonads in today's aquatic ecosystems, some comments on the ecology of cryptomonad evolution have also been included. These supplement the

An early draft of this manuscript was discussed with Lois M. Granskog and Kenneth Estep. At a later stage, Asbjorn Skogstad supported the author with useful comments on content and style. Craig Sandgren did an excellent job of improving style and language. Their inspiration to the author is gratefully acknowledged.

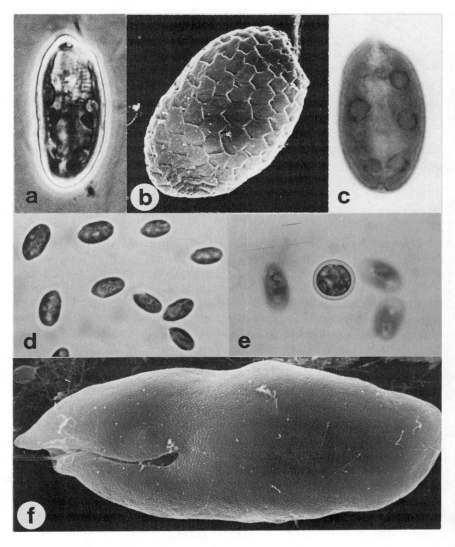

Fig. 3-1. Light and scanning electron micrographs of Cryptomonadida showing various aspects of the living and fixed cell. (a) Living cell
photographed at low primary magnification (large depth of field but
low resolution) showing transparency of cell. Organelles are visible in
the living cell if motility can be checked (*Cryptomonas tetrapyrenoidosa* ×1,490). (b) Cell surface structures of *Chroomonas* sp. as
observed by the scanning electron microscope, (×8,000). (c) Pyrenoids can be visualized under favorable conditions by careful staining
of starch grains by Lugol's iodine (*C. tetrapyrenoidosa* ×1,490). (d)
Size variation between cells is distinct even in clonal cultures (*Cryp-*

overwhelming evidence for serial endosymbiosis in this group (cf. Ludwig & Gibbs 1985).

MORPHOLOGY AND TAXONOMY

The cryptomonads are mainly free-living, highly motile flagellates; sizes range from 3 μm to more than 50 μm. Coccoid and palmelloid types have also been described, where only the swarmers exhibit the typical morphology (cf. Fritsch 1935; Bicudo 1966). Reproduction mainly takes place as an oblique binary division of the vegetative cell. Sexuality has rarely been indicated (Wawrik 1969), but resting stages have been described in different species (Ettl et al. 1967; Klaveness 1985b).

At the light microscope level, the cells are best studied alive (Fig. 3-1a) because cell shape is altered and content obscured upon fixation (there is no adequate fixative). The transparency of the living cytoplasm allows intracellular organelles to be recognized: gullet (if present) with ejectosomes (trichocysts); "Maupas's ovals," if well developed (two oval, light-breaking vesicles); a transparent nucleus; lipid vesicles; a contractile vacuole (freshwater); and the shape and size of the chloroplast(s) with its pyrenoid(s) (if present) and starch grains. Schematic drawings, based upon electron microscopy, may aid in interpretation of the light microscope image (see Gibbs 1981b, p. 76; Santore 1984).

The cryptomonads are considered difficult to discriminate at species level, and current taxonomy is inadequate (Pringsheim 1944; Gantt 1980; Oakley & Santore 1982; Santore 1984; Klaveness 1985b). The uniformity of morphological features has given rise to some very variable and widespread "species," apparently tolerating a variety of conditions (e.g., *Cryptomonas "ovata";* cf. Huber-Pestalozzi 1968; Butcher 1967). Therefore, improving taxonomic precision is a necessary prerequisite for advancing ecological knowledge of this widespread group of algal flagellates (see Klaveness 1985b). Table 3-1 presents a summary of morphological criteria that have been used or are potentially useful for species discrimination. Another source of taxonomic confusion may be inconsistent use of terminology between authors, or the lack of a precise terminology. Butcher (1967, his Fig. II) improved on the situation by naming the various views of cells; this terminology is

Caption to Fig. 3-1. (*cont.*)
tomonas pyrenoidifera ×550). (e) Resting spore of *Cryptomonas* sp. Three swimming cells out of focus (×550). (f) *Cryptomonas rostratiformis* ventral side showing papillate surface, apical rostrum, vestibulum, and furrow leading into "gullet" region (×3,200).

Table 3-1. *Criteria useful for discriminating between species of Cryptomonadida.*[a]

Cell shape	Starch grains
Cell size	Lipid vesicles
Cell ratios	Color, pigments
Flagella	Resting stages
Flagella/cell size ratio	Palmellae
Furrow/gullet	Periplast structure
Gullet/cell size ratio	Mastigonemes
Ejectosomes	Flagellar scales
"Maupas's ovals"	Flagellar root system
Chloroplast number	Thylakoids
Chloroplast shape	Nucleomorph
Chloroplast location	Habitat
Pyrenoids	

[a]See Klaveness 1985b.

adhered to here and is recommended generally (see also Klaveness 1985b).

Modern electron microscopical techniques (Fig. 3-1b,f) have not facilitated species determination so far. Morphological details of great potential value in taxonomy have been discovered, but since the original description of species are insufficient and no holotypes exist (contrary to diatoms, for example), the modern techniques call for an urgent revision where a number of "species" may be invalidated. For the general phytoplankton worker, the light microscope will continue to be the most important tool. There are still details to be discovered: for example, the pyrenoids of some species (Fig. 3-1c) or the resting stages of others (Fig. 3-1e).

Cryptomonads have been variously treated taxonomically, depending on whether they were considered as "central" or "isolated" in an evolutionary context. Botanists assign the cryptomonads to class level (Cryptophyceae; Christensen 1980) or division level (Cryptophyta; Bold & Wynne 1985), whereas in a zoological context they may be included as an order (Cryptomonadida) within the class Phytomastigophorea (e.g., Lee, Hutner, & Bovee 1985).

HABITATS

The Cryptomonadida comprise no more than about 100 species of predominantly photosynthetic flagellates, but members may be found in almost any body of natural water in the world (Klaveness 1985a). They are found in lakes of the Antarctic mainland (Parker et al. 1982), in snow of the alpine environment (Javornicky & Hindak 1970), in lakes of the tropics (cf. Lewis & Riehl 1982), in cold and temperate marine provinces, and along the coasts of the continents (Grall & Jaques 1982; Buch & Garrison 1983; Throndsen 1969, 1983; Hsiao & Pinkewycz 1983). Cryptomonads are also sometimes dominant in the open ocean (Taylor & Waters 1982), in tidal pools of highly variable salinity (Meyer & Pienaar 1984), in brackish environments (Conrad & Kufferath 1954; Hulburt 1965; Campbell 1973; Caljon 1983), and in lakes of deviant chemical composition (Kuzel-Fetzmann 1979; Dokulil 1979). Cryptomonads may occasionally form "blooms" under lacustrine (Rosén 1981) and marine conditions (Iwasaki 1979), and cryptomonads causing fish kills in a pond (Pfiester & Holt 1977) or taste/odor problems in drinking water (Palmer 1962; Collins & Kalnins 1966; Heinonen 1980) have been reported. Cryptomonads have also successfully entered the intracellular environment in some ciliates (Barber, White, & Siegelman 1969; Taylor, Blackbourn, & Blackbourn 1969) and dinoflagellates (Wilcox & Wedemayer 1984), where they play a role as functional "chloroplasts." It is, however, in lakes of the temperate region where the Cryptomonadida display the largest diversity in spatial and temporal niche occupation, and where they are found under widely different conditions (cf. Taylor et al. 1979; Ettl 1980; Heinonen 1980; Rosén 1981; Anton & Duthie 1981; Holmgren 1984).

Lack of records of cryptomonads from a geographical area generally only indicates that no investigations have taken place yet, or that destructive fixation procedures have been employed (e.g., formaldehyde). Van Meel's (1954) record of only a single species in African rift valley lakes has subsequently proved unreliable (e.g., Evans 1962; Hecky & Kling 1981). Only recently has the diversity of cryptomonads on the Australian mainland, indicated by the pioneer investigation by Playfair (1921), been adequately followed up. Lewis and co-workers have shown the importance of cryptomonads in tropical South American ecosystems (e.g., Lewis & Riehl 1982), and they have also shown that they are important in tropical lakes in general. In the marine environment, Lohmann (1922) recorded a paucity of cryptomonads in the Atlantic only where temperature was above 22°C. A more recent transect at 24°30′N by Estep and co-workers (1984) confirms the scarcity of Cryptomonadida there. It is possible that high temperature in combination with high salinity is the only combination of factors that, within a common range, may restrict the distribution or diversity of cryptomonads. Further observations, using adequate methods, in places such as the Indian Ocean or the Red Sea may reject this hypothesis. To my knowledge, members of the Cryptomonadida have not yet been recorded in hypersaline lakes or in hot springs.

SEASONAL AND SPATIAL DISTRIBUTION

Peculiar to cryptomonads in several aquatic systems (e.g., limnic, estuarine, or marine) is their permanent presence through the year. Smaller or larger population peaks follow perturbations like wind mixing or periods of precipitation and are not always related to season (see Reynolds & Reynolds 1985). They have frequently been recorded to increase when other populations were declining as if a temporary "niche" were opening. This appearance has been termed "opportunistic" (cf. Rott 1983), a strategy possible because the Cryptomonadida have been considered as distinctly "r-selected" (Sommer 1981). This descriptive terminology, however, has not contributed further to a real understanding of seasonality or lack of it among populations of cryptomonads in different localities.

In some large, nutrient-poor lakes at the temperate latitudes, photosynthetic cryptomonad flagellates may tend to develop their population maxima in the summer (e.g., Bailey-Watts & Duncan 1981). Factors such as wind exposure of the lake and the depth of the summer epilimnion in relation to the lake productivity may determine the summer density of plankton grazers. As soon as a critical grazing pressure is reached, an epilimnetic summer maximum of "grazeable" algae like cryptomonads may not persist except during and after periods of turbulence. Bimodal population maxima are therefore likely in more productive lakes at temperate latitudes (vernal–autumnal), but irregular waxing and waning of populations following periods of turbulence and/or influx of nutrients is common. Larger, highly motile species may appear and disappear from the epilimnion as a result of diel vertical migration (discussed presently).

Related factors may regulate cryptomonad seasonality in lakes under widely different climatic conditions. In Lake Kinneret, Israel, the dominating cryptomonads *Rhodomonas* and *Cryptomonas* appear in the epilimnion when autumn destratification starts, "when nutrients from the hypolimnion and metalimnion together with the algae are dispersed in the upper layers" (Pollingher 1981). In Lake Tanganyika, the relative standing crop of Cryptomonadida was large during the cool, dry season (Hecky & Kling 1981, Figs. 2, 3), but it was variable during the wet season when the lake was stratified nearer to the surface (Hecky & Fee 1981). In smaller lakes in the temperate region, the cryptomonads may seasonally exhibit a diversity of demographic strategies including stably stratified populations (e.g., Ichimura, Nagasawa, &

Tanaka 1968; Takahashi & Ichimura 1968; Baker & Brook 1971; Schindler & Holmgren 1971; Burns & Mitchell 1974; cf. also Sommer 1982), diel vertical migration (Soeder 1967; Tilzer 1973; Kalff & Welch 1974; Ilmavirta 1980, 1983; Salonen, Jones, & Arvola 1984; Arvola et al. in press), or formation of resting stages (references in Klaveness 1985b). Small, more or less humic lakes may develop populations at the interface with oxygen-free bottom water in late summer where light may be extremely low but nutrients more abundant (e.g., Takahashi & Ichimura 1968; Schindler & Holmgren 1971; Burns & Mitchell 1974). Salonen et al. (1984) have shown that diel vertical migration may extend over several meters and may traverse large temperature gradients; such behavior was probably not detected by earlier sampling procedures. The cryptomonads may therefore be able to utilize the combination of available light and nutrients in deeper waters. Finnish scientists have also shown that flagellates, and especially Cryptomonadida, increase in dominance with increasing water color but not with increasing nutrient content (Ilmavirta 1980, 1983). A similar effect with increasing water color can be read from the diagram of Rosén (1981), a regional investigation comprising 1,000 lakes throughout Sweden (surveyed August 1972).

In many lakes, the Cryptomonadida may have their maximal population development well below the surface. Nauwerck (1968) found in the extremely oligotrophic Lake Latnjajaure that the cryptomonad populations had their "optimal depth" between 15 and 23 m, deepest in April–June and shallowest in October–December. Capblancq (1972) also found a tendency toward deep distribution of cyrptomonads in his high-altitude lakes of the Pyrenees, and he could distinguish, as Nauwerck did, prominent depth and time differences among the populations of species. Accumulations of cryptomonads at depth in lakes are also well documented in more productive lakes (e.g., Braarud, Føyn, & Gran 1928; Nauwerck 1963; Rott 1983). Under low-turbulence summer conditions, the metalimnetic strata may satisfy the requirements of cryptomonad flagellates, provided there is adequate light (and, for some species, that the water's buffering capacity is sufficient to prevent a drastic increase in pH due to photosynthesis). In this metalimnetic depth stratum, cryptomonad growth may balance diel losses to grazing. By comparison, in mesotrophic and eutrophic lakes of low alkalinity, the increase in pH as a result of photosynthesis may favor the development of Cyanobacteria (cf. Shapiro 1984) in the thermocline region. These two widely different groups of planktonic organisms are both capable of utilizing low light levels as found in the thermocline region of meso-/eutrophic lakes during summer (cf. Nauwerck 1963), but some cryptomonads may be less capable of dealing with high pH (discussed subsequently). In marine provinces, cryptomonads are also present in the "deep chlorophyll maxima" (Tsuji & Adachi 1979; Furuya & Marumo 1983), but they are certainly not obligate low-light organisms and are not counted as members of the marine "knepho-plankton" (Sournia 1982).

From 250 lakes of the eastern and southeastern United States sampled in the National Eutrophication Survey, it appears that *Cryptomonas* spp. may have a preference for colder water (Taylor et al. 1979); this genus dominated primarily in spring samples. Also in Caljon's brackish-water lakes (Caljon 1983), many cryptomonads belonged to his "winter–spring group," but the seasonal dominance of the same species was different between lakes. The detailed temporal and spatial patterns exhibited are as varied as the individual lakes. *Cryptomonas ovata* was clearly a winter–spring species in Grote Geule, but fairly indifferent to season in Rode Geule (with a slight maximum in September). Wawrik (1983) found Cryptomonadida to be dominant at ice-break in 4 out of 20 ponds in Austria. Late summer/autumnal peaks are also very common (e.g., Nauwerck 1963; Elster & Motsch 1966; Reynolds 1980, 1984b). In Piburger See the cryptomonads occurred irregularly, but there were great differences between the nine taxa recorded (Rott 1983). *Cryptomonas phaseolus* occurred regularly in the late summer plankton maximum at a depth of 15–18 m, close to the upper limit of oxygen depletion, whereas other "opportunistic" species occurred at different times and depths as niches were temporarily opened for occupation.

The particular distribution of cryptomonad flagellates in the marine environment is not well described, because they frequently are lumped into groups like "monads and flagellates" or "sp. indet" in quantitative phytoplankton work. At temperate latitudes, the diversity of marine cryptomonads is astounding when adequate techniques are employed (cf. Butcher 1967; Throndsen 1969, 1983). Under favorable conditions they may be quantitatively dominant even in the open ocean (Taylor & Waters 1982). In coastal waters or under estuarine conditions, they may show a distinct seasonality. Throndsen (1969) found the total number of cryptomonad cells at his main station in the Oslofjord to be highest in the late summer, but their relative importance was highest in the early spring. The most common species, *Hemiselmis virescens,* occurred most frequently (but not exclusively) at temperatures below 10°C. In South Bay (San Francisco Bay), Cloern and co-workers (1985) found the plankton community in the spring to be dominated by smaller algae, of which *Chroomonas* and *Cryptomonas* made up a very significant fraction. But cryptomonads, especially *Chroomonas minuta,* were significantly present through the whole year.

NUTRITION

Cryptomonads have an assumed high demand for nutrients (e.g., Reinertsen 1982; Sommer 1983), but experimental corroboration comprises few suitable investigations so far. Cloern (1977) found the laboratory strain of *Cryptomonas ovata* var. *palustris* (UTEX 58) to be a weak competitor for phosphate. Further experimental investigations on the competitive abilities of cryptomonads have so far been hampered by their sensitivity to mechanical stirring (Sommer 1983, 1985). Chemostat studies of *Rhodomonas lacustris* (strain N 750301) confirms that carbon yield per unit phosphate, when limiting, is rather moderate (Olsen et al., 1986, and pers. comm.).

There are other aspects of growth and nutrition among the cryptomonads that are still less understood. Many lakes have measurable quantities of degradable dissolved organic matter. Heterotrophic growth in darkness of phototrophic freshwater species has not been documented at temperatures higher than 3–5°C, where Wright (1964) found some species to grow upon 100 mg acetate·L^{-1}. Some organic substances may stimulate growth slightly at low light levels (e.g., *Cryptomonas rostratiformis* and *C. tetrapyrenoidosa:* Klaveness unpub.). Growth of *Cryptomonas ovata* var. *palustris* (UTEX 358) in light is stimulated by addition of proteose peptone to the axenic culture (Bowen & Ward 1977). Most marine species tested seem to be stimulated in light by addition of glycerol; some are also stimulated by acetate or related organic acids (Antia 1980). *Chroomonas salina* is known to grow heterotrophically upon concentrations of glycerol not encountered in nature (see Antia 1980). Cryptomonads seem to have an obligate requirement for one or more vitamins (Provasoli & Carlucci 1974).

Vieira and Klaveness (1986) found that a cryptomonad (*Cryptomonas* cf. *tetrapyrenoidosa*) was the least versatile with regard to utilization of organic nitrogen sources among five freshwater planktonic algae tested, therefore excluding a competitive advantage of utilizing such compounds in the eutrophic lake from where it was isolated. The marine species *Chroomonas salina* and *Hemiselmis virescens* were slightly more versatile, but *Rhodomonas lens* could only use urea among several compounds tested (Antia 1980). Kuentzler (1965) found phosphatase activity to be low and constitutive (not repressible by phosphate) in two species of cryptomonads. The ecological significance of these abilities for open-water species can only be surmised at present.

The cosmopolitan distribution of the colorless marine *Leucocryptos marina* (Braarud) Butcher was noted by Braarud (1962), and its wide distribution has been corroborated from the arctic to the tropics by subsequent records (Hasle 1969; Heimdal 1983; Rojas de Mendiola 1981). The trophic role of this cryptomonad flagellate and related colorless flagellates in the marine environment has only recently been appreciated (Davis & Sieburth 1984, "Bodo sp." from Narragansett Bay), and results from experimental work are in preparation (Cynar & Sieburth unpubl.; Klaveness et al. unpubl.). Griessman's *Telonema subtilis* (Griessman 1913), perhaps related to *Cyathomonas* (Hollande & Cachon 1950), is not uncommon in some areas (Throndsen 1969), but requires further investigation. There is evidence for bacterial phagotrophy in the phototrophic *Cryptomonas ovata* (Porter et al. 1985), and this could be an important means of survival. Phagotrophy has been anticipated or documented in colorless freshwater cryptomonads such as *Katablepharis ovalis* Skuja (Skuja 1948; Pringsheim 1963; Keskitalo 1977), *Phyllomitus apiculatus* Skuja (Skuja 1964; Steinberg, Lenhart, & Klee 1983) and *Cyathomonas truncata* From. (Fritsch 1935; Schuster 1968). Skuja (1956) discusses briefly the ecology of these "Hypolimnobionten," which mainly occur in the deeper strata of lakes. Subsequent work has fully verified the general occurrence of these interesting flagellates (Pavoni 1963; Vollenweider, Munawar, & Stadelman 1974; Keskitalo 1977; Rosén 1981; Steinberg et al. 1983), but much work remains to be done to evaluate their role and influence on the function of the lacustrine ecosystems.

ENVIRONMENTAL TOLERANCES

Temperature and Light

Some cryptomonad species are capable of persisting under extremely poor light conditions (e.g., Schindler & Holmgren 1971; Morgan & Kalff 1975; Rott 1983). *Chroomonas* sp. (strain CSIRO MB-24) was capable of chromatic adaptation under simulated deep-sea conditions (Vesk & Jeffrey 1977). Reinertsen (1982) has confirmed the observation of Ahlgren (1970) that other species, such as *Rhodomonas pusilla,* are not particularly sensitive to high light intensities at adequate nutrient concentrations. Lichtle (1979, 1980) discovered that high light intensity in combination with increasing nitrogen deficiency could induce formation of resting stages in *Cryptomonas rufescens.*

Dark survival of seven marine strains was tested by Antia (1976). Most strains survived best at low temperature (2°C), but one, the Lasker strain of *Rhodomonas lens* frequently used in laboratory work, survived best at the highest temperature tested, 20°C. This strain was assumed to be a warm-water or tropical strain, as judged from the locality of isolation (Gulf Stream). Only one strain survived for more than 24 weeks in darkness, whereas *Hemiselmis virescens* (Droop, Millport-64 strain) and *Rhodomonas lens* were particularly sensitive, the latter surviving four weeks at maximum. In lakes, winter survival in darkness under ice seems to find a plausible explanation in the investigations of Morgan & Kalff (1979), where low temperature resulted in low respiration (and low grazing). A stable water column and optimal adaptation to utilize whatever light there is may be sufficient for winter survival (cf. Rodhe 1955; Wright 1964; Pechlaner 1971; Parker et al. 1982).

Bowen & Ward (1977) tried to establish optimal conditions for growth of *Cryptomonas ovata* var. *palustris* (UTEX 358, freshwater) and *Chroomonas* sp. (Provasoli, strain M20, marine). One significant result was the preference of a light/dark cycle to continuous light. A 12-h light/12-h dark cycle yielded maximal growth, which was slightly better than 14/10, but significantly better than 10/14 and continuous light. Nelson & Brand (1979) found *Chroomonas salina* (clone 3C) to divide during the dark period of a 14/10 LD cycle. Humphrey (1979) found that his strain of *Chroomonas* grew best with continuous light, compared to a 12/12 cycle in experiments done at moderate light intensities of 60–80 $\mu E \cdot m^{-2} \cdot s^{-1}$.

It is known that some planktonic algae will suffer under continuous light, but much depends on light intensity. An ecological significance of day length may be further appreciated when its variation is considered with depth. If "day" is defined as an irradiance above a threshold of 1 $\mu E \cdot m^{-2} \cdot s^{-1}$, for instance, then daybreak will occur later and night-time will occur earlier with increasing depth. Under stratified conditions, day length will act as a selective force, favoring those algal forms with the appropriate day length–light intensity response. Cryptomonads may counteract or make use of this effect by diel migration. Phytochromes and their potential role in sensing day length have not been studied in the Cryptomonadida. In a series of papers, Watanabe and co-workers (1974, 1976, 1978, 1982) studied the physiology of phototaxis in *Cryptomonas* sp. (strain IAM CR-1). Additional physiological parameters and responses of possible ecological significance have been studied as functions of light intensity by Brown & Richardson (1968), Faust & Gantt (1973), Lichtle (1979, 1980), Thinh (1983), and Raven (1984).

The long-term study of Ramberg (1979) showed that discrete populations of different species of Cryptomonadida inhabitating two small forest lakes in Sweden exhibited distinctly different temporal behavior. In his illustrative light–temperature diagrams, he showed that *Cryptochrysis* sp. grew at a narrow temperature range but tolerated a wide range in light climate. *Cryptomonas erosa,* on the other hand, grew at temperatures from less than 5°C to 25°C, but was correlated with a narrower range in light climate. Findenegg (1971) found *Cryptomonas erosa* to be dominant in lakes of the Austrian Alps under temperatures ranging from 5° to 21°C. Under such conditions, *C. erosa* was the fastest growing of the 14 different species monitored during their respective periods of dominance. Willén, Oké, & González (1980), in an admirable study of population dynamics and form variation of *Rhodomonas minuta* and *R. lens* populations in Lakes Mälaren and Vättern, found that *R. minuta* developed well at temperatures of 0.5°–1°C, but also dominated at temperatures of 16°C in turbulent early autumnal surface waters. Although *R. minuta* was present through the growing season, *R. lens* was absent in the late summer when water temperature was in excess of 16°C and blue-green algae were at a maximum.

Other studies of temperature responses of cryptomonads include the important study of Mucibabic (1956), where population growth, cell shape, and size of a colorless *Chilomonas* were recorded. Cell size (and coefficient of variation) was largest at the temperature extremes. An increase in cell size was also found by Morgan & Kalff (1975) when *Cryptomonas erosa* isolated from Lac Hertel (near Montreal, Quebec) was grown in light at low temperature (4°C). Upon subsequent storage in the dark, cell volume decreased as a result of carbohydrate reserve (starch) respiration. This work was followed by a thorough study of the interaction of light and temperature upon growth, carbon uptake, chlorophyll content, and cell volume of *C. erosa* (Morgan & Kalff 1979). Cloern (1977) studied the effect of irradiance and temperature upon growth rate, uptake of phosphate, ammonium, nitrate, and chlorophyll content in *Cryptomonas ovata* var. *palustris* (UTEX 58). Maximal growth rate and optimal light intensity as functions of temperature were modeled, and the resulting equations were implemented in a simulation model for *Cryptomonas ovata* in southern Kootenay Lake (Cloern 1978). The model reflected the summer, autumn, and early spring population peaks quite successfully in spite of the different origin of the laboratory culture.

A shift of light optima for growth with temperature in *C. erosa* was similar to that of *C. ovata* var. *palustris,* but the maximum growth rate was always higher and generally obtained at lower light intensities in *C. erosa* (Cloern 1977; Morgan & Kalff 1979). The fastest rate of growth of *C. ovata* var. *palustris* (UTEX 58) was obtained near 25°C (Bowen & Ward 1977; Cloern 1977); *C. erosa* grew fastest at 23.5°C (Morgan & Kalff 1979). It is a common experience that temperature optima found in the laboratory do not coincide well with observations on occurrence or dominance in nature; various aspects of this discrepancy are discussed by Braarud (1961), Karentz & Smayda (1984), and others.

Temperature fluctuations appear to influence cryptomonad growth. Following passage through a power plant with concomitant water temperature rise of 6°C, *Cryptomonas acuta* showed a significant reduction of productivity, in contrast to dinoflagellates and diatoms (Sellner, Kachur, & Lyons 1984). This selective loss of flagellates in power plants may be ecologically significant locally due to the large water masses involved.

pH

There are differences in cryptomonad populations among lakes with regard to pH gradients, but a number of environmental factors are correlated with pH under natural conditions. When lakes undergo acidification, there is some evidence for a shift in cryptomonad species composition. *Rhodomonas minuta* and *Katablepharis ovalis,* common in oligotrophic lakes, are absent in acid lakes (Almer et al. 1974). Several authors indicate that some cryptomonads may have distinct pH preferences – for example, *Cryptomonas cylindrica* according to Javornicky (1967), *C. curvata* according to Anton and Duthie (1981), and *C. czosnowskii* according to Javornicky (1978). Anton and Duthie (1981), on the other hand, failed to find any taxonomic or distributional significance attached to pH in the preliminary analysis of their regional data. Apparently wide tolerance of pH may reflect problems with discriminating between species, but truly tolerant species may exist.

Laboratory studies of Cryptomonadida in culture have revealed that growth of different strains is limited to restricted ranges of pH (Pringsheim 1968). Pringsheim found that different strains grew optimally at either pH 4.5–6, pH 6.5–7.2, or pH 7.5–8.5, and their preference was clearly related to locality of isolation: "Die Grenzwerte hängen etwas von der Zusammensetzung der Nährlösung ab; ... " (Pringsheim 1968, p. 371). Bowen & Ward (1977) confirmed the narrow range of preference for two more strains, one of which was the much studied *Cryptomonas ovata* var. *palustris* (UTEX 358), which had a tolerable pH range for growth of pH 4.8–6.0 under the experimental conditions employed. This strain is a descendent of Pringsheim's clone 979-3, isolated from a pond in Czechoslovakia in 1939 (Schlosser 1982), and has been used frequently in ecological research. Other scientists who keep this strain in culture have found good growth at higher pH (pers. comm., also Pringsheim's statement cited earlier). A strain (N 750301) of the common *Rhodomonas lacustris* isolated from the mesotrophic (and meromictic) lake Nordbytjern (Hongve 1974; Klaveness 1977) grew well between pH 6 and pH 8.5 in MES-TRIS buffered media with optimum around pH 7.8, but tolerated a daily pH fluctuation in the light phase in unbuffered media up to pH above 10 at a cost of growth rate (Klaveness unpubl.)

Salinity

Braarud was among the pioneers in designing careful experiments with planktonic algae aimed at resolving ecological questions. One of his studies involved the salinity response of a marine *Cryptomonas* sp. (Braarud 1951). Additional experiments have since been performed to test the generally euryhaline response of neritic (cf. Smayda 1958) species recorded in nature (e.g., Iwasaki 1979; Throndsen 1969, 1976). Apparently, some cryptomonads of marine origin have an extraordinary tolerance to variation in salinity (Meyer & Pienaar 1984). The present author has successfully obtained unialgal cultures of cryptomonads from Norwegian coastal waters by diluting seawater with a freshwater medium. The salinity tolerance of genuine freshwater strains of cryptomonads, like *Rhodomonas lacustris* (strain N 750301, Klaveness 1981), however, is very limited (unpubl.). The repeated records of *Rhodomonas lens* (first described from Austrian Alp lakes and Bohemian forest ponds: Pascher 1913) and *Rhodomonas minuta* (first described from Swedish lakes: Skuja 1948) from marine environments illustrate both the problems associated with a purely morphological species concept based on limited diacritical features, as well as the possibility of the presence of a wide range of physiological strains, as is well known from other groups of marine and freshwater phytoplankton (e.g., Brand 1982; Gallagher 1982; Francke & Coesel 1985).

BIOLOGICAL INTERACTIONS

Zooplankton Grazing

Convincing evidence has been accumulated from many independent sources indicating that flagellates and cryptomonads, in particular, are good food items for herbivorous zooplankton of various taxonomic affinities and of a large size range. Cryptomonads were among the most heavily grazed plankton by the calanoid copepod *Eudiaptomus* in Lake Erken (Nauwerck 1963). Hargrave and Geen (1970) found the ambient flagellate fraction, dominated by cryptomonads, to be removed by copepods most efficiently from mixed estuarine phytoplankton. Ferguson, Thompson, and Reynolds (1982), found cryptomonads (especially *Rhodomonas*) generally present in the gut contents of *Daphnia hyalina* and *Diaptomus gracilis* collected from their experimental enclosures in Blelham Tarn. Further experimental evidence corroborating the grazability of cryptomonads by larger zooplankton include Geller (1984), Lehman and Sandgren (1985), Knisely and Geller (1986), and references cited therein. Edmondson (1965) found the presence of cryptomonads to enhance the reproduction of planktonic rotifers. Pejler (1977) found *Rhodomonas* to be superior to *Chrysochromulina* as food for various rotifers, and Gilbert and Bogdan (1981) found a certain selectivity for flagellated cells when *Keratella* and *Polyarthra* were given a choice. Repak (1983) found that among 45 strains of axenically grown algae, only the 7 strains of Cryptomonadida were unambiguously accepted as nutritious by the ciliate *Fabrea salina*. Klaveness (1984) confirmed the superiority of cryptomonads as a food source for freshwater strains of *Coleps* sp. It is frequently inferred that flagellate biomass is "controlled" by grazing (cf. Holligan & Harbour 1977; Reynolds 1982, 1984a). Although there are counterindications (Sarnelle 1986), one can assume that grazing has a heavy impact on the planktonic cryptomonad populations and probably regulates their population dynamics.

Temporal relief from grazing may come during periods of turbulence, when the populations may be "diluted" into larger volumes of water and the water is fortified with nutrients from deeper strata. It is during and immediately after such periods of weak to moderate turbulence that planktonic Cryptomonadida may find favorable conditions and establish a population maximum (e.g., Sommer 1981; Reynolds 1984b; Fahnenstiel & Glime 1983). When the conditions again equilibrate, nutrient concentrations are reduced and grazing may take its toll. The experiments of Reynolds and co-workers (1982) suggest that other loss processes may be minor compared to grazing among the Cryptomonadida. Grazing is, however, a process by which nutrients are rapidly regenerated to support the productivity and growth of cryptomonad populations (Ilmavirta 1980; Reinertsen 1982).

Parasitism

Cryptomonads may have "enemies" of parasitic nature. A phycomycete, *Rhizophydium fugax* Canter, may infect the extracellular polysaccharide of a *Cryptomonas* sp. and eventually harm the cells (Canter 1968). An intracellular parasite of uncertain affinity that is damaging to cells of *Cryptomonas* spp. was studied with the light microscope by Ettl and Moestrup (1980). Intracellular structures reminiscent of the latter were earlier recorded by Schiller (1957) and Wawrik (1977). Caljon (1983) recorded a phycomycete, a possible member of the family Olpidiaceae, parasitic in *Chilomonas striata*. *Spiromonas perforans* fulfills part of its life cycle inside *Chilomonas* (Brugerolle & Mignot 1979). The prosthecate bacterium *Caulobacter* may modify the morphology of the cell (Klaveness 1982), and what appear to be viral infections may also occur in cryptomonads (Pienaar 1976).

THE ECOLOGY OF CRYPTOMONAD EVOLUTION

Many papers discuss the possible symbiotic origin of photosynthetic Cryptomonadida (e.g., Greenwood 1974; Taylor 1976; Dodge 1979; Gibbs & Gillott 1980; Cavalier-Smith 1982; Gibbs 1981a; Ludwig & Gibbs 1985). Much less has been said about the ecological context in which this possible evolution has taken place; that is, the ecology of cryptomonad evolution. Extending the current hypotheses regarding the origin and evolution of the Cryptomonadida beyond the cellular level (where theories have been corroborated in a remarkable manner) to the environment that made this development favorable may influence the further development of theories in a fruitful manner. This area of speculation is certainly open to the inquisitive mind.

Phagotrophy and subsequent establishment of a permanent symbiotic consortium between a colorless "host" flagellate and an early eucaryotic alga possessing phycobiliprotein pigments is a possible line of evolution leading to today's functionally well-integrated, photosynthetic and planktonic cryptomonads. Present colorless, phagotrophic Cryptomonadida such as *Cyathomonas*, are possible descendents of the ancient stock of host flagellates confined to environments where prey (bacteria, benthic microalgae?) are abundant. Did the establishment of a successful symbiosis free the consortium from the detrital and silt environment and allow penetration into the pelagic domain? The symbiontic consortiae, the planktonic photosynthetic cryptomonads of today, may still have physiological and morphological relics of this ancestry influencing their survival and success in the extant environments.

If symbiosis allowed penetration into the pelagic environment, the consortia were also exposed to different enemies. Might a lack of early coevolution explain the apparent absence or inefficient avoidance or defense mechanisms in the Cryptomonadida toward the grazing zooplankton? Does the observation that optimal population development of photosynthetic cryptomonads mainly occurs under conditions of nutrient abundance reflect their host cell's early existence in saprobic (nutrient-rich) environments? An origin of the photosynthetic Cryptomonadida from benthic stocks of saprobic phagotrophs may be the underlying reason for the apparent diversity in freshwater compared to the sea. The effect of seawater may have been a selection among stocks originating in the benthic silt where they were frequently exposed to groundwater seepage and surface runoffs. Consequently, the toleration of variable salinity and freshwater may be an ancient property, whereas the adaption of enzymes and physiological processes to higher salinities has only been mastered by a minority among the Cryptomonadida. The total diversity of Cryptomonadida in the truly marine environment seems small compared to that of brackish and freshwater environments.

CRYPTOMONAD SURVIVAL STRATEGIES

Much of the present interest in pelagic food webs of limnic and marine environments has focused on the role of flagellates. Being among the most common and widespread flagellate primary producers in freshwater and, under favorable conditions, in the sea, the Cryptomonadida deserve a concerted research effort simply to bring our knowledge up to the level of understanding comparable to dinoflagellates and diatoms. At present we can make only a few generalizations on the ecology and survival strategies of the Cryptomonadida:

1. Some smaller species, adapted to a very moderate light climate, may outgrow grazing loss in well-mixed water columns where nutrients are available.
2. Some cryptomonads seem to survive the summer in the thermocline of lakes where available nutrients allow them to outgrow the loss from grazing.
3. Other larger species, particularly when present in smaller lakes, may avoid grazing by virtue of size, and optimize use of light and nutrients by diel migration.
4. A few species may embed themselves in palmelloid mucilage layers that may be effective against some grazers, whereas others may retreat seasonally from the plankton by means of resting spores (Klaveness 1984, 1985b).

To gain insight into the various strategies at play during the growth season, our foremost requirement is to distinguish between the different waxing and waning populations in the ecosystem. This can only be done by a serious study of the taxonomy at the species level (Klaveness 1985b). Our knowledge, compared to other groups of phytoplankton, is lacking in the following areas:

1. Basic information on nutrient uptake, storage, and release as well as requirements for growth are lacking for the most common and

quantitatively important members of the Cryptomonadida. pH effects on growth are also poorly known.

2. Despite the present activity of research on cryptomonad pigments, little has been done on the process of light adaptation and its ecological relevance. Basic laboratory work needs to be done on strains of common species to supplement the numerous observations of low-light requirements from nature. Of particular importance is information concerning adaptation to low-light, complementary chromatic adaptation to in situ spectral composition, the minimum PAR required, and the influence of nutritional state on light adaptation, requirements, and tolerance.

3. The potential for heterotrophy and photoassimilation of organic compounds at very low light levels needs further testing, especially at low temperatures, where organic supplementation may be important under natural conditions. Phagotrophy in phototrophic and colorless species also requires additional investigation.

4. Numerous strains should be screened in culture for temperature tolerances in order to identify the possible existence of physiologically different strains, or "warm-water" and "cold-water" species, the existence of which is well documented in other groups of phytoplankton. Further experimental study is also needed to identify interactions between temperature and light/nutrients.

5. The nutritional value of cryptomonads for zooplankton, as compared to other common phytoplankters, should be more critically evaluated. Variation of nutritional value with growth conditions is also an interesting area of research.

6. Of great importance is that laboratory work should employ recently isolated strains from the kind of ecosystem under investigation rather than culture collection strains isolated from very different ecosystems and frequently held in culture for decades.

Progress in understanding whole-system dynamics, including the effect of turbulence, light climate, in situ nutrient regeneration, and food-web feedbacks, is hampered as long as basic knowledge of an important component is lacking. The low, but seemingly constant, background population densities of cryptomonads in most localities through the year may reflect the high affinity of herbivorous consumers for them. Their relative importance in the system cannot be gauged from their standing stock or instantaneous diversity: "communities ... should be regarded not as the menu, but as the unserved portions of the meal" (Reynolds 1980). Altogether, the irregular waxing and waning of cryptomonad populations, compared to the predictability of certain diatom and blue-green algal populations, may reflect the availability of a large register of evolutionary adapted responses in different species, all interacting through the year.

REFERENCES

Ahlgren, G. (1970). Limnological studies of lake Norrviken, a eutrophicated Swedish lake. *Schweiz. Z. Hydrol.,* 32, 353–96.

Almer, B., Dickson, W., Ekström, C., Hörnström, E., and Miller, U. (1974). Effects of acidification on Swedish lakes. *Ambio,* 3, 30–6.

Antia, N. J. (1976). Effects of temperature on the darkness survival of marine microplanktonic algae. *Microb. Ecol.,* 3, 41–54.

(1980). Nutritional physiology and biochemistry of marine cryptomonads and chrysomonads. In *Biochemistry and Physiology of Protozoa,* 2d ed., vol. 3, ed. M. Levandowsky and S. H. Hutner, pp. 67–115. New York: Academic Press.

Anton, A. and Duthie, H. C. (1981). Use of cluster analysis in the systematics of the algal genus *Cryptomonas. Can. J. Bot.,* 59, 992–1002.

Arvola, L., Salonen, K., Jones, R. I., Bergstrom, I., and Heinonen, A. (In press). A three day study of the diel behaviour of plankton in a highly humic and steeply stratified lake. *Holarct. Ecol.*

Bailey-Watts, A. E. and Duncan, P. (1981). The phytoplankton. *Monogr. Biologicae,* 44, 91–118.

Baker, A. L. and Brook, A. J. (1971). Optical density profiles as an aid to the study of microstratified phytoplankton populations in lakes. *Arch. Hydrobiol.,* 69, 214–33.

Barber, R. T., White, A. W., and Siegelman, H. W. (1969). Evidence for a cryptomonad symbiont in the ciliate, *Cyclotrichium meunieri. J. Phycol.,* 5, 86–8.

Bicudo, C. E. M. (1966). *Björnbergiella,* a new genus of Cryptophyceae from Hawaiian soil. *Phycologia,* 5, 217–21.

Bold, H. C. and Wynne, M. J. 1985. *Introduction to the Algae. Structure and Reproduction,* 2d. ed. Englewood Cliffs, NJ: Prentice-Hall.

Bowen, M. S. and Ward, H. B. (1977). Laboratory culture of *Cryptomonas ovata* and *Chroomonas* sp. (Cryptophyceae). *Microbios. Letters,* 6, 77–84.

Braarud, T. (1951). Salinity as an ecological factor in marine phytoplankton. *Physiol. Plantarum.* 4, 18–34.

(1961). Cultivation of marine organisms as a means of understanding environmental influences on populations. In *Oceanography,* ed. M. Sears, pp. 271–98. Washington: American Association for the Advancement of Science, Publ. no. 67.

(1962). Species distribution in marine phytoplankton. *J. Oceanogr. Soc. Japan,* 20, 628–49.

Braarud, T., Føyn, B., and Gran, H. H. (1928). Biologische Untersuchungen in einigen Seen des östlichen Norwegens, August–September 1927. *Avh.*

Det. Norske Vidensk. -Akad. Oslo. I. Mat. -Nat. Klasse, 1928, No. 2, 1–37.

Brand, L. E. (1982). Genetic variability and spatial patterns of genetic differentiation in the reproduction rates of the marine coccolithophores *Emiliania huxleyi* and *Gephyrocapsa oceanica. Limnol. Oceanogr.,* 27, 236–45.

Brown, T. E. and Richardson, F. L. (1968). The effect of growth environment on the physiology of algae: light intensity. *J. Phycol.,* 4, 38–54.

Brugerolle, G. and Mignot, J. P. (1979). Observations sur le cycle l'ultrastructure et la position systématique de *Spiromonas perforans* (*Bodo perforans* Hollande 1938), flagellé parasite de *Chilomonas paramaecium:* ses relations avec les dinoflagellés et sporozoaires. *Protistologica,* 15, 183–96.

Buch, K. R. and Garrison, D. L. (1983). Protists from the ice-edge region of the Weddel Sea. *Deep-Sea Res.,* 30, 1261–77.

Burns, C. W. and Mitchell, S. F. (1974). Seasonal succession and vertical distribution of phytoplankton in lake Hayes and lake Johnson, South Island, New Zealand. *N.Z. J. Mar. Freshw. Res.,* 8, 167–209.

Butcher, R. W. (1967). *An Introductory Account of the Smaller Algae of British Coastal Waters. Part IV: Cryptophyceae.* Fishery Investigations, Ser. IV, Ministry of Agriculture, Fisheries and Food, Her Majesty's Stationery Office, London.

Caljon, A. (1983). Brackish-water phytoplankton of the Flemish lowland. *Dev. in Hydrobiol.,* 18, 1–272.

Campbell, P. H. (1973). *Studies on Brackish Water Phytoplankton.* University of North Carolina, Sea Grant Program, Sea Grant publication UNC-SG-73-07.

Canter, H. M. (1968). Studies on British chytrids. XXVII. *Rhizophydium fugax* sp. nov., a parasite of planktonic cryptomonads with additional notes and records of planktonic fungi. *Trans. Br. Mycol. Soc.,* 51, 699–705.

Capblancq, J. (1972). Phytoplancton et productivité primaire de quelques lacs d'altitude dans les pyrénées. *Ann Limnol.,* 8, 231–321.

Cavalier-Smith, T. (1982). The origin of plastids. *Biol. J. Linn. Soc.,* 17, 289–306.

Christensen, T. (1980). *Algae. A Taxonomic Survey,* Fasc. 1 AiO Tryk as, Odense.

Cloern, J. E. (1977). Effects of light intensity and temperature on *Cryptomonas ovata* (Cryptophyceae) growth and uptake rates. *J. Phycol.,* 13, 389–95.
    (1978). Simulation model of *Cryptomonas ovata* population dynamics in southern Kootenay Lake, British Columbia. *Ecol. Modelling,* 4, 133–49.

Cloern, J. A., Cole, B. E., Wong, R. L. J., and Alpine, A. (1985). Temporal dynamics of estuarine phytoplankton: A case study of San Francisco Bay. *Hydrobiologia,* 129, 153–76.

Collins, R. P. and Kalnins, K. (1966). Carbonyl compounds produced by *Cryptomonas ovata* var. *palustris. J. Protozool.,* 13, 435–7.

Conrad, W. and Kufferath, H. (1954). Recherches sur les eaux saumatres des environs de Liloo. II. *Mém. Inst. Roy. Sci. Nat. Belgique,* 127.

Cosgrove, W. B. (1950). Studies on the question of chemoautotrophy in *Chilomonas paramecium. Physiol. Zool.,* 23, 73–84.

Danforth, W. F. (1967). Respiratory metabolism. In *Research in Protozoology,* vol. 1, ed. Tze-Tuan Chen, pp. 201–306. Pergamon Press.

Davis, P. G. and Sieburth, J. McN. (1984). Estuarine and oceanic microflagellate predation of actively growing bacteria: estimation by frequency of dividing-divided bacteria. *Mar. Ecol. Progr. Ser.,* 19, 237–46.

Dodge, J. D. (1979). The phytoflagellates: fine structure and phylogeny. In *Biochemistry and Physiology of Protozoa,* 2d ed., vol. 1, ed. M. Levandowsky and S. H. Hutner, pp. 7–57. New York: Academic Press.

Dokulil, M. (1979). Seasonal pattern of phytoplankton. *Monogr. Biol.,* 37, 203–31.

Edmondson, W. T. (1965). Reproductive rate of planktonic rotifers as related to food and temperature in nature. *Ecol. Monogr.,* 35, 61–111.

Elster, H-J. and Motsch, B. (1966). Untersuchungen über das Phytoplankton und die organische Urproduktion in einigen Seen des Hochscwarzwaldes, im Schleinsee und Bodensee. *Arch. Hydrobiol. Suppl.,* 28, 291–376.

Estep, K. W., Davis, P. G., Hargraves, P. E., and Sieburth, J. McN. (1984). Chloroplast containing microflagellates in natural populations of North Atlantic nannoplankton, their identification and distribution; including a description of five new species of *Chrysochromulina* (Prymnesiophyceae). *Protistologica,* 20, 613–34.

Ettl, H. (1980). Beitrag zur Kenntnis der Süsswasseralgen Dänemarks. *Botanisk Tidsskrift,* 74, 179–223.

Ettl, H. and Moestrup, Ø. (1980). Über einen intrazellularen Parasiten bei *Cryptomonas* (Cryptophyceae), I. *Pl. Syst. Evol.,* 135, 211–26.

Ettl, H., Müller, K., Neumann, K., von Stosch, H. A., and Weber, W. (1967). Vegetative Fortpflanzung, Parthenogenese und Apogamie bei Algen. In *Handbuch der Pflanzenphysiologie,* vol. 18, ed. W. Ruhland, pp. 597–776. Berlin: Springer-Verlag.

Evans, J. H. (1962). Some new records and forms of algae in central East Africa. *Hydrobiologia,* 20, 59–86.

Fahnenstiel, G. L. and Glime, J. M. (1983). Subsurface chlorophyll maximum and associated *Cyclotella* pulse in lake Superior. *Int. Revue ges. Hydrobiol.,* 68, 605–16.

Faust, M. A. and Gantt, E. (1973). Effect of light intensity and glycerol on the growth, pigment composition, and ultrastructure of *Chroomonas* sp. *J. Phycol.,* 9, 489–95.

Ferguson, A. J. D., Thompson, J. M., and Reynolds, C. S. (1982). Structure and dynamics of zooplankton communities maintained in closed systems, with special reference to the algal food supply. *J. Plank. Res.,* 4, 523–43.

Findenegg, I. (1971). Die Producktionsleistungen einiger planktischer Algenarten in ihrem naturlichen Milieu. *Arch. Hydrobiol.,* 69, 273–93.

Francke, J. A. and Coesel, P. F. M. (1985). Isozyme variation within and between Dutch populations of *Closterium ehrenbergii* and *C. moniliferum* (Chlorophyta, Conjugatophyceae). *Br. Phycol. J.* 20, 201–9.

Fritsch, F. E. (1935). *The Structure and Reproduction of the Algae,* vol. 1. Cambridge: Cambridge University Press.

Furuya, K. and Marumo, R. (1983). The structure of the phytoplankton community in the subsurface chlorophyll maxima in the western North Pacific Ocean. *J. Plank. Res.,* 5, 393–406.

Gallagher, J. (1982). Physiological variation and electrophoretic banding patterns of genetically different seasonal populations of *Skeletonema costatum* (Bacillariophyceae). *J. Phycol.,* 18, 148–62.

Gantt, E. (1980). Photosynthetic cryptophytes. In *Phytoflagellates,* ed. E. R. Cox, pp. 381–405. New York: Elsevier North-Holland.

Geller, W. (1984). A device for in situ measurement of zooplankton food selection, grazing and respiration rates. *Verh. Internat. Verein. Limnol.,* 22, 1425–31.

Gibbs, S. (1981a). The chloroplasts of some algal groups may have evolved from endosymbiotic eucaryotic algae. *Ann. N. Y. Acad. Sci.,* 361, 193–208.

(1981b). The chloroplast endoplasmic reticulum: structure, function, and evolutionary significance. *Int. Rev. Cytology,* 72, 49–99.

Gibbs, S. and Gillott, M. A. (1980). Has the chloroplast of cryptomonads evolved from an eucaryotic symbiont? In *Endocytobiology,* eds. W. Schwemmler and H. E. A. Schenk, pp. 737–43. Berlin and New York: de Gruyter.

Gilbert, J. J. and Bogdan, K. G. (1981). Selectivity of *Polyarthra* and *Kertella* for flagellate and aflagellate cells. *Verh. Internat. Verein. Limnol.,* 21, 1515–21.

Grall, J-R. and Jaques, G. (1982). Communautes phytoplanctoniques Antarctiques. In *Campagne Oceanographique Antiprod,* 2, pp. 111–129. Comité National Francaise de Recherches Antarctiques.

Greenwood, A. D. (1974). The Cryptophyta in relation to phylogeny and photosynthesis. *8th Internat. Congr. Electron Microscopy. Canberra,* Abstract no. 32, pp. 566–7.

Griessman, K. (1913). Über marine Flagellaten. *Arch. Protistenk.* 32, 1–78.

Hargrave, B. T. and Geen, G. H. (1970). Effects of copepod grazing on two natural phytoplankton populations. *J. Fish. Res. Bd. Can.,* 27, 1395–1403.

Hasle, G. R. (1969). An analysis of the phytoplankton of the Pacific Southern Ocean: Abundance, composition, and distribution during the "Brategg" expedition, 1947–1948. *Hvalrådets Skrifter,* 52. Oslo: Det Norske Vitenskaps-Akademi.

Hecky, R. E. and Fee, E. J. (1981). Primary production and rates of algal growth in Lake Tanganyika. *Limnol. Oceanogr.,* 26, 532–47.

Hecky, R. E. and Kling, H. J. (1981). The phytoplankton and protozooplankton of the euphotic zone of Lake Tanganyika: species composition, biomass, chlorophyll content, and spatio-temporal distribution. *Limnol. Oceanogr.,* 26, 548–64.

Heimdal, B. R. (1983). Phytoplankton and nutrients in the waters northwest of Spitsbergen in the autumn of 1979. *J. Plank. Res.,* 5, 901–18.

Heinonen, P. (1980). Quantity and composition of phytoplankton in Finnish inland waters. *Publ. Water Res. Inst.* (Helsinki), 37, 1–91.

Hollande, A. and Cachon, J. (1950). Structure et affinités d'un flagellé marin peu connu: *Telonema subtilis* Griesm. *Ann. des Sc. Nat. Zool.,* IIe serie, 12, 109–13.

Holligan, P. M. and Harbour, D. S. (1977). The vertical distribution and succession of phytoplankton in the western English channel in 1975 and 1976. *J. Mar. Biol. Ass. U.K.,* 57, 1075–93.

Holmgren, S. K. (1984). Experimental lake fertilization in the Kuokkel area, Northern Sweden. Phytoplankton biomass and algal composition in natural and fertilized subarctic lakes. *Int. Revue ges. Hydrobiol.,* 69, 781–817.

Hongve, D. (1974). Hydrographical features of Nordbytjernet, a manganese-rich meromictic lake in SE Norway. *Arch. Hydrobiol.,* 74, 227–46.

Hsiao, S. I. C. and Pinkewycz, N. (1983). Phytoplankton data from Frobisher Bay, 1979 to 1981. *Can. Data Rep. Fish. Aquat. Sci.,* No. 419.

Huber-Pestalozzi, G. (1968). Das Phytoplankton des Süsswassers, 2 Aufl., ed. B. Fott. *Die Binnengewässer,* 16 (3), 2–78, 311–13.

Hulburt, E. M. (1965). Flagellates from brackish waters in the vicinity of Woods Hole, Massachusetts. *J. Phycol.,* 1, 87–94.

Humphrey, G. F. (1979). Photosynthetic characteristics of algae grown under constant illumination and light-dark regimes. *J. Exp. Mar. Biol. Ecol.,* 40, 63–70.

Ichimura, S., Nagasawa, S., and Tanaka, T. (1968). On the oxygen and chlorophyll maxima found in the metalimnion of a mesotrophic lake. *Bot. Mag. Tokyo* 81, 1–18.

Ilmavirta, V. (1980). Phytoplankton in 35 Finnish brown-water lakes of different trophic status. *Dev. Hydrobiol.,* 3, 121–30.

(1983). The role of flagellated phytoplankton in chains of small brown-water lakes in southern Finland. *Ann. Bot. Fennici.* 20, 187–95.

Iwasaki, H. (1979). Physiological ecology of red tide flagellates. In *Biochemistry and Physiology of Protozoa,* 2d ed., vol. 1, eds. M. Levandowsky and S. H. Hutner, pp. 357–93. New York: Academic Press.

Javornicky, P. (1967). Some interesting algal flagellates. *Folia Geobot. Phytotax.* (Praha) 2, 43–67 + Pl. 1–9.

(1978). 6. trieda Cryptophyceae-kryptomonady. In *Sladkovodne Riasy,* ed. F. Hindak, pp. 451–65. Bratislava: Slovenske Pedagogicke Nakladatel'stvo.

Javornicky, P. and Hindak, F. (1970). *Cryptomonas frigoris* spec. nova (Cryptophyceae), the new cyst-forming flagellate from the snow of the High Tatras. *Biologia* (Bratislava), 25, 241–50.

Kalff, J. and Welch, H. E. (1974). Phytoplankton production in Char Lake, a natural polar lake, and in Meretta Lake, a polluted polar lake. *J. Fish. Res. Bd. Can.*, 31, 621–36.

Karentz, D. and Smayda, T. J. (1984). Temperature and seasonal occurrence patterns of 30 dominant phytoplankton species in Narragansett Bay over a 22-year period (1959–1980). *Mar. Ecol. Prog. Ser.*, 18, 277–93.

Keskitalo, J. (1977). The species composition and biomass of phytoplankton in the eutrophic Lake Lovojärvi, southern Finland. *Ann. Bot. Fennici.*, 14, 71–81.

Klaveness, D. (1977). Morphology, distribution and significance of the manganese-accumulating microorganism. *Metallogenium* in lakes. *Hydrobiologia*, 56, 25–33.

  (1981). *Rhodomonas lacustris* (Pascher & Ruttner) Javornicky (Cryptomonadida): Ultrastructure of the vegetative cell. *J. Protozool.*, 28, 83–90.

  (1982). The *Cryptomonas–Caulobacter* consortium: facultative ectocommensalism with possible taxonomic consequences? *Nord. J. Bot.*, 2, 183–8.

  (1984). Studies on the morphology, food selection and growth of two planktonic freshwater strains of *Coleps* sp. *Protistologica*, 20, 335–49.

  (1985a). The Cryptophyceae – geographical distribution, ecological range, and role in the aquatic food-webs. *Second Int. Phycol. Congr. Copenhagen Book of Abstracts*, p. 82.

  (1985b). Classical and modern criteria for determining species of Cryptophyceae. *Bull. Plankton Soc. Japan*, 32, 111–28.

Knisely, K. and Geller, W. (1986). Selective feeding of four zooplankton species on natural lake phytoplankton. *Oecologia* (Berlin), 69, 86–94.

Kudo, R. R. (1977). *Protozoology*, 5th ed. Springfield: Charles C. Thomas,.

Kuenzler, E. J. (1965). Glucose-6-phosphate utilization by marine algae. *J. Phycol.*, 1, 156–64.

Kuzel-Fetzmann, E. (1979). The algal vegetation of Neusiedlersee. *Monogr. Biol.*, 37, 171–202.

Lee, J. J., Hutner, S. H., and Bovee, E. C. 1985. *An Illustrated Guide to the Protozoa*. Lawrence, Kansas: Society of Protozoologists.

Lehman, J. T. and Sandgren, C. D. (1985). Species-specific rates of growth and grazing loss among freshwater algae. *Limnol. Oceanogr.*, 30, 34–46.

Lewis, W. M. and Riehl, W. (1982). Phytoplankton composition and morphology in Lake Valencia, Venezuela. *Int. Revue ges. Hydrobiol.*, 67, 297–322.

Lichtlé, C. (1979). Effects of nitrogen deficiency and light of high intensity on *Cryptomonas rufescens* (Cryptophyceae). I. Cell and photosynthetic apparatus transformations and encystment. *Protoplasma*, 101, 283–99.

  (1980). Effects of nitrogen deficiency and light of high intensity on *Cryptomonas rufescens* (Cryptophyceae). II. Excystment. *Protoplasma*, 102, 11–19.

Lohmann, H. (1922). Zentrifugenplankton und Hochseeströmung. *Int. Revue ges. Hydrobiol.*, 10, 603–82.

Ludwig, M. and Gibbs, S. (1985). DNA is present in the nucleomorph of Cryptomonads: further evidence that the chloroplast evolved from a eucaryotic endosymbiont. *Protoplasma,* 127, 9–20.

Meyer, S. R. and Pienaar, R. N. (1984). The microanatomy of *Chroomonas africana* sp. nov. (Cryptophyceae). *S. Afr. J. Bot.,* 3, 306–19.

Morgan, K. and Kalff, J. (1975). The winter dark survival of an algal flagellate – *Cryptomonas erosa* (Skuja). *Verh. Internat. Verein. Limnol.,* 19, 2734–40.

(1979). Effect of light and temperature interactions on the growth of *Cryptomonas erosa* (Cryptophyceae). *J. Phycol.,* 15, 127–34.

Mucibabic, S. (1956). Some aspects of the growth of single and mixed populations of flagellates and ciliates. *Exp. Biol.,* 33, 627–44.

Nauwerck, A. (1963). Die Beziehungen zwischen Zooplankton und Phytoplankton im See Erken. *Symb. Bot. Upsalienses,* 17 (4), 1–163.

(1968). Das Phytoplankton des Latnjajaure 1954–1965. *Schweiz. Z. Hydrol.,* 30, 188–216.

Nelson, D. M. and Brand, L. E. (1979). Cell division periodicity in 13 species of marine phytoplankton on a light:dark cycle. *J. Phycol.,* 15, 67–75.

Oakley, B. R. and Santore, U. J. (1982). Cryptophyceae: Introduction and bibliography. In *Selected Papers in Phycology,* eds. J. R. Rosowski and B. C. Parker, pp. 682–6. Kansas: Phycological Society of America.

Olsén, Y., Jensen, A., Reinertsen, H., Børsheim, K. Y., Heldal, M., and Langeland, A. (1986). Dependence of the rate of release of phosphorus by zooplankton on the P:C ratio in the food supply, as calculated by a recycling model. *Limnol. Oceanogr.,* 31, 34–44.

Palmer, C. M. (1962). *Algae in water supplies.* Washington: U.S. Dept. Health and Human Services.

Parker, B. C., Simmons, G. M., Jr., Seaburg, K. G., Cathey, D. D., and Allmutt, F. C. T. (1982). Comparative ecology of plankton communities in seven Antarctic oasis lakes. *J. Plank. Res.,* 4, 271–86.

Pascher, A. (1913). Cryptomonadinae. In *Süsswasserflora,* 2 (Flagellatae 2), pp. 96–114.

Pavoni, M. (1963). Die Bedeutung des Nannoplankton im Vergleich zum Netzplankton. *Schwiez. Z. Hydrol.,* 25, 219–341.

Pechlaner, R. (1971). Factors that control the production rate and biomass of phytoplankton in high-mountain lakes. *Mitt. Internat. Verein. Limnol.,* 19, 125–45.

Pejler, B. (1977). Experience with rotifer cultures based on *Rhodomonas. Arch. Hydrobiol. Beih. Ergebn. Limnol.,* 8, 264–6.

Pfiester, L. A. and Holt, J. R. (1977). A freshwater "red tide" in Texas. *J. Phycol.,* 13, Abstr. Suppl., no. 305, 53.

Pienaar, R. N. (1976). Virus-like particles in three species of phytoplankton from San Juan Island, Washington. *Phycologia.* 15, 185–90.

Playfair, G. I. 1921. Australian freshwater flagellates. *Proc. Linn. Soc. N.S. Wales,* 46, 99–146.

Pollingher, U. (1981). The structure and dynamics of the phytoplankton assemblages in lake Kinneret, Israel. *J. Plank. Res.,* 3, 93–105.

Porter, K. G., Sherr, E. B., Sherr, B. F., Pace, M., and Sanders, R. W. (1985). Protozoa in planktonic food webs. *J. Protozool.*, 32, 409–15.

Pringsheim, E. G. (1944). Some aspects of taxonomy in the Cryptophyceae. *New Phytol.*, 43, 143–50.

(1963). *Farblose Algen.* Stuttgart: Gustav Fischer Verlag.

(1968). Zur Kenntnis der Cryptomonaden des Süsswassers. *Nova Hedwigia*, 16, 367–401 + Tab. 148–54.

Provasoli, L. and Carlucci, A. F. (1974). Vitamins and growth regulators. In *Algal Physiology and Biochemistry*, ed. W. D. P. Stewart, pp. 741–87. Blackwell Scientific.

Ramberg, L. (1979). Relations between phytoplankton and light climate in two Swedish forest lakes. *Int. Revue ges. Hydrobiol.*, 64, 749–82.

Raven, J. A. (1984). A cost-benefit analysis of photon absorption by photosynthetic unicells. *New Phytol.*, 98, 593–625.

Reinertsen, H. (1982). The effect of nutrient addition on the phytoplankton community of an oligotrophic lake. *Holarct. Ecol.*, 5, 225–52.

Repak, A. J. (1983). Suitability of selected marine algae for growing the marine heterotrich ciliate *Fabrea salina. J. Protozool.*, 30, 52–4.

Reynolds, C. S. (1980). Phytoplankton assemblages and their periodicity in stratifying lake systems. *Holarct. Ecol.*, 3, 141–59.

(1982). Phytoplankton periodicity: its motivation, mechanisms and manipulation. *Freshwater Biol. Assoc., Ann. Rep.*, 50, 60–75.

(1984a). *The Ecology of Freshwater Phytoplankton.* Cambridge: Cambridge University Press.

(1984b). Phytoplankton periodicity: the interactions of form, function and environmental variability. *Freshwat. Biol.*, 14, 111–42.

Reynolds, C. S. and Reynolds, J. B. (1985). The atypical seasonality of phytoplankton in Crose Mere, 1972: an independent test of the hypothesis that variability in the physical environment regulates community dynamics and structure. *Br. Phycol. J.*, 20, 227–42.

Reynolds, C. S., Thompson, J. M., Ferguson, A. J. D., and Wiseman, S. W. (1982). Loss processes in the population dynamics of phytoplankton maintained in closed systems. *J. Plank. Res.*, 4, 561–600.

Rodhe, W. (1955). Can plankton production proceed during winter darkness in subarctic lakes? *Verh. Internat. Verein Limnol.*, 12, 117–22.

Rojas de Mendiola, B. (1981). Seasonal phytoplankton distribution along the Peruvian coast. In *Coastal Upwelling*, ed. F. A. Richards, pp. 348–56. Washington, D.C.: American Geophysical Union.

Rosén, G. (1981). *Tusen sjöar. Växtplanktons miljökrav.* Naturvårdsverket, Liber distribution, 162 89 Stockholm.

Rott, E. (1983). Sind die Veränderungen im Phytoplanktonbild des Pilburger Sees Auswirkungen der Tiefenwasserableitung? *Arch. Hydrobiol. Suppl.*, 67, 29–80.

Round, F. E. (1981). *The Ecology of Algae.* Cambridge: Cambridge University Press.

Salonen, K., Jones, R. I., and Arvola, L. (1984). Hypolimnetic phosphorus retrieval by diel vertical migrations of lake phytoplankton. *Freshwat. Biol.,* 14, 431–38.

Santore, U. J. (1984). Some aspects of taxonomy in the Cryptophyceae. *New Phytol.,* 98, 627–46.

Sarnelle, O. (1986). Field assessment of the quality of phytoplankton food available to *Daphnia* and *Bosmina. Hydrobiologia.* 131, 47–56.

Schiller, J. (1957). *Untersuchungen an den planktischen Protophyten des Neusiedelersees 1950–1954.* II. Teil, Burgenländisches Landesmuseum, Eisenstadt.

Schindler, D. W. and Holmgren, S. K. (1971). Primary production and phytoplankton in the Experimental Lakes Area, Northwestern Ontario, and other low-carbonate waters, and a liquid scintillation method for determining 14C activity in photosynthesis. *J. Fish. Res. Bd. Canada,* 28, 189–201.

Schlösser, U. G. (1982). Sammlung von Algenkulturen. *Ber. Deutsch. Bot. Ges.,* 95, 181–276.

Schuster, F. L., (1968). The gullet and trichocysts of *Cyathomonas truncata. Exp. Cell Res.,* 49, 277–84.

Sellner, K. G., Kachur, M. E., and Lyons, L. (1984). Alterations in the carbon fixation during power plant entrainment of estuarine phytoplankton. *Water. Air Soil Poll.,* 21, 359–74.

Shapiro, J. (1984). Blue-green dominance in lakes: the role and management significance of pH and CO_2. *Int. Revue ges. Hydrobiol.,* 69, 765–80.

Skuja, H. (1948). Taxonomie des Phytoplanktons einiger Seen in Uppland, Schweden. *Symb. Bot. Upsalienses,* 9, 1–399 + Taf. 1–39.

(1956). Taxonomische und biologische Studien über das Phytoplankton Schwedischer Binnengewässer. *Nova Acta Reg. Soc. Sci. Upsaliensis,* Ser. 4, 16 (3), 1–404 + Taf. 1–63.

(1964). Grundzüge der Algenflora und Algenvegetation der Fjeldgegenden um Abisko in Schwedisch-Lappland. *Nova Acta Reg. Soc. Sci. Upsaliensis,* Ser. 4, 18 (3), 1–465.

Smayda, T. J. (1958). Biogeographical studies of marine phytoplankton. *Oikos,* 9, 158–91.

Soeder, C. J. (1967). Tagesperiodische Wanderungen bei begeisselten Planktonalgen. *Umschau.,* 1967 (12), 338.

Sommer, U. (1981). The role of r- and K-selection in the succession of phytoplankton in Lake Constance. *Acta Oecologica: Oecol. Gener.,* 2, 327–42.

(1982). Vertical niche separation between two closely related planktonic flagellate species (*Rhodomonas lens* and *Rhodomonas minuta* v. *nannoplanctica*). *J. Plank. Res.,* 4, 137–42.

(1983). Nutrient competition between phytoplankton species in multispecies chemostat experiments. *Arch. Hydrobiol.,* 96, 399–416.

(1985). Comparison between steady state and non-steady state competition: experiments with natural phytoplankton. *Limnol. Oceanogr.,* 30, 335–46.

Sournia, A. (1982). Is there a shade flora in the marine plankton? *J. Plank. Res.,* 4, 391–9.

Steinberg, C., Lenhart, B., and Klee, R. (1983). Bemerkungen zur Ökologie eines farblosen Phytoflagellaten, *Phyllomitus apiculatus* Skuja (1948), Cryptophyceae, *Arch. Protistenk.,* 127, 307–17.

Takahashi, M. and Ichimura, S. (1968). Vertical distribution and organic matter production of photosynthetic sulfur bacteria in Japanese lakes. *Limnol. Oceanogr.,* 13, 644–55.

Taylor, F. J. R. (1976). Flagellate phylogeny: a study in conflicts. *J. Protozool.,* 23, 28–40.

Taylor, F. J. R., Blackbourn, D. J., and Blackbourn, J. (1969). Ultrastructure of the chloroplasts and associated structures within the marine ciliate *Mesodinium rubrum* (Lohmann). *Nature,* 224, 819–21.

Taylor, W. D., Hern, S. C., William, L. R., Lambou, V. W., Morris, M. K., and Morris, F. A. (1979). *Phytoplankton Water Quality Relationships in U. S. Lakes. Part 6: The Common Phytoplankton Genera from Eastern and Southeastern Lakes.* U.S. Environmental Protection Agency, Environmental Monitoring and Support Laboratory, Working Paper No. 710.

Taylor, F. J. R. and Waters, R. E. (1982). Spring phytoplankton in the subarctic North Pacific Ocean. *Mar. Biol.,* 67, 323–35.

Thinh, L-V. (1983). Effect of irradiance on the physiology and ultrastructure of the marine cryptomonad, *Cryptomonas* strain Lis (Cryptophyceae), *Phycologia,* 22, 7–11.

Throndsen, J. (1969). Flagellates of Norwegian coastal waters. *Nytt Mag. Bot.,* 16, 161–216.

——— (1976). Occurrence and productivity of small marine flagellates. *Norw. J. Bot.,* 23, 269–93.

——— (1983). *Ultra- and Nannoplankton Flagellates from Coastal Waters of Southern Honshu and Kyushu, Japan (Including Some Results from the Western Part of the Kuroshio off Honshu).* Tokyo: Working Party on Taxonomy in the Akashiwo Mondai Kenkyukai, Fishing Ground Preservation Division, Research Department, Fisheries Agency.

Tilzer, M. M. (1973). Diurnal periodicity in the phytoplankton assemblage of a high mountain lake. *Limnol. Oceanogr.,* 18, 15–30.

Tsuji, T. and Adachi, R. (1979). Distribution of nano-phytoplankton including fragile flagellates in the subtropical northwestern Philippine Sea. *J. Oceanogr. Soc. Japan,* 35, 173–8.

Van Meel, L. (1954). Le Phytoplancton. In: *Exploration Hydrobiologique du Lac Tanganika (1946–1947),* vol. IV, fasc. I. Bruxelles: A. Texte.

Vesk, M. and Jeffrey, S. W. (1977). Effect of blue-green light on photosynthetic pigments and chloroplast structure in unicellular marine algae from six classes. *J. Phycol.,* 13, 280–8.

Vieira, A. A. H. and Klaveness, D. (1986). The utilization of organic nitrogen compounds as sole nitrogen source by some freshwater phytoplankters. *Nord. J. Bot.,* 6, 93–7.

Vollenweider, R. A., Munawar, H., and Stadelmann, P. (1974). A comparative review of phytoplankton and primary production in the Laurentian great lakes. *J. Fish. Res. Bd. Can.,* 31, 739–62.

Watanabe, M. and Furuya, M. (1974). Action spectrum of phototaxis in a cryptomonad alga, *Cryptomonas* sp. *Plant Cell Physiol.,* 15, 413–20.

(1978). Phototactic responses of cell population to repeated pulses of yellow light in a phytoflagellate *Cryptomonas* sp. *Plant Physiol.,* 61, 816–18.

(1982). Phototactic behavior of individual cells of *Cryptomonas* sp. in response to continuous and intermittent light stimuli. *Photochem. Photobiol.,* 35, 559–63.

Watanabe, M., Miyoshi, Y., and Furuya, M. (1976). Phototaxis in *Cryptomonas* sp. under condition suppressing photosynthesis. *Plant Cell Physiol.,* 17, 683–90.

Wawrik, F. (1969). Sexualität bei *Cryptomonas* sp. und *Chlorogonium maximum. Nova Hedwigia,* 18, 283–92.

(1977). Neue und seltene "μ"-und Nannoalgen aus Teichen des Waldviertels II. *Arch. Protistenk.,* 119, 407–17.

(1983). Planktonaspekte in Teichen des nordlichen und nordöstlichen Waldviertels während und nach dem Eisbrechen 1982. *Arch. Protistenk.,* 127, 271–82.

Wilcox, L. W. and Wedemayer, G. J. (1984). *Gymnodinium acidotum* Nygaard (Pyrrophyta), a dinoflagellate with an endosymbiotic cryptomonad. *J. Phycol.,* 20, 236–42.

Willén, E., Oké, M., and González, F. (1980). *Rhodomonas minuta* and *Rhodomonas lens* (Cryptophyceae) – Aspects on form-variation and ecology in Lakes Mälaren and Vättern, Central Sweden. *Acta Phytogeogr. Suec.,* 68, 163–72.

Wright, R. T. (1964). Dynamics of a phytoplankton community in an ice-covered lake. *Limnol. Oceanogr.,* 9, 163–78.

Chapter 4

FRESHWATER ARMORED DINOFLAGELLATES: GROWTH, REPRODUCTION STRATEGIES, AND POPULATION DYNAMICS

UTSA POLLINGHER

INTRODUCTION

The freshwater dinoflagellates comprise only 220 species (Bourrelly 1970). They are unicellular algae with a life cycle that allows them to inhabit, alternatively, the plankton (as a motile vegetative cell) and the benthos (as a nonmotile cyst). By far the best-known dinoflagellates ecologically are large-celled species of *Peridinium* and *Ceratium*.

The dinoflagellates populate various types of natural and artificial ecosystems (Pollingher 1986a). Most of the morphological and physiological features of dinoflagellates do not permit them to compete successfully with nannoplanktonic algae. In spite of this, the common species form water blooms during the summer in temperate water bodies and in winter–spring in subtropical Lake Kinneret. This is possible due to strategies developed and used by these algae to counterbalance their handicaps. Their blooms give a fishy smell and taste to the water, and are thus of both practical and ecological concern.

The aim of this chapter is to discuss the important factors affecting dinoflagellate growth, population dynamics, and the strategies that they have developed for survival and blooming under conditions unfavorable for other algae. Emphasis will primarily be placed on two well-studied annual blooms: *Peridinium* spp. in Lake Kinneret, Israel and *Ceratium hirundinella* in Esthwaite Water in England.

DINOFLAGELLATES AS ORGANISMS

Dinoflagellate Cells

Most dinoflagellates are biflagellate motile organisms. Some nonmotile coccoid and amoeboid forms also occur, but in very low numbers. In addition to the predominant pigmented species, there are some colorless forms.

The dinoflagellates possess a distinctive covering of the vegetative cell called an amphiesma (Loeblich 1970). It consists of several layers of membranes that sometimes enclose cellulosic (or polyglucan) plates. Two types of dinoflagellates can be distinguished morphologically: *thecate* and *athecate* or naked. The thecate (armored) species have walls divided into plates arranged in specific series. Athecate forms have a cell covering of membranes only. The dinoflagellates possess a pair of unequal, laterally inserted, heterodynamic flagella that have independent beating patterns (Jahn, Harmon, & Landman 1963). The beat of the transversal flagellum provides both rotational motion and the forward propulsion of the cell (LeBlond & Taylor 1976). The combination of the beating of the two flagella maintains the position of the cell in the water. The flagellar motility of dinoflagellates does not require a high rate of energy expenditure (Raven & Beardall 1981).

The nucleus in dinoflagellates is relatively large, containing a high number of permanently condensed chromosomes (up to 500). The food reserve in the vegetative cells is starch and lipid material in varying proportions, depending on the stage of the life cycle (Messer & Ben-Shaul 1970; Chapman, Dodge, & Heaney 1985).

Vertical Migration

The motile dinoflagellates are able to perform discernible diel vertical migration: They ascend toward the surface during the light period and descend and/or disperse at night. Dinoflagellates are positively phototactic, and there is a relationship between vertical migration and circadian rhythms in photoresponses. Cells show maximum responsiveness to light in the first half of the day, and the photoresponsiveness declines by the end of the day (Forward 1976). Patterns of vertical migration seem to be altered by gradients of light and temperature, by limitation of nutrients, and by population age (Berman & Rodhe 1971; Pollingher & Berman 1975; Heaney 1976; George & Heaney 1978; Harris, Heaney, & Talling 1979; Heaney & Talling 1980a; Heaney & Eppley 1981).

Table 4-1. *Surface: volume ratio (SA/VOL) of algae of various shapes and sizes in Lake Kinneret.*

Algae	Radius (μm)	SA/VOL	Shape
Chroococcus minutus	3	1	sph
Chroococcus turgidus	5	0.60	sph
Microcystis aeruginosa	1.5	2	sph
Microcystis flos-aquae	3.5	0.86	sph
Melosira granulata	7.5 × 18	0.27	cyl
Cyclotella meneghiniana	9 × 10	0.42	cyl
Cyclotella sp.	3 × 4	0.67	cyl
Chrysochromulina	3 × 6	1.52	sph
Scenedesmus quadricauda	6 × 12	2.4	cyl
Tetraedron minimum	4	1.5	cube
Rhodomonas minuta	6 × 12	2.25	cyl
Peridinium cinctum	24	0.12	sph
Peridinium cinctum	30	0.10	sph

sph = sphere; cyl = cylinder.

General Size Range and Cell Size Variability

Armored dinoflagellates are relatively large organisms (*Peridinium* diameter 30–70 μm; *C. hirundinella* length 90–450 μm, Huber-Pestalozzi 1950; *C. furcoides* length 162–322 μm, Hickel 1985a), and therefore they have a low surface-area-to-volume ratio (SA/VOL). The SA/VOL ratio provides information about the exchange potential of a given biomass unit. This ratio decreases with increasing size. At a given size, a spherical shape provides the smallest SA/VOL, which is essentially the case with the armored dinoflagellates. The SA/VOL of *P. cinctum* in Lake Kinneret is 0.10–0.12; that of the cyanobacterium *Chroococcus* spp. is 0.60–1.00 (Table 4-1). Thus dinoflagellates cannot compete effectively with small algae that appear at an early stage of seasonal succession (Pollingher 1981).

It is known from cultures started from a single cell that size variation is a natural feature of *Peridinium* populations as a function of both cell nutrition and stages in the cell cycle. In Lake Kinneret, at the beginning of the bloom, all cells of *Peridinium cinctum* f. *westii* are approximately the same size. Very soon a difference in cell age appears, because only a small percentage of the population divides daily and the old cells are larger than the young ones. Mean *Peridinium* cell size has also been demonstrated to vary from year to year in the lake as a function of phosphorus availability, which affects population growth rate and, hence, the distribution of cell sizes (Serruya & Pollingher 1977). Stosch (1973) has shown that cells of *Woloszynskia apiculata* cultivated at different light intensities have different sizes and shapes.

Great morphological variability also exists in *C. hirundinella* with respect to both local variations and seasonal ones. Most of the investigations of this variability were based on qualitative net samples (Wesenberg-Lund 1908; Lemmermann 1910; Krause 1911–12; Langhans 1925; Entz 1927; Pearsall 1929; Huber-Pestalozzi 1950; Florin 1957; Komarovsky 1959; Daily 1960). The relative distribution of form types and the abundance of *C. hirundinella* in Lake Erken were studied by Dottne-Lindgren and Ekbohm (1975) using an integrated water sample and statistical analysis. The whole *Ceratium* population was divided into three different form types. The authors found that the total length of cells was the most variable parameter and could give the most reliable indication of a possible variation in the environment. They concluded that long individuals appear in early summer, disappear in June (probably dying off), and reappear in mid-October. These cells are succeeded by a "summer form" population consisting of smaller individuals originating from resting cysts. The decreasing length of organisms can perhaps also be explained by the increase of temperature, which increases the division rate, and thus could be a change parallel to that documented for *Peridinium cinctum.*

Seasonal length variation was described for a second *Ceratium* species, *C. furcoides,* from Plußsee by Hickel (1985a). At the beginning of the bloom, the cells were short (170 μm); with increasing cell density, cell length increased (211 μm). The maximum mean cell length was observed at the end of the bloom (269 μm) in cells that did not develop into cysts. With decreasing temperature, cells became shorter. Since sexual reproduction was found in situ in *C. furcoides,* Hickel (1985a) proposed that the large cells at the end of the bloom may be planozygotes that were not transformed into hypnozygotes. Hickel also analyzed the *Ceratium* population in Plußsee during two seasonal cycles and found that the dominant organism was *C. furcoides,* but it was accompanied by *C. furcoides* f. *gracile* and by *C. hirundinella* Figs. 4-1, 4-2).

In Windermere and Esthwaite water, *C. hirundinella* is the numerically dominant species, although in some years (e.g., 1982) *C. furcoides* constitutes approximately 30% of *Ceratium* cells in Esthwaite Water at the time of maximum population density. In Blelham Tarn, *C. furcoides* is often the dominant species, with *C. hirundinella* contributing less than 3% of the *Ceratium* cells in 1982 (Canter & Heaney 1984). It thus seems that in many cases, form variation may be accompanied also by species diversity.

Fig. 4-1. *Ceratium furcoides,* dorsal (left) and ventral view. Scale bar: 20 μm. (Courtesy of Dr. B. Hickel.)

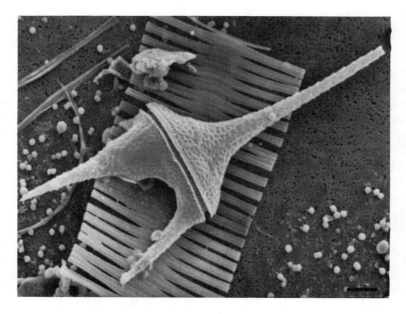

Fig. 4-2. *Ceratium hirundinella.* (Courtesy Dr. B. Hickel.)

The number of horns on *Ceratium* cells also varies with environmental factors. In cultures of *C. hirundinella* isolated from different lakes, the cells usually have two posterior horns (Bruno 1975), even when the cultures were started with three posterior horned cells. It is possible to bring about the production of a third posterior horn by varying temperature and/or trace metals. At 21–25°C all cells in Bruno's cultures had two posterior horns, but at 15°C a large percentage of cells had three posterior horns. In the small population of *C. hirundinella* in Lake Kinneret, different form types have also been observed (Komarovsky 1959).

SEASONAL AND SPATIAL DISTRIBUTION OF DINOFLAGELLATES

Most of the common species of dinoflagellates are not perennial forms. They persist as benthic stages during winter in water bodies located in temperate zones and in summer in those located in the subtropical warm belt (Lake Kinneret). In Lake Kinneret, vegetative cells of *P. cinctum* have disappeared almost completely during the summer in each of the last 10 years. The bloom is generally initiated from cysts resuspended from the sediments. Young cells that emerge from cysts (recognized because they preserve red storage bodies) are recorded at the littoral stations from November until February (temperatures 20–14°C, respectively). The first vegetative cells are found on the mud-water interface where the high concentration of nutrients allows them to store phosphorus inside their cells. This mechanism explains why the early stage of the bloom does not depend entirely on the nutrient concentrations of the lake water.

A similar situation was observed in the development of *C. hirundinella* in Esthwaite Water, a small productive lake. Heaney, Chapman, & Morison (1983) showed evidence for the major role of benthic cysts in providing the inoculum for the development of the *Ceratium* water bloom. In winter, less than 1 cell/100 mL of water was recorded. The bloom began in March, and the March–July period was characterized by nearly exponential growth, followed by a stationary phase during July–September, and ending with a rapid decline in late September or early October (Heaney & Talling 1980a).

The water bloom of *Peridinium* is a constant yearly event in Lake Kinneret, and this may be explained by the geographical location of the lake. In subtropical regions, light is available throughout the year, and the seasonal succession is strongly influenced by winds and rains (Pollingher 1986b). Precipitation and wind direction display a regular seasonal pattern in the lake region. Due to the effect of turbulence on

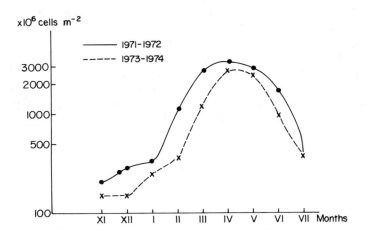

Fig. 4-3. Growth curve for *Peridinium* in Lake Kinneret (monthly averages).

the division rate of *Peridinium* (discussed later), the development of the bloom is directly correlated with the timing and duration of the mixing period in the lake. *Peridinium* appears in greatest numbers at the end of the mixing period (Table 4-2).

In some lakes located in temperate zones, dinoflagellate water blooms are also a regular yearly event (Pollingher 1986a). In other lakes, the succession pattern of algae changes nearly every year. In Plußsee, for instance, the phytoplankton succession pattern has been variable. In summer 1969, *Anabaena flos-aquae* dominated. In summer 1974, a bloom of *Oscillatoria redeckei* developed and was followed by *Cryptomonas, Ceratium,* and *Oscillatoria* again. In summer 1976, *Oscillatoria* and *Peridinium elpatiewsky* developed, and in summer of 1981 and 1982 water blooms of *C. furcoides* and *C. hirundinella* developed (Hickel 1978, 1985a). Such variations are probably correlated with year to year changes in the environmental conditions.

Normal dinoflagellate population growth occurs through asexual division. Dinoflagellate specific growth rate is low and the generation time is long compared with those of other algae. The development of the bloom resembles the growth curve of a batch culture in Lake Kinneret (Fig. 4-3). A short lag phase occurs in November–December, followed by a relatively long exponential phase until March–April, then a stationary phase of about one month, and then a drastic decrease. Such a sequence describes the monthly averages. However, a different picture is obtained when weekly fluctuations are followed (Fig. 4-4).

Table 4-2. The timing of the mixed period (marked with asterisks) and the fluctuations of the number of Peridinium cells ($\times 10^6 \cdot m^{-2}$, monthly averages) in the trophogenic layer of Lake Kinneret.

	1696	1970	1971	1972	1973	1974	1975	1976	1977	1978	1980	1981	1982	1983	1984	1985
Jan.	146*	1320	460*	356*	140*	175*	33*	25*	101*	170*	22*	23*	105*	31*	220	232*
Feb.	417	2040*	877*	1210	1473	400*	52*	258*	655	418*	56*	45*	89*	32*	1470*	330
Mar.	2040	4780	2345	3193	3053	1343	788	740	1761	874	108*	197	247	421	1604	2000
Apr.	4505	4870	3705	3430	4027	2793	2033	2683	2068	1675	1880	1858	1150	1590	1873	3356
May	3700	2885	3850	2990	1270	3148	2912	2158	2542	2369	2954	2338	2067	2616	1109	3000
Jun.	1640	2017	2015	1787	1315	1087	1274	1362	1391	1187	1560	682	909	1025	1785	625

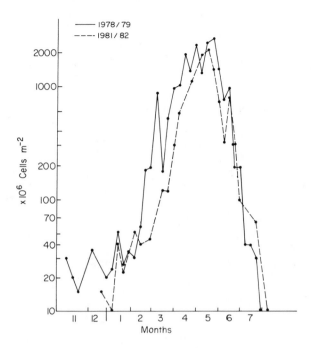

Fig. 4-4. Growth curve of *Peridinium* in Lake Kinneret (weekly variations).

Short-term fluctuations in the overall pattern of growth are probably losses due to high turbulence or mortality. In Lake Kinneret, the end of the dinoflagellate water bloom may be due to three factors: (a) the onset in June of daily westerly winds (from noon to the next early morning) that generate turbulence; (b) currents that force the algae to be exposed to damaging radiation and high temperature; and (c) very low concentrations of nutrients.

The spatial distribution of dinoflagellate cells in the lake is the result of the interaction of three different components: circadian vertical migration, turbulent mixing resulting from high wind stress, and water displacement. The first of these, circadian vertical distribution, results mainly from active swimming and is correlated with light irradiance and chemical composition of the water during quiescent periods. *Peridinium* performs an active upward movement from early morning until afternoon, and a downward movement from late afternoon until midnight (Fig. 4-5). On a windy night, the vertical distribution during the same months is very different. High wind intensity pushes the *Per-*

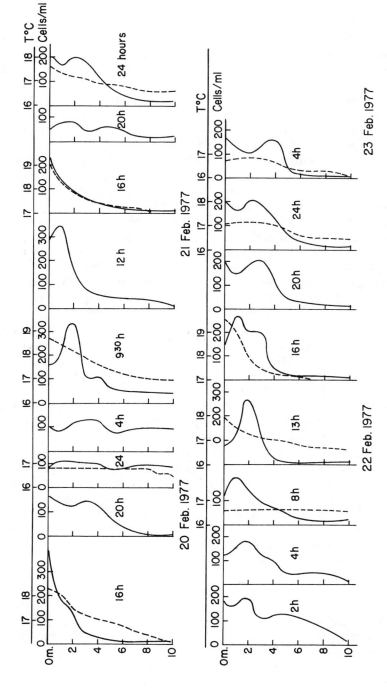

Fig. 4-5. Diel vertical distribution of *Peridinium* cells and water temperature in Lake Kinneret during four consecutive days.

idinium population downward, where it aggregates at the thermocline but never past it (Fig. 4-6). Berman and Rodhe (1971) described the diurnal migrations of *Peridinium* in the mid- and late phases of the bloom period. The vertical distribution of *Peridinium* cells during the bloom season in the early morning was described by Pollingher and Berman (1975).

The vertical distribution of *C. hirundinella* in a small lake (Esthwaite Water, England) as correlated with controlling factors has been very well documented (Talling 1971; Heaney 1976; George & Heaney 1978: Heaney & Talling 1980a,b). The diel pattern of *Ceratium* distribution over a seasonal growth cycle in the same water body was studied by Frempong (1984). Reynolds (1984) studied the vertical distribution of *C. hirundinella* in Crose Mere in calm conditions and during strong winds, and concluded that in the latter conditions the *Ceratium* cells were unable to regulate their vertical distribution within the epilimnion, but were able to resist elimination from it.

Extreme horizontal patchiness is a regular feature of dinoflagellate blooms (Berman & Rodhe 1971; Serruya 1971; Pollingher & Berman 1975; George & Heaney 1978; Heaney & Talling 1980a). Horizontal distribution patterns can be influenced by many physical, chemical, and biological factors, but the dominant factor influencing gross heterogeneity in most lakes is the circulation pattern induced by wind (George & Edwards 1976). Serruya (1975) has described the physical mechanisms caused by wind stress, which lead to patchy distribution of *Peridinium* in Lake Kinneret during the bloom period. At the end of the bloom, these mechanisms force the algae to be exposed to damaging radiation. Figure 4-7 represents the patchy distribution of *Peridinium* on Apr. 11, 1977, a cloudy, calm day preceded by a quiet day. The average number of cells (0–3 m depth) at different stations varied from 220 to 1,070 mL^{-1}. The maximum number of cells found in a patch was 5,000 mL^{-1}. In Esthwaite Water, a light northern wind transports large surface populations of *Ceratium* to the southern end of the lake within a few hours. At other occasions, patches resulted from a combined action of horizontal transport and upwelling of deeper water rich in *Ceratium* (Heaney & Talling 1980a).

GROWTH AND GROWTH-LIMITING FACTORS

The most common bloom forming armored dinoflagellates belong to the genera *Ceratium, Peridinium,* and *Peridiniopsis.* These genera are apparently cosmopolitan; they prefer hard water with a high concentration of calcium, and they tolerate a wide range of environmental conditions.

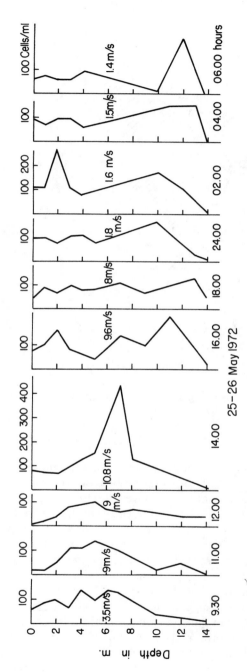

Fig. 4-6. Diel vertical distribution of *Peridinium* cells during both a windy night and a night when Lake Kinneret was stratified.

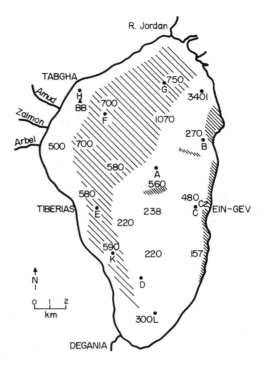

Fig. 4-7. Patchy near-surface distribution of *Peridinium* in Lake Kinneret on 11 Apr. 1977.

Tolerance Limits

Light. Dinoflagellates seem to prefer abundant light and are frequently found in the upper strata of water bodies, but their light requirements appear to be quite flexible. In cultures, the highest growth rates of *Peridinium cinctum* f. *westii* under intermittent light were found at intensities between 20 and 2,000 $\mu E \cdot m^2 \cdot s^{-1}$ at 20°C (Berman & Dubinsky 1985). The highest yield of *Peridinium* was found at about 130 $\mu E \cdot m^{-2} \cdot s^{-1}$ (Lindstrom 1984). This alga requires a minimum of about 10 to more than 125 $\mu E \cdot m^{-2} \cdot s^{-1}$ to produce high biomass. *Peridinium* is able to survive in complete darkness for seven to nine days (Pollingher 1986a), which would suggest that, if cells are carried down from the photic zone, they can survive for some time before returning to the upper layer. The motile cells of *C. hirundinella* in Esthwaite Water (England) tend to aggregate at depth with a relative irradiance level of 140 $\mu E \cdot m^{-2} \cdot s^{-1}$ (Harris et al. 1979). The preferred irradiance level in the lake corresponds to that at which net photosynthesis was maximal.

Salinity. Freshwater dinoflagellates are usually stenohaline. Most are halophobic and will not tolerate more than 100 mg chloride·L^{-1} (Holl 1928). *Peridinium,* however, bloomed in Lake Kinneret in 1964 before the diversion of salty springs when the water chloride content was 345 mg $Cl·L^{-1}$, and it also blooms today at 220 mg $Cl·L^{-1}$. *Peridiniopsis borgei* appears to be an exceptional species with broad salinity tolerance. It has been blooming for more than 60 years in Hortyteich, Hungary, a freshwater pond (Bourrelly 1968). It also blooms in summer in the small, brackish Lake Holmsjon, Finland (Lindholm, Weppling, & Jensen 1985), and it dominates the plankton for the entire year in Lake Kalkbrottsdammen, Sweden, an abandoned chalk mine with 220 mg $Cl·L^{-1}$ (Cronberg 1981).

Oxygen. Most dinoflagellates prefer well-oxygenated water bodies; they avoid eutrophic systems that experience periodic oxygen depletion. For this reason they are not found in ponds supplied with sewage water. A lack of O_2 leads to shedding of the flagella, immobility, and ecdysis of *Peridinium* in Lake Kinneret. *Peridinium* does not reach the anoxic hypolimnion in Lake Kinneret; *Ceratium* rarely penetrates into the anoxic hypolimnion of Esthwaite Water (Harris et al. 1979).

pH. Among 40 species of dinoflagellates studied by Holl (1928), 13 were oxyphile, and 24 alkalinophile. *Peridinium* tolerates a wide range of pH, although its growth is reduced below pH 6, and highest numbers and growth rate were found at a pH above 8 (Lindstrom 1984). *Peridinium* blooms in Lake Kinneret at a pH range of 8 to 9 (Berman & Dubinsky 1985). *C. hirundinella* can tolerate a wide range of pH under experimental conditions but the optimal levels are 7.0–7.5 (Bruno 1975). Hickel (1985b) describes a water bloom of *Peridinium pusillum* (reaching 3,000 cells·mL^{-1}) in a mineral acidotrophic pond with a pH below 4 in northern Germany, so tolerances are quite species specific.

Temperature. Most dinoflagellates are not strong stenotherms, but they do demonstrate maximum population development during the warmest time of the year in the temperate zones. Only a few apparently persist perennially, in that they can always be found in the plankton of some lakes *(P. bipes, P. willei).* Two distinct thermal optima may exist for *C. hirundinella* in temperate zones: 12–13°C (Hutchinson 1967; Heaney 1976) and 16–23°C. In Lake Kinneret, *Ceratium* disappears when the temperature becomes higher than 25°C. In lakes located in temperate zones, it was found that the *Ceratium* population reacted developmentally only to an increase in temperature, not to a decrease (Wesenberg-Lund 1908; Dottne-Lindgren & Eckbohm 1975). In cultures, *C. hirundinella* grew most rapidly at 21°C (Bruno & McLaughlin 1977).

Turbulence. It is well known that agitation inhibits the growth of marine dinoflagellates in culture (Tuttle & Loeblich 1975; Galleron 1976; White 1976). A high rate of agitation (more than 125 rotations·min^{-1}) causes death and disintegration of cells (White 1976). In Lake Kinneret after strong storms, dead cells of dinoflagellates are revealed. Pollingher and Zemel (1981) have shown that batch cultures of *Peridinium,* shaken at a speed of 100 rotations·min^{-1} only during the light period (14 h) had a death rate of 25.8% after 45 days of agitation. Cultures shaken continuously had a death rate of 62.5%, whereas unshaken cultures maintained in the same conditions had a death rate of only 9.5%. The cells that remained alive and were removed from the shaker, after 23 days of agitation, were not affected. They resumed their growth at a high rate of division. The theca of *Ceratium* seems to be less resistant to stress than that of *Peridinium.*

Growth Dynamics

The division of dinoflagellates in situ is phased; every 24 h a portion of the cells divides at a fairly precise time of the night. The cell division timing is optimally phased during the dark period that follows the period of energy accumulation in the cell (during the day). The division of *Peridinium* in Lake Kinneret takes place at night between 0100 and 0700 h, with a peak between 0200–0400 h (Pollingher & Serruya 1976). Nuclear division starts at 2300 h and is completed by 0200 h, when the cells contain two nuclei. Cytokinesis begins only after nuclear division is completed. The duration of the complete division process seems to extend over a 4–5 h period (Pollingher & Zemel 1981), and a similar period is needed for division of *C. hirundinella* (Entz 1931). Phased division is strongly synchronized in Lake Kinneret for *Peridinium* and *Ceratium* and also for *Ceratium* in Esthwaite Water, occurring between 0130 and 0430 h (Heller 1977; Frempong 1982). In other temperate water bodies, however, the division is less synchronized. In Lake Constance, for instance, synchronization was not as tight as reported in England (Sommer, Wedemeyer, & Lowsky 1984), and in lakes located in northern Germany division of *Peridinium* and *Ceratium* species in July–August is delayed until 0200–0300 h and cells in division are still evident at 0900–1000 h (Pollingher & Hickel unpublished data).

The percentage of cells in division in freshwater dinoflagellates varies from 1 to 40%. The maximum value was reported for *Peridinium* in Lake Kinneret, and a maximum of 33% was reported for *C. hirundinella* in Hungary (Entz 1931). As a consequence, the dinoflagellates have a low specific growth rate and a long generation time. The average specific growth rate of *Peridinium* in cultures using a variety of media was 0.06 days^{-1} with a generation time of 11 days. The maximum specific growth rate of *Peridinium* grown on Lindstrom's medium number 16×2 was 0.19 days^{-1}, and the average rate was 0.11 days^{-1} (Pollingher unpublished data). Higher values were obtained for *C. hirundinella* in cultures: 0.25 days^{-1} by Bruno & McLaughlin (1977), and 0.21 by Jaworsky (in Reynolds 1984). The specific growth rate and doubling time of various dinoflagellate species in situ are shown in Table 4-3.

Population division rate (DR) is affected by environmental factors. Although division occurs in complete darkness, the dividing cells are found generally only from the surface to 7 m. It is likely that the low concentration of oxygen below 10 m in the lake prevents the oxidative processes associated with division. Small variations in temperature do not affect the DR of the *Peridinium* population in Lake Kinneret, but on very cold nights the DR is very low. Elbrächter (1973) described a similar behavior in a population of *Ceratium* from the coastal waters of the Kiel Bay, and Leedale (1959), studying the cell division of euglenoids in cultures, found that raising or lowering the temperature by 5°C resulted in a drop of DR.

Weekly in situ measurements of the DR of *Peridinium* in 1973–4 showed that the seasonal sudden increase in DR was not related to temperature (Pollingher & Serruya 1976), but was correlated with wind intensity (Serruya, Serruya, & Pollingher 1978). In January–February, when the easterly wind reached a daily average of 8 m·s^{-1}, the DR never exceeded 10%. In early March, the wind velocity dropped below 2 m·s^{-1} and remained at this low value for nearly one month. The DR then rose to 32% and remained between 30% to 40% during this quiescent period. In early April, with the onset of the westerly winds, the DR dropped back to its winter values. On Mar. 28, 1973, the daily wind average was 1.5 m·s^{-1}, and the DR that night was 21.2%; the next day the daily wind average was 5.4 m·s^{-1}, and the DR dropped to 13.7%. A comparison of the fluctuations in the wind velocity during 24 h and the timing and rate of division of *Peridinium* during a period of four years showed that the intensity of the wind blowing during the day does not affect the timing and rate of division (Pollingher & Zemel 1981). But, a decisive role is played by wind extending from 1800 to 0200 h, that is, during the synthetic and nuclear division period of *Peridinium* as observed in situ (see Table II, Pollingher & Zemel 1981). After 0200 h, strong winds do not affect the DR.

Table 4-3. *Specific rates of daily population increase and doubling time of dinoflagellates in various water bodies.*

Species	Water body	Specific growth rate (ln units day^{-1})	Doubling time (days)	Division rate (%)	Source	Observations
P. cinctum	Kinneret (Israel)	0.22–0.02	3.15–34.5	1–40	Pollingher (unpubl.)	—
P. borgei	Kinneret (Israel)	0.21–0.02	3.15–34.5			—
P. elpatiewsky	Kinneret (Israel)	0.25–0.05	2.7–13.8			—
P. cinctum	Gr. Ploner See (FRG)			19.3		Jul
P. bipes	Ob. Ausgraben See (FRG)			8.8	Pollingher & Hickel (unpubl.)	Jul
P. willei	Schohsee (FRG)			23.0		Jul
P. willei	Schluensee (FRG)			20.0		Jul
C. hirundinella	Lagymanyos (Hungary)			1–33	Entz 1931	—
C. hirundinella	Esthwaite Water (UK)	0.092–0.016	7.5–43	2–10	Heller 1977; Heaney & Talling 1980b; Frempong 1982; Heaney et al. 1983	—
C. hirundinella	Crosmere (UK)		≈5		Reynolds 1976	—
C. hirundinella	Constance (FRG)	0.27–0.12	2.6–5.8		Sommer et al. 1984	—
C. hirundinella	Lambert Quarry Lake (Tennessee, USA)		7–15	4–10	Mueller-Elser & Smith 1985	—
Ceratium spp.	Plußsee (FRG)	0.24–0.13	2.9–5.3	4–20	Pollingher & Hickel (unpubl.)	Jul–Aug
Ceratium	Schohsee (FRG)			6–11		Jul–Aug
Ceratium	Schluensee (FRG)			10–15		Jul
Ceratium	Erken (Sweden)		2.9–4.5		Dottne-Lindgren & Ekbohm 1975	—

Nutrients, chiefly the concentration of phosphorus, may play an important role in dinoflagellate growth. In Crose Mere, *Ceratium* dominates in August (ca. 400 cells mL^{-1}) when lowest levels of dissolved P and nitrate are recorded (Reynolds & Reynolds 1985). *Peridinium* in Lake Kinneret blooms when the concentrations of phosphorus are also very low. During 1973–6, major environmental changes occurred in Lake Kinneret (Serruya & Pollingher 1977); consequently, the concentration of dissolved phosphorus increased. The new environmental conditions allowed greater development of all other algal groups, which in turn delayed the appearance of *Peridinium* and the intensity of the bloom (Pollingher 1986b). A drastic decrease in the *Peridinium* population occurred, from a semiannual average cell density of $2,160 \times 10^6$ cells·m^{-2} in 1972 to $1,180 \times 10^6$ cells·m^{-2} in 1975, and from a semiannual average of dinoflagellate biomass of 131 g·m^{-2} (1972) to 41 g·m^{-2} (1975). Lake Balaton provides another example. In the less eutrophic central and northern part of the lake, *C. hirundinella* is dominant in summer; in the rest of the lake where the concentration of nutrients is higher, blue-green algae prevail. The abundance of *Ceratium* decreases with the increase of nutrients (Padisak 1985).

Loss Processes

Losses occur during the entire persistence of dinoflagellate populations, but usually they increase at the end of the water bloom. We discuss here some of the best-established factors regulating dinoflagellate populations through the loss of vegetative cells.

Physiological Death. Weekly sampling of *Peridinium* during the Lake Kinneret bloom periods (Fig. 4-4) showed that decrease of the population is 8–10% during exponential growth and 40–80% at the decline of the bloom. At the end of the bloom, most of the population dies off and most of the cells are decomposed in the water column. The decay of *Peridinium* is a rapid process (Serruya, Pollingher, & Gophen 1975). The theca disintegrates into individual plates, which are then quickly destroyed. This may explain why only the inner membranes of the cells, and not thecae or plates, are found in the sediments. During the process of decay, the thecae of dead *Peridinium* are surrounded by a great number of bacteria and phagotrophic colorless flagellates. The breakdown of *Peridinium* biomass was studied on lyophilized cells under various conditions in situ or in the laboratory by Hertzig, Dubinsky, and Berman (1981) and by Sherr, Sherr, and Berman (1982). They found that degradation depended on a specialized bacterial population and was accelerated in the presence of protozoans and by turbulence.

Assuming grazing of *Peridinium* to be constant and insignificant, and assuming a very low sedimentation rate, because of the active swimming of the cells, then losses due to physiological death in situ, expressed in the same rate terms as growth, vary from $K = -0.002$ to $K = -0.10$ days^{-1}. Sommer et al. (1984) found rates of decrease of the *Ceratium* population in Lake Constance during its decline phase of -0.09 and -0.17 days^{-1} and, during encystment, loss rates of -0.56 and -0.36 in 1979 and 1980, respectively (Sommer 1981).

The enumeration of dead cells in batch cultures of *Peridinium* has shown that the incidence of dead cells varied between 1.5 and 3.0% of the population during the exponential growth and stationary phase, but during the decline phase the incidence increased to 35–62% of the population. The daily death rate of *Peridinium* in cultures varied between $K = -0.02$ and $K = -0.07$ days^{-1} (Pollingher unpublished data).

Grazing. Large dinoflagellates such as *Peridinium* and *Ceratium* species are not grazed by filter-feeding zooplankton. Only the rotifers *Hydatina* sp., *Asplanchna* sp., *A. priodonta,* and *A. herricki* feed on *Ceratium* and *Peridinium* (Barrois 1894; Entz 1927; Gophen 1973; Gilyarov 1977). *Peridinium* in Lake Kinneret is grazed and digested by the cichlid fish *Tilapia galilaea (Sarotherodon galilaeus)* and in smaller quantities by *T. aurea* (Leventer 1972; Spataru 1976; Spataru & Zorn 1978; Landau 1979; Gophen 1980; Gophen, Drenner, and Vinyard 1983). At the end of the bloom in Lake Kinneret, the ciliate *Bursaria truncatella* has been seen to have many *Peridinium* cells inside it (Pollingher 1986a). Small species, such as *Peridiniopsis cunningtonii,* are ingested by littoral protozoa, such as *Ophrydium* (30 cells of *P. cunningtonii* were found in one living organism), *Lacrymaria,* and *Euplotes* (Pollingher unpublished data). However, the fish *Tilapia galilaea* and *T. aurea* are the principal consumers of *Peridinium* in Lake Kinneret. The quantity of dinoflagellate biomass used by these fish per month represents about 14% of the dinoflagellate standing stock of the lake (25,000 tons w.w., Pollingher & Serruya 1976).

Parasitism. At the end of the bloom in Lake Kinneret a very small number of dead cells of *Peridinium* are infected by chytrids. Sommer et al. (1984) found 10% of the live population of *C. hirundinella* with attached chytrid zoosporangia in Lake Constance. Canter (1968, 1979) described cells of *Ceratium* and *Peridinium* from Esthwaite Water infected by the chytrid *Amphicypellus elegans,* whereas cysts of *Ceratium* were parasitized by *Rhizophydium nobile.* Canter and Heaney (1984) reported the presence of two more biflagellate fungi (*Lageni-*

dium sp. and *Aphanomycopsis cryptica*) and a monadlike protozoan that live within cells of *Ceratium* from Windermere, Esthwaite Water, and Blelham Tarn. In Esthwaite Water in October 1982, 87% of *Ceratium* cells were parasited by *A. cryptica,* which must have partially depleted the recruitment of cysts to the sediment. In July 1983, 80% of *Ceratium* cells were affected by the fungus, and only in August did the *Ceratium* population recover. Moderate infections of *A. cryptica* were also observed in Windermere and Blelham Tarn in 1983. The same authors also mention the presence of several undescribed organisms that kill *Ceratium* cysts present in the sediments. In Lake Kinneret, cysts of *Peridinium* or *Ceratium* attacked by fungal parasites have not yet been recorded.

Other Losses. The vegetative division of dinoflagellates entails higher energy losses than many other planktonic algae. In some species of *Ceratium* spp. or *Gonyaulax apiculata* (Hickel & Pollingher in press), the daughter cells retain parts of the old theca and regenerate the missing part. In *Peridinium* from Lake Kinneret the theca is shedded completely and both daughter cells have to synthesize a new one (Pollingher & Serruya 1976). Drastic changes in the environmental conditions, such as oxygen deprivation, osmotic pressure changes (by adding distilled water to a culture), or temperature increases, lead to theca shedding and ecdysis by *Peridinium.* Crisculo, Dubinsky, and Aaronson (1981), working with cultures of *Peridinium,* have found that imbalance of nutrients determines theca shedding. Undoubtedly this loss of metabolized carbon must contribute to the low growth rates demonstrated by large dinoflagellates.

Dinoflagellates versus Nannoplankton: Nutrition, Growth, Kinetics, Competition

Dinoflagellates are clearly much larger than competing nannoplanktonic algae, and they certainly have lower growth rates (as discussed previously). However, they are apparently able to outgrow smaller cells during the stratified season in temperate zones and establish predictable blooms in some lakes. The dinoflagellate competitive strategy can therefore perhaps best be discussed as a comparison of dinoflagellates to nannoplankton with respect to nutrition, motility, growth, and grazing losses. The most obvious "costs" conveyed by large cell size and generally compact shapes of dinoflagellates are high half-saturation constants for nutrient uptake and low specific growth and productivity. The apparent benefits of the particular morphology of dinoflagellate cells include the ability to migrate vertically, low self-shading, potential low specific phosphorus requirements, an ability for "luxury consumption" of phosphorus, and low rates of cell loss due to sinking and grazing. Each of these aspects will be discussed here.

Dinoflagellate Advantages. Due to their size, most of the dinoflagellates cannot be grazed by herbivorous zooplankton, as has already been discussed. Furthermore, flagellates have a lower sinking rate in comparison with other algae (Reynolds 1984). Motility provides an advantage to dinoflagellates by enabling the cells to return to the surface when turbulence carries them out of the photic zone. The cells can thus avoid unfavorable light and temperature conditions and can aggregate at optimal depth during calm weather. The vertical stratification of dinoflagellates in Lake Kinneret appears before the onset of the thermal and chemical stratification (Pollingher & Berman 1975).

Dinoflagellates have a low self-shading effect measured as the extinction coefficient (ϵ_s) per unit concentration of biomass taken as chlorophyl *a* Talling (1971) found ϵ_s in ln units $mg^{-1} \cdot m^{-2}$ Chl *a* of 0.01 for a dense poulation of *C. hirundinella* in Windermere waters and 0.02 for *Asterionella,* which is assumed representative for phytoplankton. For comparison, the vertical extinction coefficient for *Peridinium* in Lake Kinneret was 0.006 (Berman 1976). In Talling's (1971) opinion, the low self-shading effect may be connected with the large size of the cells, between which more light passes unintercepted (sieve effect) than would occur with the same biomass distributed in small cells.

During the peak of the *Peridinium* bloom, dense patches of cells aggregate at or close to the surface, and the effect of self-shading by such patches was observed (Pollingher & Berman 1982). To determine if the deep cells were potentially active, Pollingher & Berman (1982) took some samples from 1-m depth and others from 10-m depth. Both were incubated in situ at 0-, 1-, 5-, and 10-m depths. The deeper samples had higher chlorophyll cellular content, and their photosynthetic activity on exposure to elevated light levels resembled that of the near-surface cells. This indicates that the organisms can quickly adapt to higher irradance and confirms that most of the cells throughout the water column are potentially active, an important adaptation for vertically migrating species that would ameliorate any self-shading effects.

Peridinium has an advantage associated with the capacity for "luxury consumption" of phosphorus (Serruya, et al. 1974; Serruya & Berman 1975). In conditions of excess ambient orthophosphate (at the beginning of the bloom development), *Peridinium* takes up phosphate and stores it in an internal pool of polyphosphates organized in aggregated morphological structures (Elgavish et al. 1980). *Peridinium* sampled in early March had about three times more intracellular phosphorus than cells sampled in April–June (Wynne et al. 1982). *Peridinium* cells grown in a culture medium containing 1500 mg P L^{-1}, when washed and transferred to a medium without phosphorus, continued to grow through three to four divisions (Serruya & Berman 1975).

Stumm & Morgan (1970) calculated that 1 mg P is sufficient to synthesize 114 mg of algal biomass dry weight in one cycle of transformation. In Lake Kinneret during winter–spring, the uptake of one atom of phosphorus by *Peridinium* corresponds to the uptake of 235 to 645 atoms of carbon (Serruya et al. 1974). Therefore, this species requires a very low P:C ratio for persistence, as compared to other algae.

Due to their motility, dinoflagellate cells can screen the whole water column for nutrients, and the activity of the transverse flagellum leads to an increase in the flow of water near the cell surface. Mitchell & Galland (1981) studied the photosynthesis and vertical distribution of *Gymnodinium* spp. in Lake Mahinerangi (a power supply reservoir). They observed that the dinoflagellates were concentrated near the bottom of the euphotic zone during mid-morning sampling (when the lake was weakly stratified). Sampling carried out in the early morning showed that the cells located at 12 m depth contained a higher concentration of P ($14.5 \pm 0.4 \times 10^{-5}$ μg·cell^{-1}) than those located at the surface ($6.8 \pm 1.0 \times 10^{-5}$ μg·cell^{-1}). Cells sampled at midday at the same depths contained similar concentrations of P (8.3×10^{-5} μg· cell^{-1}). The authors concluded that these results support the hypothesis of Eppley, Holm-Hansen, and Strickland (1968) that vertical migration allows dinoflagellates to assimilate nutrients from the whole water column.

Dinoflagellate Disadvantages. Nutrient uptake parameters play an important role in algal resource competition. They reflect the uptake efficiency of nutrients. In Lake Kinneret, dinoflagellate half-saturation constants (K_s) for N and P are higher, and the maximum uptake velocities (V_m, expressed per unit chlorophyll) are lower than for nannoplankton (Table 4-4). The nannoplankton in Lake Kinneret have greater affinity and faster chlorophyll specific uptake rates for phosphorus than does *Peridinium*. Thus the dinoflagellates do not have a competitive advantage in exploiting this nutrient (Berman & Dubinsky 1985). In cultures, *Ceratium* is able to use both inorganic and organic sources of phosphorus with better growth resulting from the organic source (Bruno 1975). *Peridinium* from Lake Kinneret grows in cultures on compounds such as glucose-6-phosphate, glycerophosphate, and ATP as the only phosphorus source (Serruya & Berman 1975).

Table 4-4. *Half-saturation constants (K_s) and maximum uptake velocity (V_m) for nitrogen and phosphorus by P. cinctum in comparison to nannoplankton in Lake Kinneret.*

Algae	K_s	V_m
P. cinctum	27 μM NH$_4$-N	
Nannoplankton	2.5 μM NH$_4$-N	
P. cinctum	28.9 μM NO$_3$	
P. cinctum	0.18–0.31 μM P	3.8–7.6 $\times$ 10^{-6} μmol P·μg chl a^{-1}·min^{-1}
Nannoplankton	0.03 μM P	90 $\times$ 10^{-6} μmol P·μg chl a^{-1}·min^{-1}

Sources: McCarthy, Wynne, and Berman (1982); Sherr et al. (1982); Berman et al. (1984); Berman and Dubinsky (1985).

Peridinium in Lake Kinneret and in cultures prefers ammonium over nitrate (McCarthy et al. 1982). It is clear from Table 4-4 that *Peridinium* has lower affinities for NH$_4$ than do the nannoplanktonic algae. *C. hirundinella* in cultures prefers NaNO$_3$ and urea. It seems that NH$_4$ has a tendency to be toxic to it in high levels (Bruno 1975).

The specific photosynthetic carbon uptake of dinoflagellates is low in comparison to nannoplanktonic species (Table 4-5; Pollingher & Berman 1982). Also, carbon production per carbon biomass unit of dinoflagellates is at least 10-fold lower than that of nannoplankton. Pollingher and Berman (1982) have shown that the highest photosynthetic activity and intracellular concentration of chlorophyll in *Peridinium* are associated with cells at the beginning of the growth season and start to decline as the numbers of cells increase. The highest assimilation numbers are also found in young cells, despite differences in light and temperature regimes. A comparison of the cellular chlorophyll content and photosynthetic activities measured in *Peridinium* of Lake Kinneret with some observations in other lakes or in the laboratory in cultures with other species *(Ceratium)* shows close agreement between results (Table 4-6). They conform to predicted values for cells of this size (i.e., Reynolds 1984, Fig. 9).

REPRODUCTIVE STRATEGIES

Life Cycle and Special Cells

Cysts. The dinoflagellate's most important survival adaptation is its life cycle, which involves an annual alternation of a motile vegetative cell with a nonmotile resistant cyst. Many dinoflagellates spend more

Table 4-5. *Range of photosynthetic characteristics of dinoflagellates compared to nannoplankton.*

Algae	Carbon assimilated (pg C·cell^{-1}·h^{-1})	Carbon assimilated per unit carbon biomass	Renewal time (h)	Site	Reference
Dinoflagellates					
P. cinctum f. *westii*	80–210	0.01–0.19	31.5–158	Lake Kinneret	Pollingher & Berman 1982
P. cinctum f. *westii*	108–250	0.01–0.1	67–143	Laboratory	Pollingher & Berman 1982
P. cinctum f. *westii*	187–500			Laboratory	Rodhe 1978[a]
Peridinium cinctum	126–231			Laboratory	Prezelin & Sweeney 1979
Ceratium hirundinella		1.16			Gutelmacher 1975[b]
Nannoplankton cells					
Chroococcus spp.	3–17	0.09–0.77	2.7–11.3	Lake Kinneret	Pollingher & Berman 1982
Chroococcus limneticus	2.2	13.6	16	Castle Lake	Gutelmacher 1975[b] Stull Amezaga, and Goldman 1973[b]
Synedra rumpens	4–8	0.17–0.30	3.3–6	Lake Kinneret	Pollingher & Berman 1982
S. radians	0.68		56	Castle Lake	Stull et al. 1973[b]
Nannoplanktonic Chlorophyta	2–9		2.8–10	Lake Kinneret	Pollingher & Berman 1982
Oocystis naegeli	4.7		2.7	Castle Lake	Stull et al. 1973[b]

[a]Calculated from original data.
[b]Autoradioactive method.

Table 4-6. *Cellular chlorophyll content and photosynthetic activities measured in Peridinium and Ceratium.*

Parameter	Algae	Values	Site	Reference
Chlorophyll		pg chl·cell^{-1}		
	P. cinctum f. *westii*	61–288	Lake Kinneret	Pollingher & Berman 1982
	P. cinctum f. *westii*	51–250	Laboratory	Pollingher & Berman 1982
	P. cinctum f. *westii*	225	Lake Kinneret	Berman 1979
	Peridinium cinctum	61–113	In culture	Prezelin & Sweeney 1979
	Ceratium hirundinella	250	Esthwaite Water	Talling 1971
	Ceratium hirundinella	84–280	Esthwaite Water	Harris et al. 1979
Assimilation Number		µg C·µg chl^{-1}·h^{-1}		
	P. cinctum f. *westii*	0.7–3.4	Lake Kinneret	Pollingher & Berman 1982
	P. cinctum f. *westii*	0.8–3.8	Laboratory	Pollingher & Berman 1982
	P. cinctum f. *westii*	2.5	In culture	Rodhe 1978
	Ceratium hirundinella	2.6–15	Esthwaite Water	Harris et al. 1979

[a]Calculated assuming a cellular chlorophyll content of ≈ 0.3% of wet weight (Berman 1978).

time as resting cysts than as motile planktonic cells (i.e., *Ceratium* spp. in temperate zones). The cysts are able to withstand various adverse environmental conditions and are dispersed by birds (Krupa 1981). The cyst is an adaptive form because encystment and excystment especially occur under definite environmental conditions.

The shape, size, and ultrastructure (Bibby & Dodge 1972; Dürr 1979; Chapman, Dodge, & Heaney 1982) of the cyst are different from that of the planktonic cell, and the cysts need a period of dormancy. In cysts and recently excysted cells, large lipid reserves are found (Chapman, Livingstone, & Dodge 1981; Heaney et al. 1983). In cells of *Ceratium* in stationary phase of growth, lipid concentration increases (Chapman et al. 1985), accounting for 34% of cell dry weight (Cranwell 1976). Chapman et al. (1985) suggested that starch is the main short-term energy store for *C. hirundinella,* and that lipids may be more important for long-term storage such as within the cyst.

We do not yet have any evidence that the cysts of *Peridinium* in Lake Kinneret are the result of sexual fusion, but cysts obtained in cultures as a result of induced copulation are similar to those recorded in nature. The processes of encystment and excystment and the ultrastructure of *C. hirundinella* cysts from Esthwaite Water have been studied in detail, and the authors (Chapman et al. 1981; Heaney et al. 1983) have not yet found any proof that the cyst is connected with sexual reproduction. Cyst formation is probably variable because von Stosch (1973) showed that hypnozygotes represent only one of many different but presumably effective forms of zygotes produced by dinoflagellates. Some species form temporary cysts; others complete the sexual cycle in a motile state without cyst formation. In *Woloszynskia apiculata,* copulating gametes are still able to revert to vegetative cells when conditions are favorable.

Sexual Reproduction. In the last 20 years sexual reproduction has been recorded and described in clonal cultures (von Stosch 1964, 1972, 1973; von Stosch, Theil, & Happach-Kassan 1981) and N-deficient heterothallic cultures of freshwater dinoflagellates (Pfiester 1975, 1976, 1977; Pfiester & Skvarla 1979, 1980; Happach-Kassan 1980; Spector, Pfiester, & Triemer 1981; Sako et al. 1984). The vegetative cells of dinoflagellates are haploid. Gametes are formed by a longitudinal division of the peridinoid dinoflagellate, whereas asexual division in *Peridinium* spp. is characterized by transverse fission, and in *Ceratium* by oblique fission. The gametes are smaller than the vegetative cells and remain motile during fusion. Gamete fusion may involve equal isogamy (most *Peridinium* spp.) or anisogamy (*Ceratium* species) and may be homothallic or heterothallic. After fusion the zygote is larger

Table 4-7. *Variation in details of vegetative and sexual reproduction of dinoflagellates.*

Species	Source	Type of vegetative reproduction	Mating compatibility	Gamete fusion	Zygote morphology	Meiosis and meiotic products	Reference
Ceratium cornutum	Culture clones	MBF	He	A	RC	2 consecutive steps of division of single germling swarmer	von Stosch 1964, 1972; Happach-Kassan 1980; von Stosch et al. 1981
Gymnodinium excavatum	Culture	MBF	Ho	I	RC	2 consecutive steps of division of single hypnozygote germling	von Stosch 1972
G. paradoxum var. apiculatum mihi	Culture	Cyst, 4 or 8 zoids	He	I	RC	1 division inside hypnozygote, 2 swarmers escape, undergoing 2nd division in temporary cysts	von Stosch 1972
G. pseudopalustre	Culture	Motile + motionless zoosporangia	Ho	I formed by depauperating div.	RC	Swarmer released from hypnozygote serves as a meiocyte & undergoes two meiotic cell divisions to yield 4 haploid flagellates	von Stosch 1973
Woloszynskia apiculata	Culture	Motile + binary fission, motionless zoosporangia	He	I	RC	1 meiotic division in hypnozygote, 2 swarmers undergo 2nd meiotic division in a separate zoosporangium, or alternatively, 2nd meiotic division also inside the spore to yield 4 swarmers	von Stosch 1973
C. furcoides	Plußsee	MBF	He	A	C	—	Hickel 1984

Species	Condition					Meiosis	Reference
Peridinium cinctum f. *ovoplanum*	N-deficient cultures	MBF	Ho	I	RC	Meiosis prior to emergence of a single cell from hypnozygote	Pfiester 1975; Spector et al. 1981
P. willei	N-deficient cultures	MBF	Ho	I	RC	Meiosis prior to emergence of a single cell from hypnozygote	Pfiester 1976
P. gatunense	N-deficient cultures	MBF	Ho	I	RC	Meiosis prior to emergence of 2 cells from hypnozygote	Pfiester 1977
P. limbatum	N-deficient cultures	MBF	—	—	RC	—	Pfiester & Skvarla 1980
P. volzii	N-deficient cultures	MBF	He	I	RC	Meiosis prior to emergence of a single cell	Pfiester & Skvarla 1979
P. cunningtonii	(N + P)-deficient cultures	—	—	I	RC	—	Sako et al. 1984
P. cinctum f. *westii*	Old cultures	MBF	Ho	I	—	Sporangia with 4 small cells	Pollingher (unpubl.)
P. cinctum f. *westii*	Kinneret	MBF	—	—	C?	—	Pollingher (unpubl.)
C. hirundinella	Esthwaite Water	MBF	—	—	C?	—	Heller 1977; Heaney & Talling 1980b; Chapman et al. 1981, 1982; Frempong 1982; Heaney et al. 1983
C. hirundinella	Lagymanyos	MBF	—	—	C?	—	Entz 1931

MBF = motile binary fission; He = heterothallic; Ho = homothallic; A = anisogamous; I = isogamous; RC = resting cyst; C = cyst than the gametes and may remain motile as a planozygote, with two longitudinal flagella. After some days, the planozygote is transformed into a resting cyst or a hypnozygote that is athecate but possesses three distinct walls. Meiosis in some species occurs while the zygote is motile. In others both meiotic divisions may occur in the nonmotile hypnozygote. Alternatively, the first division may occur in the hypnozygote, from which two swarmers escape, undergoing the second division in temporary cysts. In many cases meiosis occurs prior to emergence from the hypnozygote. There are many variations in details of the vegetative and sexual reproduction; they are shown in Table 4-7. The variation in cell size resulting from sexual reproduction is shown in Table 4-8.

As an example, a detailed description of the sexual reproduction of *Ceratium cornutum* in clonal cultures was given by Happach-Kassan (1980). *C. cornutum* is a dioecious dinoflagellate with anisogamic copulation. The difference between the female gametes and the vegetative cells is not very evident. The male gametes are usually (but not necessarily) smaller and have a lighter color than the female gametes. A few days after bringing together both clones and changing the culture conditions (decrease of temperature), a mass production of gametes occurred. Copulation could be induced only by decrease of temperature with or without limitation of nutrients, and low temperature led to a better synchronization of gamete formation than did high temperature. One-third of the population underwent copulation. The resulting pairs of gametes swam for 4–5 h. Then they settled on the bottom of the vessel, and the female gamete incorporated the male one. Copulation took 1½ h, and the resulting planozygote had two longitudinal flagella. The planozygote was transformed into a hypnozygote after three to four weeks of increasing size without cell division; its nucleus contained 118–132 chromosomes. Subsequent hypnozygote formation depended on temperature, and the hypnozygote remained dormant for three to four months at 2°–4°C. Germination was induced by raising the temperature to 21°C under a light/dark regime of 14:10 h. The hypnozygote liberated one meiocyte, which underwent meiosis in a motile phase in two steps. The daughter cells resulting from the first meiosis underwent a second division after one to two days. In *C. cornutum* the most important trigger for zygote formation was temperature and not nutrient deficiency, as in many *Peridinium* species.

The only case of sexual reproduction observed in situ was described for *C. forcoides* by Hickel (1985a) in Plußsee (Fig. 4-8). In this case, at the end of the bloom, the planozygotes constituted more than 20% of

Table 4-8. Diversity in cell size of dinoflagellates in cultures.

Species	Veg. cell (μm)	Gamete (μm)	Planozygote (μm)	Hypnozygote (μm)	Reference
P. cinctum f. westii	54 × 48	38 × 38	60 × 66		Pollingher (unpubl.)
P. cinctum f. ovoplanum	48 × 48 40 × 30	42 × 40 25	40–75 D		Spector et al. 1981
P. volzii	38–42	25–28	60–62		Pfiester & Skvarla 1979
Wolosyznskia apiculata	25–46 L 18–36 W	21–25 L 14–17 W	47 × 34	41–49 L 31–36 W	von Stosch 1973

L = long; W = wide; D = diameter.

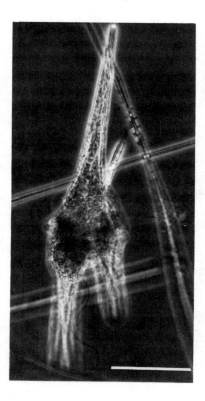

Fig. 4-8. Sexual reproduction of *Ceratium furcoides* by anisogamic copulation. Late copulation stage showing the remains of the male gamete attached to the ventral part of the female gamete, which has longitudinal flagella. Scale bar: 5 μm. (Courtesy Dr. B. Hickel.)

the population. The lack of observations of sexual reproduction in wild populations may be because it occurs in only a few cells at any given time, or because the sexual process is a short-lived phenomenon in *Peridinium* species and thus is difficult to detect. We do not know how often or when sexual reproduction occurs in situ. In Lake Kinneret, planozygotes of *Peridinium* were observed only once or twice. We have observed stages of sexual reproduction in old cultures of *Peridinium,* and once a tetrad (a sporangia with four spores) was recorded. The sexual stages were recorded concomitantly with vegetative division. The ratios between these two types of reproduction were 5.2% of cells in vegetative division and .14% in sexual stages.

Reproductive Dynamics during Population Cycles

In temperate zones, encystment occurs in fall at the end of the stratification period, and the cyst is an "overwintering" form (Entz 1925, 1926; Hickel 1978, 1984, 1985a; Chapman et al. 1982; Heaney et al. 1983; and others). Encystment is a regular event in September–October. A high percentage of the population of *C. hirundinella* is typically transformed into cysts in Esthwaite Water. In 1981 only 10% of the population encysted, and in 1980 normal encystment was not observed (Chapman et al. 1982; Heaney et al. 1983). Hickel (1985a) described mass encystment of *C. furcoides* at the end of August following a bloom in 1981 in Plußsee. Mass encystment of *C. hirundinella* was also described in Lake Vechten (Netherlands) (Steenbergen 1982). The cysts were also recorded in both epiphyton and epipelon.

In subtropical Lake Kinneret, the cysts of *Peridinium, Ceratium,* and *Peridiniopsis* are "oversummering" forms, and encystment occurs at the beginning of the stratified period (Pollingher 1986b). At the end of the bloom (May–June), the bulk of the *Peridinium* population dies off, and only 1% is transformed into cysts.

Temperature plays an important role in the process of excystment and encystment (Huber & Nipkow 1923). Excystment occurs in temperate zones when the temperature of the water increases and in Lake Kinneret when it decreases (Pollingher & Serruya 1976). The result of excystment in *Peridinium* and *Ceratium* is a single cell (Eren 1969; Chapman et al. 1981), which in *Peridinium* preserves both red bodies that belonged to the cyst. Excystment of *Ceratium* in temperate zones occurs from February to April ($+5°C$) when the "preceratium" stages can be seen in the water (Heaney et al. 1983). Germination of *Peridinium* in Lake Kinneret occurs from November to February when temperatures vary between 14° and 20°C. Encystment of *Peridinium* sampled from Lake Kinneret during the peak or the end of the bloom proceeded in the laboratory by keeping the algae in a closed vessel, whereas cells sampled during their growth phase did not encyst. However, this method did not yield cysts with *Peridinium* maintained in cultures (Pollingher, unpublished data). Eren (1969) induced encystment of *Peridinium* from the lake in the laboratory by exposing the cells to 28°C.

CONCLUSIONS: THE DINOFLAGELLATE STRATEGY

Most dinoflagellate morphological and ecophysiological features do not favor them in competition with nannoplanktonic species. In spite of this, dinoflagellates are sometimes abundant and can form water blooms in various types of freshwater ecosystems. This may be explained by ecophysiological characteristics and by reproductive strategies that enable them to develop and survive in conditions that are unfavorable for other algae. The following characteristics of dinoflagellates seem to be most critical.

Most planktonic organisms need turbulence for survival: It allows them to maintain themselves in the water column, and the stimulation of input of nutrients is often correlated with turbulence. Dinoflagellates also need turbulence during a given short period (i.e., for resuspension of cysts which constitute the inoculum for initial population development). However, the vegetative cells apparently prefer quiet windless weather. Due to their motility, dinoflagellates do not need turbulence for persistence in the water column. They are able to process the whole water column for nutrients via vertical migration, and losses by sedimentation are low in comparison with nonflagellate algae. Due to their capacity for luxury consumption of phosphorus, they can divide two to three times when phosphorus is unavailable in the water. A combination of luxury consumption, vertical migration, and long generation times gives them a strong advantage under conditions of extreme nutrient depletion, even though their actual nutrient affinities are quite low.

Due to the relatively large size of dinoflagellate cells, they are not grazed by herbivorous zooplankton, so grazing losses are low. So too is the physiological death rate of dinoflagellates during their exponential growth and stationary phase, so long as the weather remains calm.

The dinoflagellate reproductive strategies enable them to encyst when the environmental conditions become unfavorable for the vegetative cell, to spend sometimes a long period as a benthic dormant cyst, and to excyst when the conditions again become favorable. It is not yet known how widespread sexual reproduction of dinoflagellates is in nature, but sexual reproduction allows for genetic recombination and also involves the production of a resistant cyst stage. Evidence of the high adaptive abilities of dinoflagellates is their persistent abundance since the Jurassic (Tappan & Loeblich 1971).

Large-celled dinoflagellates do not dominate during early stages of seasonal succession. They appear during the period of transition between mixing and stratification, and reach their maximum in temperate zones during the period of thermal stratification and nutrient impoverishment. Their cysts are overwintering forms. In the subtropical Lake Kinneret, the dinoflagellates appear during destratification of the lake, bloom at the end of this period when the nutrient concentration (especially phosphorus) is low, and disappear at the onset of stratification. The cysts in this case are oversummering forms.

Dinoflagellates develop a population with a low specific productivity and a long generation time; thus they immobilize large amounts of nutrients and slow down the overall activity of the ecosystem (Serruya, Gophen, & Pollingher 1980).

REFERENCES

Barrois, T. (1894). Contribution a l'étude de quelques lacs de Syrie. *Rev. Biol. Nord. France,* 6, 224–314.

Berman, T. (1976). Light penetrance in Lake Kinneret. *Hydrobiologia,* 49, 41–8.

(1978). General biochemical features. In *Lake Kinneret,* ed. C. Serruya (Monographiae Biologicae 32), pp. 269–70. The Hague: Dr. W. Junk.

Berman, T. and Dubinsky, Z. (1985). The autoecology of *Peridinium cinctum* fa. *westii* from Lake Kinneret. *Verh. Internat. Verein. Limnol.,* 22, 2850–4.

Berman, T. and Rodhe, W. (1971). Distribution and migration of *Peridinium* in Lake Kinneret. *Mitt. Internat. Verein. Limnol.,* 19, 266–76.

Berman, T., Sherr, B. F., Sherr, E., Wynne, D., and McCarthy, J. J. (1984). The characteristics of ammonium and nitrate uptake by phytoplankton in Lake Kinneret. *Limnol. Oceanogr.,* 29, 287–97.

Bibby, B. T. and Dodge, J. D. (1972). The encystment of a freshwater dinoflagellate: a light and electron microscopical study. *Br. Phycol. J.,* 7, 85–100.

Bourrelly, P. (1968). Note sur *Peridiniopsis borgei* Lemm. *Phykos,* 7, 1–2.

(1970). *Les Algues d'Eau Douce.* Vol. III. *Les Eugléniens, Peridiniens et Cryptomonadines.* Paris: N. Boubée.

Bruno, S. F. (1975). Cultural studies on the ecology of *Ceratium hirundinella* (O.F.M.) Bergh, with notes on cyclomorphosis. Ph.D. Thesis, Fordham Univ.

Bruno. S. F. and McLaughlin, J. J. A. (1977). The nutrition of the freshwater dinoflagellate *Ceratium hirundinella. J. Protozool.,* 24, 548–52.

Canter, H. M. (1968). Studies on British chytrids. XXVIII. *Rhizophydium nobile* sp. nov., parasitic on the resting spore of *Ceratium hirundinella* O. F. Mull. from the plankton. *Proc. Linn. Soc. Lond.,* 192, 197–201.

(1979). Fungal and protozoan parasites and their importance in the ecology of the phytoplankton. *Rep. Freshwat. Biol. Ass.,* 47, 43–50.

Canter, H. M. and Heaney, S. I. (1984). Observations on zoosporic fungi of *Ceratium* spp. in lakes of the English lake district; importance for phytoplankton population dynamics. *New Phytol.,* 97, 601–12.

Chapman, D. V., Dodge, J. D., and Heaney, S. I. (1982). Cyst formation in the freshwater dinoflagellate *Ceratium hirundinella. J. Phycol.,* 18, 121–9.

(1985). Seasonal and diel changes in ultrastructure in the dinoflagellate *Ceratium hirundinella. J. Plank. Res.,* 7, 263–78.

Chapman, D. V., Livingstone, D., and Dodge, J. D. (1981). An electron microscope study of the excystment and early development of the dinoflagellate *Ceratium hirundinella. Br. Phycol. J.,* 16, 182–94.

Cranwell, P. A. (1976). Decomposition of aquatic biota and sediment formation: organic compounds in detritus resulting from microbial attack on the alga *Ceratium hirundinella*. *Freshwat. Biol.*, 6, 41–8.

Crisculo, C. M., Dubinsky, Z., and Aaronson, S. (1981). Skeleton shedding in *Peridinium cinctum* from Lake Kinnert – a unique phytoplankton response to nutrient imbalance. In *Developments in Arid Zone Ecology and Environmental Quality*, ed. H. Shuval, pp. 169–76. Philadelphia: Balaban ISS.

Cronberg, G. (1981). SEM and ecological investigations of *Peridiniopsis borgei* Lemm., *P. polonicum* (Wolosz.) Bourr. from Sweden and *P. palustre* var. *raciborskii* (Wolosz.) Lef. from Brazil. Preprint. Hexrose Conf. on Modern and Fossil Dinoflagellates, Tubingen, FRG.

Daily, W. A. (1960). Forms of *Ceratium hirundinella* (O. F. Muller) Schrank in lakes and ponds of Indiana. *Proc. Indiana Acad. Sci.*, 70, 213–5.

Dottne-Lindgren, A. and Ekbohm, G. (1975). *Ceratium hirundinella* in Lake Erken: Horizontal distribution and form variation. *Int. Revue ges. Hydrobiol.*, 60, 115–44.

Dürr, G. (1979). Electron microscope studies of the theca of dinoflagellates. II. The cyst of *Peridinium cinctum*. *Arch. Protistenk.*, 122, 121–39.

Elbrächter, M. (1973). Population dynamics of *Ceratium* in coastal waters of Kiel Bay. *Oikos, Suppl.*, 15, 43–8.

Elgavish, A., Elgavish, G. A., Halmann, M., and Berman, T. (1980). Phosphorus utilization and storage in batch cultures of the dinoflagellate *Peridinium cinctum* f. *westii. J. Phycol.*, 16, 626–33.

Entz, G. (1925). Über Cysten und Encystierung der Süsswasser Ceratien. *Arch. Protistenk.*, 51, 131–83.

(1926). Beitrage zur kenntnis der Peridinen. Zur Morphologie und Biologie von *Peridinium borgei*. *Arch. Protistenk.*, 56, 397–446.

(1927). Beitrage zur kenntnis der Peridinen. II. resp. VII. Studien aus Süsswasser Geratien (Morphologie, Variation, Biologie). *Arch. Protistenk.*, 58, 344–440.

(1931). Analyse des Wachstums und der Teilung einer Population sowie eines Individuums des Protisten *Ceratium hirundinella* unter den naturlichen Verhaltnissen. *Arch. Protistenk.*, 74, 310–61.

Eppley, R. W., Holm-Hansen, O., and Strickland, J. D. H. (1968). Some observations on the vertical migration of dinoflagellates. *J. Phycol.*, 4, 333–40.

Eren, J. (1969). Studies on development cycle of *Peridinium cinctum* f. *westii. Verh. Internat. Verein. Limnol.*, 17, 1013–6.

Florin, M. B. 1957). Plankton of fresh and brackish waters in the Sodertalje area. *Acta Phytogeogr. Seuc.*, 37, 1–144.

Forward, R. B., Jr. (1976). Light and diurnal vertical migration: photobehaviour and photophysiology of plankton. *Photochem. Photobiol. Rev.*, 1, 157–209.

Frempong, E. (1982). The space-time resolution of phased cell division in natural populations of the freshwater dinoflagellate *Ceratium hirundinella*. *Int. Revue ges. Hydrobiol.*, 67, 323–39.

(1984). A seasonal sequence of diel distribution patterns for the planktonic dinoflagellate *C. hirundinella* in a eutrophic lake. *Freshwat. Biol.,* 14, 401–21.

Galleron, C. J. (1976). Synchronization of the marine dinoflagellate *Amphidinium carterae* in dense cultures. *J. Phycol.,* 12, 69–73.

George, D. G. and Edwards, R. W. (1976). The effect of wind on the distribution of chlorophyll *a* and crustacean plankton in a shallow eutrophic reservoir. *J. Appl. Ecol.,* 13, 667–90.

George, D. G. and Heaney, S. I. (1978). Factors affecting the spatial distribution of phytoplankton in a small productive lake. *J. Ecol.,* 66, 133–55.

Gilyarov, A. (1977). Observations on food consumption of rotifers of the genus *Asplanchna. Zoologhiceskii J.,* 56, 1874–6.

Gophen, M. (1973). Zooplankton in Lake Kinneret. In *Lake Kinneret Data Record,* ed. T. Berman (N.C.R.D. 13–73), pp. 61–7. Tabgha: Israel Scientific Research Conference.

(1980). Food sources, feeding behaviour and growth rates of *Sarotherodon galilaeus* fingerlings. *Aquaculture,* 20, 101–15.

Gophen, M., Drenner, R. W. and Vinyard, G. L. (1983). Cichlid stocking and the decline of the Galilee St. Peter's fish *(Sarotherodon galilaeus)* in Lake Kinneret, Israel. *Can. J. Fish Aquat. Sci.,* 40, 983–6.

Gutelmacher, B. L. (1975). Relative significance of some species of algae in plankton primary production. *Arch. Hydrobiol.,* 75, 318–28.

Happach-Kassan, C. (1980). Beobachtungen zur Entwicklungsgeschichte der Dinophycee *Ceratium cornutum:* Sexualität, Gamie und Meiose. Ph.D. Dissertation, Philipps Univ., Marburg.

Harris, G. P., Heaney, S. I., and Talling, J. F. (1979). Physiological and environmental constraints in the ecology of the planktonic dinoflagellate *Ceratium hirundinella. Freshwat. Biol.,* 9, 413–28.

Heaney, S. I. (1976). Temporal and spatial distribution of the dinoflagellate *Ceratium hirundinella* O. F. Muller within a small productive lake. *Freshwat. Biol.,* 6, 531–42.

Heaney, S. I., Chapman, D. V., and Morison, H. R. (1983). The role of the cyst stage in the seasonal growth of the dinoflagellate *Ceratium hirundinella* within a small productive lake. *Br. Phycol. J.,* 18, 47–59.

Heaney, S. I. and Eppley, R. W. (1981). Light, temperature and nitrogen as interacting factors affecting diel vertical migrations of dinoflagellates in culture. *J. Plankton Res.,* 3, 331–44.

Heaney, S. I. and Talling, J. F. (1980a). Dynamic aspects of dinoflagellate distribution patterns in a small productive lake. *J. Ecol.,* 68, 75–94.

(1980b). *Ceratium hirundinella* – ecology of a complex, mobile and successful plant. *Rep. Freshwat. Biol. Ass.,* 48, 27–39.

Heller, M. D. (1977). The phased division of the freshwater dinoflagellate *Ceratium hirundinella* and its use as a method of assessing growth in natural populations. *Freshwat. Biol.,* 7, 527–33.

Hertzig, R., Dubinsky, Z., and Berman, T. (1981). Breakdown of *Peridinium* biomass in Lake Kinneret. In *Developments in Arid Zone Ecology and*

Environmental Quality, ed. H. Shuval, pp. 179–85. Philadelphia: Balaban ISS.

Hickel, B. (1978). Phytoplankton population dynamics in Plußsee (East Holstein, Germany). *Verh. Ges. Okol., Kiel,* 1977(1978), 119–26.

—— (1984). Evidence for sexual reproduction in the freshwater dinoflagellate *Ceratium furcoides.* In *Max-Planck Gesellschaft Jahrbuch 1984,* pp. 319–23. Gottingen: Vanderhoeck & Ruprecht.

—— (1985a). The population structure of *Ceratium* in a small eutrophic lake. *Verh. Internat. Verein. Limnol.,* 22, 2845–9.

—— (1985b). *Peridinium pusillum* blooms in a highly acidic pond. Abstr., 3rd Int. Conf. Modern & Fossil Dinoflagellates, Aug. 1985, Egham, U.K., p. 15.

Hickel, B. and Pollingher U. (In press). On the morphology and ecology of *Gonyaulax apiculata* Penard (Entz) from the Selenter See (West Germany). *Arch. Hydrobiol. Suppl. (Algol. Stud.).*

Holl, K. (1928). Oekologie der Peridineen. *Pflantzen Forschung,* 11, 1–105.

Huber-Pestalozzi, G. (1950). Das Phytoplankton des Süsswassers: Systematik und Biologie. Cryptophyceen, Chloromonadinen, Peridineen. In *Die Binnengewasser,* 16, Tiel 3. Stuttgart: E. Schweizerbart' sche Verlags.

Huber, G. and Nipkow, F. (1923). Experimentelle Untersuchungen über Entwicklung und Formbildung von *Ceratium hirundinella* O. F. Muller. *Flora. Jena.,* 116, 114–215.

Hutchinson, G. E. (1967). *A Treatise on Limnology,* Vol. II. New York: Wiley.

Jahn, T. L., Harmon, W. M., and Landman, M. (1963). Mechanisms of locomotion in flagellates. I. *Ceratium. J. Protozool.,* 10, 358–63.

Komarovsky, B. (1959). The plankton of Lake Tiberias. *Bull. Sea Fish. Res. Stn., Haifa,* 25, 1–94.

Krause, F. (1911/12). Studien über die Formveränderung von *Ceratium hirundinella* O. F. Mull. als Anpassungserscheinung an die Schwebefähigkeit. *Int. Revue ges. Hydrobiol.,* 3, 1–32.

Krupa, D. (1981). *Ceratium hirundinella* (O. F. Muller) Bergh. in two trophically different lakes. II. Development and morphological variation of active forms and cysts. *Ekol. Pol.,* 29, 571–83.

Landau, R. (1979). Growth and population studies on *Tilapia galilaea* in Lake Kinneret. *Freshwat. Biol.,* 9, 23–32.

Langhans, V. H. (1925). Gemischte Populationen von *Ceratium hirundinella* (O.F.M.) Schrank und ihre Deutung. *Arch. Protistenk.,* 52, 586–602.

LeBlond, P. H. and Taylor, F. J. R. (1976). The propulsive mechanism of dinoflagellate transverse flagellum reconsidered. *BioSystem,* 8, 33–9.

Leedale, G. F. (1959). The time-scale of mitosis in Euglenineae. *Arch. Mikrobiol.,* 32, 352–60.

Lemmermann, E. (1910). *Kryptogamenflora der Mark Brandenburg. Algen I,* Bd. II. Leipzig: Gebruden Borntraeger.

Leventer, H. (1972). Eutrophication control of Tsalmon reservoir by the cichlid fish *Tilapia aurea.* In *Proceedings of the 6th International Water Pollution Research,* June 18–23, 1972, pp. A/8/15/1–9. Pergamon Press Paper No. 15.

Lindholm, T., Weppling, K., and Jensen, H. S. (1985). Stratification and primary production in a small brackish lake studied by close interval siphon sampling. *Verh. Internat. Verein. Limnol.*, 22, 2190–4.

Lindstrom, K. (1984). Effect of temperature, light and pH on growth, photosynthesis and respiration of the dinoflagellate *Peridinium cinctum* fa. *westii* in laboratory cultures. *J. Phycol.*, 20, 212–20.

Loeblich, A. R. III. (1970). The amphiesma or dinoflagellate cell covering. In *Proceedings of the North American Paleontology Convention*, Chicago, IL, ed, E. L. Yochelson, pp. 867–926. Lawrence, Kansas: Allen Press.

McCarthy, J. J., Wynne, D., and Berman, T. (1982). The uptake of dissolved nitrogenous nutrients by Lake Kinneret (Israel) microplankton. *Limnol. Oceanogr.*, 27, 673–80.

Messer, G., and Ben-Shaul, Y. (1970). Ultrastructure of aging cells of *Peridinium westii*. In *Proceedings of the VIIth International Congress of Electron Microscopy*, pp. 411–13. Grenoble: Societe Francais de Microscopie Electronique.

Mitchell, S. F. and Galland, A. N. (1981). Phytoplankton photosynthesis, eutrophication and vertical migration of dinoflagellates in a New Zealand reservoir. *Verh. Internat. Verein. Limnol.*, 21, 1017–20.

Mueller-Elser, M. and Smith, W. O. (1985). Phased cell division and growth rate of a planktonic dinoflagellate *Ceratium hirundinella* in relation to environmental variables. *Arch. Hydrobiol.*, 104, 477–91.

Padisak, J. (1985). Population dynamics of the freshwater dinoflagellate *Ceratium hirundinella* in the largest shallow lake of Central Europe, Lake Balaton, Hungary. *Freshwat. Biol.*, 15, 43–52.

Pearsall, W. H. (1929). Form variation in *Ceratium hirundinella* O. F. M. *Proc. Leeds Philos. Soc.*, 1, 432–9.

Pfiester, L. A. (1975). Sexual reproduction of *Peridinium cinctum* f. *ovoplanum* (Dinophyceae). *J. Phycol.*, 11, 259–65.

(1976). Sexual reproduction of *Peridinium willei* (Dinophyceae). *J. Phycol.*, 12, 234–8.

(1977). Sexual reproduction of *Peridinium gatunense* (Dinophyceae). *J. Phycol.*, 12, 92–5.

Pfiester, L. A. and Skvarla, J. J. (1979). Heterothallism and thecal development in the sexual life history of *Peridinium volzii* (Dinophyceae). *Phycologia*, 18, 13–18.

(1980). Comparative ultrastructure of vegetative and sexual theca of *Peridinium limbatum* and *Peridinium cinctum* (Dinophyceae). *Amer. J. Bot.*, 67, 955–8.

Pollingher, U. (1981). The structure and dynamics of the phytoplankton assemblages in Lake Kinneret, Israel. *J. Plankton Res.*, 3, 93–105.

(1986a). Ecology: (b) Freshwater ecosystems. In *The Biology of Dinoflagellates*, ed. F. J. R. Taylor. Oxford: Blackwell.

(1986b). Phytoplankton periodicity in a subtropical lake (Lake Kinneret, Israel). *Hydrobiologia*, 138, 127–38.

Pollingher, U. and Berman, T. (1975). Temporal and spatial patterns of dinoflagellate blooms in Lake Kinneret, Israel (1969–1974). *Verh. Internat. Verein. Limnol.,* 19, 1370–82.

(1982). Relative contribution of net and nanophytoplankton to primary production in Lake Kinneret (Israel). *Arch. Hydrobiol.,* 96, 33–46.

Pollingher, U. and Serruya, C. (1976). Phased division of *Peridinium cinctum* f. *westii* and the development of the bloom in Lake Kinneret (Israel). *J. Phycol.,* 12, 162–70.

Pollingher, U. and Zemel, E. (1981). *In situ* and experimental evidence of the influence of turbulence on cell division processes of *Peridinium cinctum* f. *westii* (Lemm.) Lefevre. *Br. Phycol. J.,* 16, 281–7.

Prezelin, B. B. and Sweeney, B. M (1979). Photoadaptation of photosynthesis in two bloom-forming dinoflagellates. In *Toxic Dinoflagellate Blooms,* ed. D. L. Taylor and N. H. Seliger, pp. 101–6. Amsterdam: Elsevier/North-Holland.

Raven. J. A. and Beardall, J. (1981). Respiration and photorespiration. *Can. Bull. Fish, Aqaut. Sci.,* 210, 55–82.

Reynolds, C. S. (1976). Succession and vertical distribution of phytoplankton in response to thermal stratification in a lowland mere, with special reference to nutrient availability. *J. Ecol.,* 64, 529–51.

(1984). *The Ecology of Freshwater Phytoplankton.* Cambridge: Cambridge University Press.

Reynolds, C. S. and Reynolds, J. B. (1985). The atypical seasonality of phytoplankton in Crose Mere 1972: an independent test of the hypothesis that variability in the physical environment regulates community dynamics and structure. *Br. Phycol. J.,* 20, 227–42.

Rodhe, W. (1978). Growth characteristics. In *Lake Kinneret,* ed. C. Serruya (Monographiae Biologicae 32), pp. 275–83. The Hague: Dr. W. Junk.

Sako, Y., Ishida, Y., Kadota, H., and Hata, Y. (1984). Sexual reproduction and cyst formation in the freshwater dinoflagellate *Peridinium cunningtonii. Bull. Jap. Soc. Sci. Fish.,* 50, 743–50.

Serruya, C. and Berman, T. (1975). Phosphorus, nitrogen and the growth of algae in Lake Kinneret. *J. Phycol.,* 11, 155–62.

Serruya, C., Edelstein, M., Pollingher, U., and Serruya, S. (1974). Lake Kinneret sediments: nutrient composition of the pore water and mud water exchanges. *Limnol. Oceanogr.,* 19, 489–508.

Serruya, C., Gophen, M., and Pollingher, U. (1980). Lake Kinneret: carbon flow patterns and ecosystem management. *Arch. Hydrobiol.,* 88, 265–302.

Serruya, C. and Pollingher, U. (1977). Lowering of water level and algal biomass in Lake Kinneret. *Hydrobiologia,* 54, 73–80.

Serruya, C., Pollingher, U., and Gophen, M. (1975). N and P distribution in Lake Kinneret (Israel) with emphasis on dissolved organic nitrogen. *Oikos,* 26, 1–8.

Serruya, C., Serruya, S., and Pollingher, U. (1978). Wind, phosphorus release and division rate of *Peridinium* in Lake Kinneret. *Verh. Internat. Verein. Limnol.,* 20, 1096–1102.

Serruya, S. (1971). Kinneret hydromechanics, 1968–1971. Limnological Laboratory, Tiberias, Progress Report No. 5.

— (1975). Wind, water temperature and motions in Lake Kinneret: general pattern. *Verh. Internat. Verein. Limnol.,* 19, 73–87.

Sherr, B. F., Sherr, E. B., and Berman, T. (1982). Decomposition of organic detritus: a selective role for microflagellate protozoa. *Limnol. Oceanogr.,* 27, 765–9.

Sommer, U. (1981). The Role of r- and K selection in the succession of phytoplankton in Lake Constance. *Acta Oecol., Oecol. Gener.,* 2, 327–42.

Sommer, U., Wedemeyer, C., and Lowsky, B. (1984). Comparison of potential growth rates of *Ceratium hirundinella* with observed population density changes. *Hydrobiologia,* 109, 159–64.

Spataru, P. (1976). The feeding habits of *Tilapia galilaea* (Artedi) in Lake Kinneret (Israel). *Aquaculture,* 9, 47–59.

Spataru, P. and Zorn, M. (1978). Food and feeding habits of *Tilapia aurea* (Steindachner) (Cichlidae) in Lake Kinneret (Israel). *Aquaculture,* 13, 67–79.

Spector, D. L., Pfiester, L. A., and Triemer, R. E. (1981). Ultrastructure of the dinoflagellate *Peridinium cinctum* fa. *ovoplanum.* II. Light and electron microscopic observations on fertilization. *Amer. J. Bot.,* 68, 34–43.

Steenbergen, C. L. M. (1982). Phytoplankton periodicity and sediment trap recoveries. Limnological Institute, Nieuwersluis, Progress Report 1982, pp. 33–5.

von Stosch, H. A. (1964). Zum Problem der sexuellen Fortpflanzung in der Peridineengattung *Ceratium. Helgolander wiss. Meeresunters.,* 10, 140–52.

— (1972). La signification cytologique de la "cyclose nucleaire" dans le cycle de vie des dinoflagellés. *Mem. Soc. bot. Fr.,* 1972, 201–22.

— (1973). Observations on vegetative reproduction and sexual life cycles of two freshwater dinoflagellates, *Gymnodinium pseudopalustre* Schiller and *Woloszynskia apiculata* sp. nov. *Br. Phycol. J.,* 8, 105–34.

von Stosch, H. A., Theil, G., and Happach-Kassan, C. (1981). Analysis of ordered tetrads in *Ceratium cornutum* Clap. & Lachm., a freshwater dinoflagellate. Preprint. Hexrose Conf. on Modern and Fossil Dinoflagellates, Tubingen, FRG.

Stull, E. A., Amezaga, E., and Goldman, C. R. (1973). The contribution of individual species of algae to primary productivity of Castle Lake, California. *Verh. Internat. Verein. Limnol.,* 18, 1776–83.

Stumm, W. and Morgan, J. J. (1970). *Aquatic Chemistry.* New York: Wiley.

Talling, J. F. (1971). The underwater light climate as a controlling factor in the production ecology of freshwater phytoplankton. *Mitt. Internat. Verein. Limnol.,* 19, 214–43.

Tappan, H. and Loeblich, A. R., Jr. (1971). Geobiological implications of fossil phytoplankton evolution in time and space. In *Symposium on the Palynology of the Late Cretaceous and Early Tertiary,* ed. R. M. Kosanke and A. T. Cross, pp. 247–340. Geol. Soc. Amer. Spec. Pap. 127.

Tuttle, R. C. and Loeblich, A. R. (1975). An optimal growth medium for the dinoflagellate *Crypthecodinium cohnii, Phycologia,* 14, 1–8.

Wesenberg-Lund, C. (1908). *Plankton Investigations of the Danish Lakes. General part. The Baltic freshwater plankton, its origin and variation.* Copenhagen: Nordish Forlag.

White, A. W. (1976). Growth inhibition caused by turbulence in the toxic marine dinoflagellate *Gonyaulax excavata. J. Fish. Res. Bd. Can.,* 33, 2598–602.

Wynne, D., Patni, N. J., Aaronson, S., and Berman, T. (1982). The relationship between nutrient status and chemical composition of *Peridinium cinctum* during the bloom in Lake Kinneret. *J. Plank. Res.,* 4, 125–36.

Chapter 5

ECOLOGY OF FRESHWATER PLANKTONIC GREEN ALGAE

CHRISTINE M. HAPPEY-WOOD

INTRODUCTION

The green algae, or Chlorophyta, compose the largest and most varied phylum of algae. They are characterized by their grassy green color, attributable to the pigments chlorophylls *a* and *b*, α, β, and γ carotenes, and several xanthophylls (Goodwin 1974, 1980), and also by their storage of plant starch (a β 1,4-linked polyglucan). This starch is frequently associated with one or several pyrenoids, conspicuous storage regions within the chloroplast. Among the algae, the Chlorophyta are the most closely related to the higher plants based on their similar photosynthetic pigments, storage of higher-plant starch, and the fine-structural organization of the chloroplast with the photosynthetic lamellae stacked together to give grana and intergranal regions (e.g., Dodge 1973; Bisulputra 1974).

This chapter is concerned with the ecology and biology of representative green algae characteristic of phytoplankton populations of freshwaters. Definitions of phytoplankton as an ecological community are many and varied in the literature, and frequently the algae are described as photosynthetic cells floating in an aqueous medium (e.g., Hensen 1887; Morris 1980). This implies that the organisms are consistently buoyant or possess some active mechanism or cellular characteristic that enables them to remain in suspension. However, planktonic algae may not reside in open water habitats for their entire life cycle. Some may spend resting periods in benthic situations, and others spend periods of dispersal in soil particles or the aeroflora. Thus, in discussing the ecology of planktonic green algae here, we will regard them as the community of microscopic plants existing in suspension in aquatic environments that is liable to passive movements by winds and currents (modified from Reynolds 1984b).

Phytoplankton populations are characterized by diverse species assemblages with representatives from all the major algal phyla apart from the macroscopic Phaeophyta and Rhodophyta. Green algae will almost invariably be found in all freshwaters, even if present in only small numbers. However, discussions here will be focused on chlorophytes forming dominant populations in planktonic communities by virtue either of their cell numbers or cell volumes. This chapter discusses the types of chlorophytes common in freshwater phytoplankton populations, their distribution, and factors affecting their growth in naturally occurring communities. These factors may be considered to be of three types related to, first, environmental attributes such as water turbulence or nutrient status, second, attributes inherent to the cells themselves (that is, morphological, physiological, or genetic),and third, attributes related to other living organisms within the plankton. When these are all considered in conjunction with the growth characteristics, reproductive mechanisms, and life cycles, it is possible to identify a series of survival strategies that have evolved among planktonic green algae. Because there is such a wide variety of morphological types with a range of inherent biological characteristics, the ecology of green algae will be considered in five biological groups: microalgae (i.e., those smaller than 20 μm in largest diameter), flagellates (i.e., those flagellate unicells larger than 20 μm and colonial Volvocales), Chlorococcales, Desmidiales, and filamentous green algae.

TYPES OF PLANKTONIC GREEN ALGAE

The green algae include a greater diversity of cellular organization, morphological structure, and reproductive processes than are found in any other algal division (Bold et al. 1978). This is reflected in the diversity of classification systems suggested by different authors, which are summarized in an appendix by Bold and Wynne (1985). If the diatoms are excluded from the Chrysophyta, more orders of the Chlorophyta are represented in phytoplankton than in any other alga phyla (Reynolds 1984a).

Although types of planktonic chlorophytes may be considered in taxonomic groupings, such as Volvocales, Chlorococcales, Zygenematales, and so on, it is of more interest and probably greater ecological significance to group them biologically – by cell organization (either unicellular, colonial, or filamentous) and cellular activity (either nonmotile or motile through the use of flagella or secretion of mucilage). This results in three categories of organization: unicellular, colonial, and filamentous, within which exist motile and nonmotile representatives exhibiting a great range in cell size, morphology, and simple multicellular arrangement (Table 5-1).

Many workers have differentiated groups of planktonic algae by cell dimensions. In the early twentieth century, Lohman (1903) divided the phytoplankton into two major categories on the basis of whether the algae were retained by or passed through a fine-mesh net. The fraction retained by a fabric of mesh aperture 20–90 μm was referred to as *netplankton,* and the fraction that passed through was termed *nanno-plankton* (now generally accepted as nanoplankton, Sournia 1980). More recently, a category of even smaller size has been distinguished, the picoplankton, including algae within the size range of cell diameters 0.2–2 μm (Sieburth, Smetacek, & Lenz 1978; Sournia 1980). Algae falling into all three of these categories of plankton are found among the chlorophytes (Table 5-1).

DISTRIBUTION IN NATURALLY OCCURRING PLANKTONIC COMMUNITIES

Inoculum

Before a phytoplankton population can develop, a source of actively growing cells (an inoculum) is essential. In the case of green algae, several sources of inoculae may be identified. Unlike other algal phyla, some chlorophytes are common in the air flora and were recognized as such in the mid-nineteenth century (Ehrenberg 1844). The major airborne algae are *Chlorella, Chlorococcum,* and, to a lesser extent, *Chlamydomonas,* all of which are far more common in aerosols than are members of the Cyanophyta or Chrysophyta (Brown, Larson, & Bold 1976).

Colonization of standing water is extremely rapid by smaller-sized members of the Chlorococcales. When small sterilized experimental pond systems (volume 30 L, surface area 0.25 m²) were expose outside to natural irradiance and potential airborne inoculae, growth of small green algae was rapid. Colonization by *Scenedesmus acuminatus, Chlamydomonas sphaerella,* and *Chlorella pyrenoidosa* was recorded after seven days (Fig. 5-1). The experiment involved three sterile treatments: (a) distilled water, (b) nutrient addition (Bristols solution: Bold 1942), and (c) added nutrients with shading by a cover of nylon mesh. Population densities of all three green algae were greatest in the treatment with added nutrients, and growth rates were lower when the tanks were shaded. Both the patterns of growth and relative contribution of the three green algae differed between the three treatments. *S.*

Table 5-1. *Categories of representative green algae represented in freshwater plankton.*

Morphology:	Unicellular		Colonial		Filamentous
Activity:	Motile (by flagella)	Nonmotile	Motile (by flagella)	Nonmotile	Motile (by mucilage secretion)[b]
Size:	*Chloromonas* (2 μm)	*Chlorella* (1–2 μm)	*Gonium* (4–32 cells) (ca. 20 μm)	*Ankistrodesmus* Cell no. variable (2–50 μm)	Desmids
	Chlamydomonas (2–100 μm)	*Ankistrodesmus* (2 × 10 μm)	*Eudoria* 16–32 cells (ca. 20–30 μm)	*Kirchneriella* 4–16 cells (2–20 μm)	*Desmidium* many cells, (ca. 120 μm per cell)
	Carteria (10–100 μm)	*Coccomyxa*	*Pandorina* 16–32 cells (ca. 25 μm)	*Scenedesmus*[b] 2–4 cells (4–50 μm)	
	Chlorogonium (5–100 μm)	*Stichococcus*	*Volvox* 1–50 × 10³ cells (ca. 10–15 μm)		

Chlorella (10–15 μm)	*Dictyospherium* 8–32 cells (4–10 μm)	Zygnemataceae
Chlorococcum (10–20 μm)	*Pediastrum*[+] 8–64 cells (5–40 μm)	*Mougeotia* many cells
Desmids[a] (30–300 μm) *Closterium* *Cosmarium* *Euastrum* *Xanthidnum*		*Zygnema* many cells *Spirogyra* many cells

[a] Desmids secrete mucilagenous trails that may result in movements of cells over surfaces, but not active movement when suspended in water.

[b] Limited movement of certain Chlorococcalean algae has been observed on surfaces, e.g., *Pediastrum*, *Scenedesmus* (Brandeham, personal communication), but not when the cells are suspended in water.

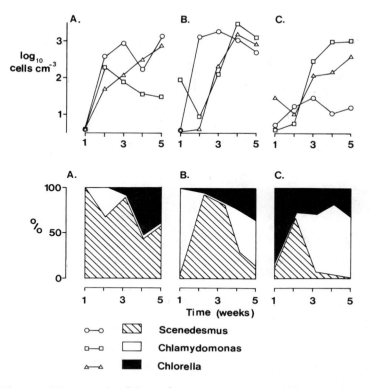

Fig. 5-1. The growth of *Scenedesmus acuminatus, Chlamydomonas sphaerella,* and *Chlorella pyrenoidosa* and the percentage contribution of these three algae to the total community population density for a five-week period in three experimental treatments exposed to air. Treatment A, distilled water; B, with added nutrients; C, with added nutrients and shading.

acuminatus was dominant initially in distilled water, but after four weeks *Chlorella pyrenoidosa* was of equal numerical importance in the plankton population. Under conditions of natural irradiance with added nutrients, *Chlamydomonas sphaerella* was the major component of the plankton for the first two weeks, to be replaced by *S. acuminatus.* After four weeks, *C. pyrenoidosa* and *C. sphaerella* accounted almost equally for a total of 80% of the algal numbers, with *Scenedesmus* of lesser importance.

In the third treatment with added nutrients and shading, *Chlorella* was the most numerous alga after seven days, growth of *Scenedesmus* then took place, but to a lesser extent and over a shorter time period than in the two unshaded treatments. After three weeks, *Chlamydomonas* had increased exponentially to become the dominant planktonic alga. Thus, without the influence of potential algal inocula within the water column or benthic niches, and lacking influence of transport into standing water by other biological vectors such as animals, growth of planktonic populations of algae dominated by small chlorophytes may be rapid. The dominant species varied with nutrient content of the water exposed in the experimental pond systems, the presence of shading, and the time course of the experiment.

On many occasions, planktonic populations are thought to be derived from a small population of algae present in the water column, probably in numbers too low to enable detection by routine counting methods. This was certainly true of the common planktonic diatom *Asterionella formosa* (Lund 1949), and in the case of desmids it is likely that a few cells may remain in the plankton throughout the year. Resting stages and thick-walled sexual spores that remain viable in the sediment may also be an important inoculum, but there appears to be little evidence for such with reference to the green algae, apart from *Eudorina elegans* (Reynolds 1986).

The majority of algal species are characteristic of a particular community within a water body. Planktonic species, apart from senescent or resting cells, are absent from benthic situations and, in general, the converse is true. However, in the case of planktonic green algae the development of populations derived from other ecological niches in a lake or pond have been described. Small chlamydomonads have been found in the benthic communities of epipelon and epiphyton (e.g., *Chlamydcapsa bacillus* and *Chlamydomonas perpusilla*) that also grew in the phytoplankton of the overlying water (Happey-Wood 1978). Thus small green algae persistent in benthic communities may act as a source of inoculum to the planktonic community. In comparing epipelic and planktonic algal populations of two small water bodies, Moss and Karim (1969) found that the larger-celled chlorophytes *Staurastrum validus* and *Staurastrum* cf. *gracile* var. *nanum* were the only algae to grow actively in both algal niches and may have originated from metaphyton. It is likely that exchange of actively growing species may be of ecological significance only in shallow or well-mixed environments, and may be more significant for the more cosmopolitan green algae of small cell size.

Biographical and Seasonal Distribution

Green phytoplanktonic algae exhibit a widespread distribution in terms of latitude, and the predominant types vary greatly with both the nutrient status of the water and the physical characteristics of the water body, such as its morphometry, which will affect the water movements, currents, and retention time.

At high latitudes the small-sized cell components of the phytoplankton are more common. Rodhe (1955) described populations of *Chlorella pyrenoidosa, Stichococcus atomus,* and *Coccomyxa coccoides* together with *Chlamydomonas* sp. and other delicate flagellates growing during spring under considerable cover of ice and snow in Swedish Lapland. Skuja (1964) found nonflagellate *Chlorella* and *Stichococcus* common in the deep waters of lakes in northern Sweden. Similarly in Alaska, Alexander and Barsdate (1971) recorded the growth of *Chlamydomonas* and *Chlorella* prior to ice melt. In reviewing the limnology of the Arctic, Hobbie (1973) found that chlorophytes featured in the spring plankton growing under ice cover, but their contribution to planktonic biomass was small in comparison to chrysosphytes, cryptophytes, dinoflagellates, and small centric diatoms. Tundra ponds are characterized by an extremely short growing season of about 100 days, and the algae are subjected to stress during the long period of ice cover when these shallow ponds may freeze solid. It is thought that many algae may form physiological resting stages in which ice formation within the cells is prevented by the accumulation of low-molecular-weight solutes such as polyhydric alcohols or amino acids and their derivatives, because no morphological obvious resting stages or sexual spores have been observed (Sheath 1986). Chlorophytes accounted for 89 out of a total of 251 taxa identified by Sheath (1986) in his review, and in terms of biomass they were a major class of primary producers together with chrysophytes, diatoms, and blue-green algae. Despite the large numbers of species of green planktonic algae in this severe habitat, the major peaks of planktonic biomass were dominated by *Rhodomonas minuta* and flagellate chrysophytes.

The predominant primary producers of the Antarctic, where the flora is generally less diverse than the Arctic (Heywood 1977, 1978), are mats of filamentous blue-green algae. However, the phytoplankton of Algal Lake on Ross Island was made up largely of *Chlamydomonas intermedia, C. subcaudata,* and a *Brachiomonas* sp. That of Skuja Lake (also on Ross Island) contained these same flagellates, but the algal bloom developing late in the season was dominated by *Ankistrodesmus falcatus* var. *acicularis* (Goldman, Mason, & Wood 1972). More recently, Heywood, Dartnall, and Priddle (1980) noted that the filamentous green algae *Spirogyra, Mongeotia,* and *Zygenema* form dense growths during summer in several lakes surveyed by the British Antarctic Survey. In the four perennially ice-covered lakes of southern Victorialand, Seaburg, Kaspar, and Parker (1983) recorded a total of eight taxa of green planktonic algae, compared with six filamentous blue-green algae, one member of the xanthophyceae, two chrysophytes, two diatoms, one cryptomonad, and four flagellates of uncertain affinity. Six of the chlorophytes were unicellular ultraplankton (five flagellates, including two species of *Chlamydomonas, Chloromonas, Polytomella, Thorakomonas* plus *Chlorella*) and there were also several larger chlorophytes, including a coccoid unicellular chlorophyte, and a colonial *Westella* sp.

In alpine lakes, Kosswig (1967) found members of the Chlorococcales common during summer, and desmids were present in the phytoplankton of Vorderer Finterstater See throughout the year (Pechlaner 1971). Population densities were low in spring after ice melt, with the population evenly distributed with depth. During summer the large cells were in greater numbers in the deeper strata of the lake but were again homogeneously distributed in October. Under ice cover, the major proportion of desmids were recorded from the 5-m depth sample, and it was suggested that this crop of desmids, persistent during the winter, was persistent on layers of dense cold water.

In general, unicellular green algae are a common component of the smaller-sized cell fractions of phytoplankton in oligotrophic lakes. Ostrofsky and Duthie (1975) described *Sphaerocystis* and *Gleocystis* as dominant in some lakes in the eastern Canadian Shield. In Lake Matamec, Quebec, *Chlamydomonas, Paramastix,* and *Oocystis* were components of the ultraplankton (cell diameter less than 64 μm, Ross & Duthie 1981). The phytoplankton of Lake Tahoe, Nevada/California, in the United States, a deep ultraoligotrophic system, is dominated by "heavily pigmented ultraplankton" (Tilzer, Paerl, & Goldman 1977). A large proportion of algae making up this fraction of the phytoplankton are likely to be small chlorophytes (Goldman, personal communication).

Members of the Chlorophyta contributed a small proportion of the total phytoplankton of Lake Ontario (Skreinivasa & Nalewajko 1975). These were predominantly chlorococcalean algae; only two flagellates, *Chlamydomonas* and *Eudorina elegans,* were recorded. The more recent review of Munawar and Munawar (1986) confirms diatoms as being the dominant algae in all five Laurentian Great Lakes. However, these investigators confirmed the presence and importance of the smaller-sized cell categories of the phytoplankton, and they concluded that plankton within the size category 5–20 μm were the most important in all five lakes in terms of efficiency of carbon fixation in relation to algal biomass. Chlorophytes did not feature in the list of common planktonic algae in oligotrophic Lake Superior, and the largest number of chlorophyte species was compiled for western Lake Erie. Chlorophyte algae dominate the phytoplankton of Lake Ontario and eastern Lake Erie in summer and fall, and their biomass approaches 40% of the total plankton during summer in Lake Ontario. In considering phytoplankton succession, 26 taxa of chlorophytes were present in spring with the greatest diversity (eight species) in Georgia Bay of Lake Huron. Thirty-seven species of green algae were found during summer, and 29 in autumn, with the largest number of species in Lake Ontario during both these seasons.

The well-mixed, shallow large lakes of central Canada, Winnipeg and Southern Indian Lake, had contrasting populations of chlorophytes. In Lake Winnipeg, large-sized forms such as *Closterium, Staurastrum, Pediastrum, Coelastrum,* and *Dictyosphaerium* were present during the ice-free season, apart from spring, and total chlorophytes accounted for between 5% and 10% of the plankton biomass (Hecky, Kling, & Brunskill 1986). A high diversity of small-celled chlorophytes was found in Southern Indian Lake, but these chlorophytes were of insignificant biomass.

Two deep oligotrophic lakes, Taupo and Waikaremoana, in the center of North Island, New Zealand, support contrasting planktonic populations (Cassie 1978; White et al., 1980). The most important contributors to the total algal biomass of Lake Taupo in summer were small-celled coccoid and ovoid chlorococcalean algae, whereas diatoms were dominant in the winter (Vincent 1983). *Sphaerocystis schroeteri* was dominant in Lake Waitaremoana from spring until summer, with maximum population density in the metalimnion. *Chlamydomonas* and *Staurastrum* were also common during this period.

In the contrasting linked lakes Sorrell and Crescent on the Tasmanian Central Plateau, members of the Chloroccales together with filamentous green algae were a major component, both in terms of cell volume and percentage composition. This was particularly so in the summer period (November–March) in the less productive Lake Sorrell. Chlorococcales were present in Lake Crescent, but were of minor importance in the biomass, which was, on average, an order of magnitude greater than in Lake Sorrell. Desmids were relatively more important in Lake Crescent (Cheng & Tyler 1973).

Green algae featured widely in the characterization of freshwater phytoplankton from temperate situations given by Reynolds (1980). He proposed 14 species assemblages of freshwater phytoplankton, and later extended this to lakes ranging from very low nutrient concentrations to higher trophic situations (Reynolds 1982, 1984b). Of these groupings, 12 contain chlorophytes; this is more than any of the other major algal phyla or classes, including diatoms (Fig. 5-2). Within these

	VERNAL PERIOD	SUMMER PERIOD	
		EARLY-	LATE-
OLIGOTROPHIC	Cyclotella spp. Rhizosolonenia spp.	Peridinium willei Ceratium hirundinella Gomphosphaeria Staurodesmus	
MESOTROPHIC	Asterionella formosa Cyclotella spp. Melosira italica Synedra spp. *Ankistrodesmus* *Chlorella* Cryptomonas / Rhodomonas - - - - - - - - ➔	Dinobryon Mallomonas *Sphaerocystis* *Gemicellicystis* *Coenocococcus* *Oocystis* *lactustris*	*Pandorina* Ceratium *Gomphosphaeria* Asterionella Tabellaria flocculosa Fragilaria crotonensis *Cosmarium* spp. Staurastrum spp.
		Oscillatoria arghadii / rubescens - ➤	
EUTROPHIC	Asterionella Fragillaria Stephanodiscus spp. *Ankistrodesmus* *Elakatothrix* *Scenedesmus* *Tetrastrum* Cryptomonas / Rhodomonas - - - - - - - - - - - - - - - - - - - ➤	*Eudorina* *Pandorina* *Volvox* Tribonema e.g. *Ankyra* Chromulina Monodus	Aphanizomenon Anabaena spp. Ceratium Gloeotrichia Microcystis Asterionella Fragilaria Melosira granulata *Closterium* spp. Staurastrum
HYPERTROPHIC	Diatoma Stephanodiscus spp. Synedra spp. *Ankistrodesmus* *Crucigena* *Scenedesmus* *Tetrastrum* Cryptomonas / Rhodomonas - ➤ Oscillatoria arghadii / redekei - - - - - - - - - - - - - - - - - ➤	*Pediastrum* *Coelastrum* *Oocystis borgei* - - - - - - - - - - ➤ - ➤	Aphanothece Aphanothece

Fig. 5-2. The assemblages of freshwater plankton from temperate latitudes showing the importance of green algae (italics). Modified from Reynolds (1984); used with permission.

phytoplankton assemblages, three are vernal, four occupy the early summer period, and the remainder are either mid- or late summer in seasonality. It is generally accepted that growth of planktonic chlorophytes follows the vernal populations of diatoms, and this is more pronounced where nutrients are in plentiful supply (e.g., Hutchinson 1967). However, reference to Reynolds's analyses of planktonic algae (Fig. 5-2) indicates that the situation is not so straightforward; green algae may be important components of phytoplankton at any time from spring through to autumn, depending on the situation.

Subtropical polymictic Lake Kinneret; is characterized by a Pyrrophyta–Chlorophyta assemblage (Pollingher 1986; Pollingher, Chapter 4 this volume). Biomass of green algae was found to be high in winter, summer, and autumn. Nannoplankton were maximum in the September–October period, and, after the overturn in October or November, they formed the dominant algal group with high species diversity but low biomass. Two desmids formed significant populations in the lake: *Cosmarium laeve,* which achieved maximum numbers of up to 500 cells cm^{-2} in September–October, and *Closterium aciculare,* which had a pulse of abundance in November–December. *Cosmarium laeve* grew during a period of low nutrient availability and was thought to obtain nutrients via excretion from zooplankton abundant at that time. Other chlorophytes, particularly *Tetraedron minimum* and, to a lesser extent, *Coelastrum,* were recorded during most of the year over many years of observation.

Tropical lakes are subject to less seasonal change than either temperate or high-latitude lakes (Hutchinson 1967). As a result, seasonal periodicity of the phytoplankton is minimized (Beadle 1981). But annual patterns of phytoplankton seasonality are somewhat more pronounced in larger and deeper tropical lakes (Beadle 1981), where they are governed by either hydrological or hydrographic features (Talling 1986). In the shallow Lake George, the Chlorococcalean algae *Pediastrum, Scenedesmus,* and *Kirchneriella* accounted for less than 5% of the total phytoplankton (Burgis et al. 1973). Unlike the dominant bluegreen algae in this lake, the chlorophytes were more or less evenly distributed with depth throughout the day, exhibiting little diurnal variation in depth distribution. A variety of green algae, particularly *Volvox* spp. and *Pediastrum,* were characteristic of the phytoplankton populations that developed in the Old Aswan reservoir during reimpoundment (Talling 1986). Appreciable populations of *Chlamydomonas* sp. developed in a small montane reservoir in Malawi during the short growing seasons of dryer seasons (Moss 1970). In the very large deep African rift lake, Lake Tanganyika, chlorophytes were present throughout the year. They accounted for almost the entire phytoplankton volume in August–September following the major diatom maximum and mixing of the water column. During February, March, and April, green algae made up about 50% of the biomass, decreasing to about 20% during the later period of stratification (Talling 1986). The shallower, more rapidly mixed Lake Victoria supported a relatively constant population of desmids and a total of 13 species of green algae, although the dominant algae were diatoms, particularly *Melosira nilsensis,* and blue-green algae.

Two small rift valley lakes in Kenya, Naivasha and Oloidien, were studied by Kalff and Watson (1986). They found 68 species of green algae in the phytoplankton populations. In Lake Naivasha, large-celled chlorophytes such as *Oocystis lacustris, Cosmarium pseudoproturberans* var. *alpinum, Cosmarium* sp., and *Botryococcus* were dominant from January onward, whereas the nannochlorophytes, particularly species of *Cosmarium* with cells less than 35 μm, became the major component of the plankton after February. These small-celled green algae similarly obtained maximum populations after February in Lake Oloidon, and they remained dominant for the rest of the year. Although the diversity of chlorophytes, a total of 38 species, was less in Lake Oloidon, their contribution was more than 10% of the total plankton, reaching about 40% in November–December.

Filamentous green algae accounted for the greatest phytoplankton biomass in Lake Titicaca (Richerson et al. 1986). The most numerous species were *Planctonema lauterbornii, Mougeotia viridis,* and *Gloeotilopsis planktonica* (which was probably the same alga as *Ulothrix planktonica* described previously). These algae were persistent throughout the year, with *Gloetilopsis planktonica* the dominant early in the stratification season, and *Planctonema lauterbornii* occupying a similar niche late in October–November toward the end of the period of stratification.

Descriptions of phytoplankton from river habitats are rare in comparison with studies of lakes. Rivers with slow-moving currents may be compared with lakes with short water retention times, and in these or rivers with managed flow regimes, phytoplankton development may occur. Diatoms have been described as the algae present in greatest population densities, but Swale (1969) found chlorophytes as a major component of English river plankton. In several English rivers, *Ankistrodesmus* was the most common genus together with *Scenedesmus* and *Chlamydomonas.* The chlorococcalean algae were persistent from late spring until autumn, but growth of *Chlamydomonas* was more sporadic. Green algae showed the greatest species diversity in the phytoplankton of the river Avon, although population densities of diatoms were greater (Aykulu 1978). Eighteen representatives of the Chlorococcales were identified, but they were less abundant than *Chlamydomonas.* Studies on the River Thames (Lack 1971) revealed the presence of considerable phytoplankton populations in which green algae, particularly *Ankistrodesmus falcatus* and *Scenedesmus acuminatus* formed a peak density during the summer months. On other occasions mixed populations of green algae were recorded, including flagellates such as *Chlamydomonas* sp., *Gonium* sp., and *Pandorina* sp. In the tributary River Kennet, chlorophytes were more common in the winter period.

Variations in Distribution within a Lake

Discussions of the ecology of phytoplankton frequently are represented by observations at specific sites or even a single sampling station within a given water body. It must be emphasized that variations in phytoplankton distribution exist within the dynamic three-dimensional environment of a water column. These may exist under mixed water conditions in addition to situations of stable thermal stratification. Thus it is important to consider the spatial distribution of phytoplankton, both in terms of vertical distribution down a depth profile and in a horizontal plane. In relation to planktonic chlorophytes, both flagellates and nonmotile algal species may exhibit discontinuities in such distribution.

Vertical heterogeneity in distribution of phytoplankton occurs during more stable water conditions developing during conditions of thermal stratification. Generally, flagellate green algae are considered to be at a selective advantage during such periods, and their distribution is concentrated in narrower depth zones than nonmotile members of the Chlorococcales (Fig. 5-3). In Blelham Tarn, *Eudorina* populations were concentrated in the surface 2 m and *Volvox* populations in Crosemere exhibited a similarly very narrow depth distribution. However,

Fig. 5-3. The depth distriubtion of nonmotile (unshaded) and flagellate green algae during conditions of thermal stratification during summer in phytoplankton of three lakes: (a) *Ankyra judayi* in Blelham Tarn, enclosure A, in 1978; (b) *Sphaerocystis schroeteri* in the same enclosure in 1977; (c) *Oocystis* sp. in Abbot's Pool in 1967; (d) *Eudorina elegans* in Blelham Tarn, enclosure B in 1977; (e) *Eudorina elegans* in Crosemere in 1977; (f) *Volvox aureus* in Crosemere in 1977; (g) *Pandorina morum* in Abbot's Pool in 1967; (h) *Chlamydomonas monadina* in Abbot's Pool in 1967. All data are expressed as individuals·mL^{-1}. Based on data in Reynolds (1976, 1984), Reynolds original, and Happey-Wood (1976); used with permission.

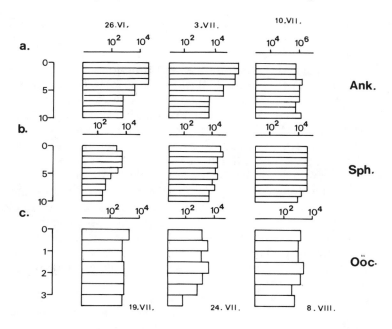

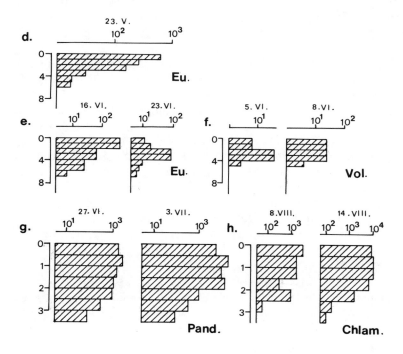

the nonmotile Chlorococcalean algae may also grow during conditions of stratification; *Ankyra* and *Sphaerocystis* were more or less evenly distributed within the surface 6 m, the depth interval corresponding to the euphotic zone, with small numbers of individuals recorded down to a depth of 10 m (Reynolds & Butterwick 1979; Reynolds & Wiseman 1982; Reynolds 1984a). A similar situation was found in the shallow eutrophic Abbot's Pool (Happey-Wood 1976a,b). Motile *Pandorina morum* and nonmotile *Oocystis* sp. were major components of the phytoplankton, with the nonmotile alga exhibiting a more homogeneous distribution with depth (Fig. 5-3).

Vertical migration of flagellates, particularly the larger colonial members of the Volvocales such as *Pandorina morum,* has been documented (Happey-Wood 1976a) and is more pronounced in shallow situations or under very calm conditions and stable stratification. That vertical redistribution of green flagellate cells is a real phenomenon associated with movement of motile algal cells has been confirmed by statistical analysis of depth distribution of flagellates in comparison with that of nonmotile cells over a 24-h period (Happey-Wood 1976a). Vertical migration of green algal flagellates under these stratified conditions may be extremely rapid, even for a small unicellular alga such as *Chloromonas,* as has been shown by Simla (1988) in a sheltered meromictic humic lake in Finland. Rapid changes in vertical distribution of such motile cells may be complicated by horizontal movement (Tyler in press), and thus the three-dimensional, patchy structure of flagellate populations must be borne in mind.

Horizontal variations in planktonic distribution may be relatively small-scale in comparison to the size of lakes (e.g., Richerson, Armstrong, & Goldman 1970; Richards & Happey-Wood 1979). This is true both for motile and nonmotile chlorophytes. The population density of *Chlorella vulgaris* and *Chlamydomonas moewussii* exhibited changes along a transect sampled at 1-m intervals in the surface waters of Llyn Padarn, North Wales, U.K. (Fig. 5-4). Such small-scale horizontal variations in phytoplankton numbers may reflect differences in water quality that are reflected in growth responses of the algae, or may reflect surface water movement in Langmuir cells (Langmuir 1938; Smith 1975) induced by direct wind stress. The contrasting horizontal distributions of these two phytoplankters in this case are more likely

Fig. 5-4. The horizontal distribution of *Chlorella* and *Chlamydomonas* along a 64-m transect in surface waters of Llyn Padarn. (A) population density of these two algae; (B) running average of five consecutive samples of population density; (C) pattern analysis expressed as (mean squared − mean)/(mean squared).

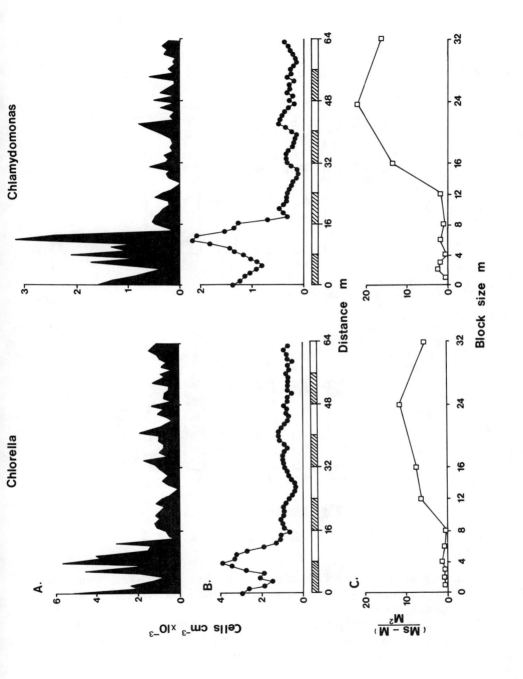

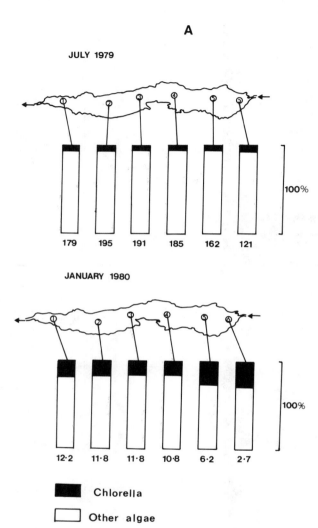

A

JULY 1979

JANUARY 1980

■ Chlorella

□ Other algae

to be autogenic (a reflection of the organisms themselves) because they have contrasting distributions within the same transect. Also, more extensive studies of pattern in distribution of diatoms has failed so far to detect any allogenic feature that might reflect or influence the regular patchy distribution of *Asterionella formosa* and *Tabellaria flocculosa* (Richards & Happey-Wood unpubl.).

Heterogeneous horizontal distribution may be of a larger scale than discontinuities in depth distribution or small-scale horizontal patches,

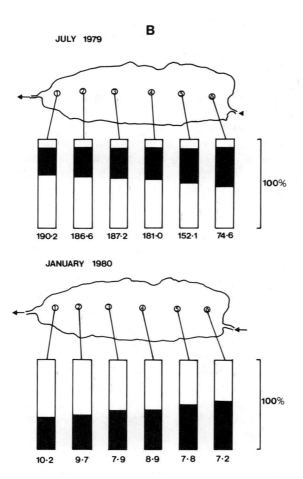

Fig. 5-5. Whole lake surveys of the phytoplankton in Llyn Padarn (A) and Cwellyn (B). Histograms show the percentage of algal biomass contributed by *Chlorella minutissima* (shaded) compared with the other phytoplankton species, and numbers indicate the total population density of phytoplankton in thousands of cells·mL^{-1} for Llyn Padarn and hundreds of cells·mL^{-1} for Cwellyn. (Original data of Haycroft.)

and they may relate to a lake as a whole. Wind action over a lake surface may result in increased phytoplankton density along the lee shore (George & Edwards 1976; George & Heaney 1978; Heaney & Talling 1980). Gradients of population density have been found down the length of a lake, and they presumably reflect the retention time for water in the basin plus the inherent generation time of the organisms, which will vary with environmental variables such as temperature and irradiance. Large-scale horizontal gradients of phytoplankton population density may also differ between species within a lake basin (Fig. 5-5). Population densities of larger planktonic species such as *Asterionella formosa* Hass. increase down lake with distance from the inflow. However, the pattern of distribution of *Chlorella minutissima*, an organism of size characteristic of picoplankton, showed a completely different pattern of population density within the same lake, Llyn Padarn (Fig. 5-5a). The greatest proportion of the population attributable to *Chlorella pyrenoidosa* was found in the region of the inflow, particularly during the winter period. In January, this very small alga accounted for 30% of the algae in the phytoplankton of Llyn Padarn. The same distribution patterns were found in Cwellyn (Fig. 5-5b). The small green algae *Ankistrodesmus falcatus* and *Chlorella minutissima* accounted for all the phytoplankton at the inflow end of Cwellyn in January and, though numbers were less, they were the major phytoplankters in the water column along the entire length of the lake. In July, *Chlorella minutissima* contributed 48% of the plankton at the inflow end, decreasing along the length of the lake to 30% of the total phytoplankton at the outflow end. Thus, small microchlorophytes, such as *Chlorella* and *Ankistrodesmus*, are able to outcompete large species of phytoplankton in the early stages of phytoplankton development at the inflow end of lakes of low or moderate nutrient status. This is likely to be a reflection of their inherently rapid growth rate, because no significant populations of these algae were characteristic of the inflowing river waters.

ECOLOGY OF MORPHOLOGICAL GROUPS OF CHLOROPHYTA

Micro-Green Algae

In general, these small-celled algae (0.2–20 μm) feature far less commonly in the literature discussing phytoplankton ecology than do larger green algae, such as members of the Chlorococcales, flagellate colonial organisms, and desmids. This may reflect greater problems encountered in dealing with small-celled and often motile organisms on a quantitative basis.

Specific methodology aids the recognition of these smaller-celled chlorophytes, and may be necessary for taxonomic determination. Culturing aids both the estimation of numbers of small-celled individuals such as *Chlorella* spp. (Happey 1970) and is essential for study of life cycles necessary for correct taxonomic designation (e.g., Starr 1955). More recently, the advent of fluorescence microscopy has enabled the enumeration of the smallest-celled algae, and such picoplankton have been found to be common in freshwater plankton (Craig 1984; Caron, Pick, & Lean 1985; Happey-Wood et al. 1988).

In oligotrophic situations micro–green algae are the major photosynthetic component in numerical terms. In the softwater mountain lakes of North Wales, Priddle and Happey-Wood (1983) described populations of *Chlamydomonas microsphaerella* Pasher and Jahoda, *Chlorella minutissima* Fott and Nováková, *Chlorella vulgaris* Beijerinck, *Chlorella* sp., and the very small-celled *Scenedesmus graheinsii* (Heynig) Fott (Fig. 5-6). These small algae accounted for more than 90% of the algae counted in the phytoplankton samples over a period of a year (Priddle & Happey-Wood 1983). In Llyn Llydaw, *Ankistrodesmus falcatus* was persistent and present throughout the year with population densities between 1000 and 10,000 cells·mL^{-1} on several occasions. Both species of *Chlorella* achieved similar numbers of cells·mL^{-1} but were not recorded for the entire 12 months. In neighboring Lake Cwellyn, a similar pattern of growth of these micro-Chlorophyta was observed, but *Chlorella minutissima* was present on more occasions that in Lake Llydaw. Considering growth of this group of small green algae over the season, there appear to be three periods of continued exponential increase in numbers; the first in March–April, the second in midsummer, and the third in October.

Such micro-Chlorophyta are characterized by small size and thus may have been thought to contribute relatively little to the overall phytoplankton biomass. However, the standing crops of larger planktonic algae, mainly small numbers of dinoflagellates, cryptomonads, and chrysophytes, were extremely low in these Welsh lakes. When the population structure of the phytoplankton is considered in terms of biomass rather than cell numbers, the micro-Chlorophyta contributed more than 75% of the phytoplankton biomass expressed as cell volumes (Fig. 5-7). They must provide a large proportion of the fixed carbon in the open water of such lakes. Ong (1977) found that organisms of size less than 20 μm frequently accounted for more than half the planktonic primary production, and more recently Kennaway (Kennaway & Happey-Wood 1988) differentiated phytoplankton carbon fixation into three size categories and found that up to 80% of the fixed carbon was retained in samples passing a 3-μm pore-sized filter.

In Welsh lakes of moderate or mesotrophic status, planktonic small green algae feature less prominently in the planktonic biomass and account on average for 45% of the number of phytoplankters (Priddle

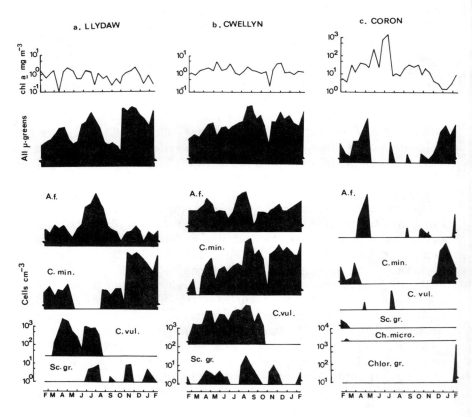

Fig. 5-6. Chlorophyll *a* concentration, population densities of total small green algae, *Ankistrodesmus falcatus* (A.f.), *Chlorella minutissima* (C.min.), *Chlorella vulgaris* (C.vul.), *Scenedesmus grahensii* (Sc.gr.), *Chlamydomonas microsphaerella* (Ch.micro.), and *Chloromonas grovei* (Chlor.gr.) for three Welsh lakes: L. Llydaw, L. Cwellyn, and L. Coron. Redrawn from Priddle & Happey-Wood (1983); used with permission.

& Happey-Wood 1983). Algal species were the same as in the two very oligotrophic mountain lakes, Llydaw and Cwellyn. However, in shallow eutrophic lakes on Anglesey, differing in maximum inorganic phosphorus levels by an order of magnitude compared with the mountain situations, small-celled chlorophytes were less frequent. The three nonmotile species common to the mountain lakes were recorded in Llyn Coron, and small-celled flagellates, *Chlamydomonas microsphaerella* and *Chloromonas grovei*, were also recorded. Despite their rela-

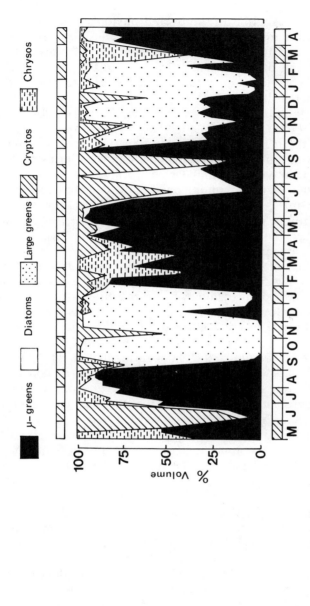

Fig. 5-7. Relative contribution of algal groups, small green algae, diatoms, cryptomonads, and chrysophytes in terms of cell volume for Llyn Cwellyn.

tively small contribution to the overall phytoplankton of 5% of the annual total, the maximum population densities were similar, being between 1000 and 10,000 cells·mL^{-1}. The major difference in the population dynamics of these small-celled algae in the eutrophic shallow lakes was the short duration of the populations and very rapid increase in population density (Fig. 5-6).

Chlorella minutissima has been found in population densities greater than 100,000 cells·mL^{-1} in Abbot's Pool, a shallow pool characterized by thermal and chemical stratification, and extremely high nutrient concentrations (Happey-Wood 1978). This alga was found to be present in the water column, even if only at a restricted depth, throughout the year. Maximum population densities of more than 100,000 cells·mL^{-1} occupied the depth of the developing chemocline in spring and were recorded throughout the water column in winter. However, when the dynamics of the population are considered, the periods of exponential increase of *Chlorella minutissima* coincided with the periods of time between the decline and subsequent growth of larger algae (Fig. 5-8). Thus microchlorophytes were capable of very rapid growth on eight occasions when competition from other plankton species was diminished. To a lesser extent, this "opportunistic" type of growth pattern was identified in the phytoplankton dynamics of Llyn Coron (Fig. 5-4); growth of *Ankistrodesmus falcatus* followed the decline of *Thalassiosira,* the major spring diatom, growth of *Chlorella minutissima* was subsequent to the decrease in population of *Cyclotella,* and the major growth of *Chl. vulgaris* was after summer growth of *Anabaena flos-aquae* and before the main growth of *Microcystis aeruginosa* (Priddle & Happey-Wood 1983).

The population dynamics of microchlorophytes contrasts greatly in environments of low nutrient status and eutrophic situations. In nutrient-poor, oligotrophic waters, micro–green algae are often the major component of phytoplankton populations. This is true both in numerical and biomass terms and also in relation to the amount of organic carbon fixed by photosynthesis. In such situations, the success of micro–green algae may be accounted for partly by morphological features of the cells, such as their small size. Small cells have a higher surface-area-to-volume ratio than larger cells and thus a larger capacity for both nonactivated and activated nutrient–solute influx. In terms of unit volume of cell, there is more area of membrane per cell, thus photosynthesis will be enhanced by higher diffusion rates of carbon dioxide and nutrients (Raven 1986). Similarly, light absorption per unit volume of cell may be greater than in larger cells (Reynolds, Chapter 10 this volume). Thus small-sized chlorophyte cells may be at a selective advantage in clear water oligotrophic situations where essential nutrients, such as nitrogen and phosphorus, may be scarce. In

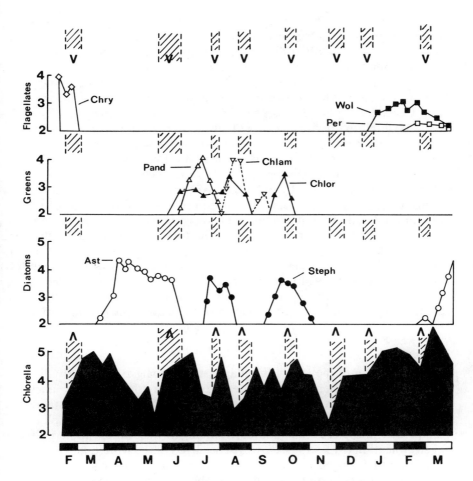

Fig. 5-8. The mean population density of eight depth samples of *Chlorella minutissima* and major phytoplankton species for Abbot's Pool. Flagellates: Wol, *Woloszynschia tenuissimum;* Per, *Peridinium* sp.; Green algae: Pand, *Pandorina morum;* Chlam, *Chlamydomonas monadina;* Chlor, *Chlorococcum minutum;* Diatoms: Ast, *Asterionella formosa;* Steph, *Stephanodiscus astraea.* Periods of exponential increase in *Chlorella* are emphasized by shading. Redrawn from Happey-Wood (1985); used with permission.

eutrophic situations, their "opportunistic" growth pulses are short-lived, but their characteristic small size may be important because they grow in conditions of low nutrient availability in relation to the larger euplankton. Their small size and efficient nutrient uptake mechanisms enable them to take advantage of nutrients in concentrations too low or in a form unavailable for larger cells. On the other hand, they may be capable of efficiently utilizing the nutrients and organic substances leaked from senescing cells of larger algae.

The small-celled or microchlorophytes grade in size from the lower end of nannoplankton, down to the size range defined for picoplankton. Certainly, *Chloromonas grovei* and *Chlorella minutissima* (Figs. 5-6, 5-8) fall into the size category of picoplankton. More recent studies with fluorescence microscopy have found small chlorophytes as a common component of the picoplankton of freshwaters, following size fractionation of "whole water samples" using nucleopore filters. Craig (1984), in studying the phytoplankton of Little Round Pond, a small meromictic lake in Ontario, found large numbers of picoplankton, and the carbon fixation ascribable to organisms retained by a 3-μm nucleopore filter accounted for up to 75% of the total primary productivity. The picoplankton constituted 50% of the algal biomass for much of the autumn. Thus more extension studies of these, the smallest-sized green algae, are required to gain an understanding of their ecological role in the complex interrelationships of pico- and nannoplankton with inorganic nutrients and other components of the planktonic community.

Flagellates–Volvocales

Flagellate green algae (excluding flagellate microalgae previously discussed) are characteristic components of the phytoplankton of lakes and pools of relatively high nutrient status during periods of stable thermal stratification. Growth of chlamydomonads in a eutrophic, temperate, dimictic lake in southern Michigan was limited to periods of stratification (Moss 1972). Growth occurred in the winter, and a second, more prolonged period of development was recorded from July until the water column mixed in November (Fig. 5-9). The population density was less during winter, presumably a reflection of lower growth rates in the cold water temperature linked with the shorter days of February. Under ice, the population was concentrated in the surface 2 m in the coldest water, whereas in summer the greatest population densities in late July and October were recorded toward the lower part of the epilimnion. After mixing of the water column, the chlamydomonads declined. Thus certain flagellate chlorophytes are capable of growth in conditions of inverse stratification under ice, where motility is essential to maintain the cells in the euphotic zone because the water surface is isolated completely from wind stress and, hence, water turbulence.

Members of the Volvocales, particularly colonial organisms, are

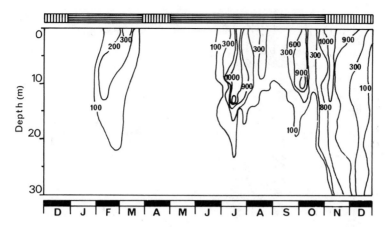

Fig. 5-9. Depth–time distribution of green flagellates from Gull Lake, Michigan. Periods of stratification are indicated by horizontal shading and water column mixing by vertical shading along the top of the figure. Redrawn from Moss (1972); used with permission.

characteristic components of summer plankton, growing after the demise of the spring diatom bloom. The phytoplankton of Crose Mere, a small eutrophic lake maintained by nutrient-rich ground water (Reynolds 1976), exemplifies this well. Following the onset of thermal stratification in late May, *Volvox aureus* exhibited a brief phase of abundance in early June (Fig. 5-10a). The population occupied the upper 4-m depth zone of the lake above the thermocline, and from 5–10 June development was localized in the surface 2 m of the water column. This was followed by redistribution of the population, which subsequently declined. *Eudorina elegans* increased to reach its population maximum several weeks later than *Volvox aureus* (Fig. 5-10b), but the pattern of growth and distribution of the population within the depth profile paralleled that of the larger colonial green alga. However, persistence of *Eudorina* in the epilimnion was some three weeks longer than that of *Volvox*.

Growth of large colonial Volvocales is frequently accompanied by considerable utilization of available nitrogen from lake water. Reynolds (1976) found significant relationships, by calculation of partial correlation coefficients, between growth of *Volvox aureus* and the availability of nitrate ($r = 0.860$, $P < 0.05$) and ammonium ($r = 0.818$). Growth of *Eudorina* was possibly related to nitrate availability ($r = 0.380$, $P < 0.1$). Since the growth of *Eudorina* followed that of *Volvox*, and by this time some decrease in nitrate concentrations had

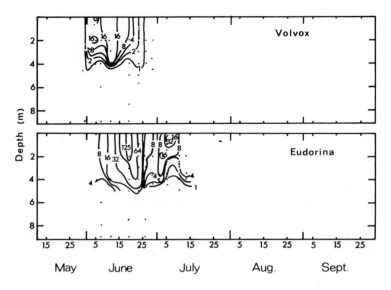

Fig. 5-10. Depth–time distribution of *Volvox aureus* and *Endorina elegans* in Crose Mere. Isopleths in colonies·mL^{-1}. Redrawn from Reynolds (1976); used with permission.

taken place, this lower level of correlation is not surprising. During the four weeks of growth of *Pandorina morum* in Abbot's Pool, the nitrate–nitrogen concentrations throughout the water column decreased to the level of detection (Happey-Wood 1976b). Thus although high levels of available nitrogen, particularly nitrate, appear necessary for the development of the colonial flagellate green algae, their growth may result in considerable depletion of this essential nutrient, which may contribute to the decline of these algae.

Deprivation of nitrogen in cultures of *Chlamydomonas*, either by dilution of the medium with distilled water or by growing cultures under nitrogen starvation, resulted in induction of gamete formation (Kochert 1982). It appears that many other volvocalean algae become sexually competent when starved of nitrogen. Thus, in naturally occurring phytoplankton, the environmental conditions, particularly decreasing availability of nitrogen for growth during decline of flagellate populations may induce sexual reproduction. Although flagellate population may be responsible in part for self-limitation of growth by nutrient depletion, the resistant zygospores resulting from sexual reproduction likely under these conditions provide a strategy for survival.

In addition to the presence of adequate nutrients in the water column, calm water conditions characteristic of periods of stable thermal stratification have been shown to be essential for colonial Volvocales to achieve maximum rates of population increase. Rates of change of population density of *Volvox aureus* were found to respond to fluctuations in the absolute depth of water column mixing and its relation to light penetration (Reynolds 1983; Reynolds et al. 1983). Increase in population density of colonial Volvocales is a complex process and involves a 16- or 32-fold increase in individuals in the case of *Eudorina* and the release of daughter colonies by *Volvox*. Once the rate of release of daughter colonies is reduced or daughter colonies are lost from the euphotic zone by greater water turbulence, the population as a whole declines.

The ability of colonies and unicellular flagellates to move within the water column is of great strategic importance to the survival of these algae. Active migration confers the ability of movement to environments suitable for optimum growth, either in terms of irradiance or nutrients. Uptake of nutrients by migration of green flagellates has not been demonstrated experimentally, although it is likely because downward movement to the thermocline in shallow eutrophic systems takes the organisms into regions of higher nutrient concentrations, and nutrient transport upward through a water column has been confirmed for the cryptomonad *Cryptomonas marssonii* (Salonen, Jones, & Alvola 1984; Klaveness, Chapter 3 this volume).

Moss (1969) suggested that downwardly directed convection currents resulting in deeper water mixing in the epilimnion of Abbot's Pool resulted in redistribution of *Pandorina morum* with a greater proportion of the population occupying deeper layers of the epilimnion, although the surface aggregation of more than half the population was maintained. This indicated that perhaps a proportion of the population was more actively motile or phototactic. Vertical migration rhythms are not as well documented for green flagellates as for other algae capable of swimming, such as dinoflagellates (e.g., Berman & Rodhe 1971; Talling 1971; Heaney & Talling 1980). Vertical upward movement of *Pandorina morum* at 06.30 h and downward migration between 16.00 and 20.00 h was demonstrated from statistically significant changes in population densities over a depth of 2 m in a shallow eutrophic pool (Happey-Wood 1976a). Similar to the observations of Moss (1969) on this alga, only the epilimnetic portion of the population was exhibiting active vertical movement. Thus, colonies below the epilimnion and euphotic zone appeared unable to migrate back into conditions suitable for optimum growth.

The most important advantage of movement to flagellates lies in the reduction in loss to the population by sinking out of the euphotic zone during calm water conditions. This may be illustrated by the growth of *Chlamydomonas monadina* in Abbot's Pool. The chlamydomonad was present in the water column during August under conditions of thermal stratification and during September after the autumn overturn (Happey-Wood 1976b). In the summer period the apparent generation time for the population resident in the euphotic zone was 1.1 days. However, when the water column became mixed and cells were carried in water turbulence throughout the depth profile, the apparent generation time increased to 6.2 days. Although water temperature had decreased by about 2°C, nutrient concentrations had increased markedly, and thus it was unlikely that these factors were having great constraints on the development of *Chlamydomonas*. This flagellate was evenly distributed throughout the water column and thus not able to maintain itself in the euphotic zone by active movement. The change in light climate experienced by the chlamydomonad, from the organism maintaining itself in an optically shallow epilimnion to experiencing an optically deep system in which the cells would spend long periods in darkness or at irradiances where the cells were below their compensation point, resulted in a sixfold increase in apparent generation time and decline of the population.

Large size may be an important characteristic in aiding survival of the colonial green flagellates because they are present in the water column when pressure from grazing by filter-feeding zooplankton is likely to be high. In the case of *Volvox,* daughter colonies are likely to be too large to be ingested, but this does not apply to *Eudorina.* In one cycle of division of 5–6.2 days, a 16- or 32-fold increase in colonies is produced by *Eudorina* (Reynolds & Rogers 1983), and for the first 24 h after release from the mother colony when the daughter colonies are less than 50 μm in diameter they may be liable to grazing pressure. However, this time period represents less than 20% of the generation time, and thus compared with smaller-sized algae, size confers a selective advantage against grazing for about 80% of the life of *Eudorina.*

In summary, then, there is some understanding of the population dynamics of flagellate green algae. Their dominance in a phytoplankton population is generally short-lived. Growth occurs under conditions of stratification, usually in summer when nutrient levels are still moderately high. Their ability to move enables them to overcome transport of cells down through the water column by water turbulence and loss from the actively growing population by sinking. Movement enables them to exist at an optimum irradiance by maintaining the photosynthetically active individuals in an optically shallow environment, and may enable them to obtain otherwise scarce or limiting nutrients by vertical migration through a greater volume of water to deeper regions of higher nutrient content. However, the advantages of motility may be easily outweighed by increased water movements due to cooling and wind stress, and thus the ability to move is of greatest advantage only under conditions of very restricted water turbulence. Size may reduce or exclude grazing pressure in the case of colonial Volvocales, but size variation during reproduction cycles makes these forms temporarily susceptible to grazing losses.

Nonmotile Algae: Chlorococcales

The major seasonal growth period characteristic of nonmotile green algae generally follows the decline of diatom populations in the spring. These green algae are negatively buoyant and have no inherent feature such as the possession of flagella to enable them to move actively. Thus they are dependent entirely on water turbulence to maintain the individuals in suspension. Under conditions of stable thermal stratification, cells or colonies will tend to sediment down through the water column from the euphotic zone. Only when the rate of growth of the survivors in the euphotic region of the epilimnion is greater than the rate at which the population is depleted by sedimentation through the water column will growth be expressed as net population increase in the phytoplankton. Thus growth of these algae, which mainly occurs during the early stages of thermal stratification, depends on the changing balance between photosynthesis expressed as cellular increase and loss processes such as respiration and sedimentation of the population. Net algal growth depends on a suitable light climate and adequate nutrients.

Studies on the population dynamics of phytoplankton in Abbot's Pool illustrate the effect of water mixing on the growth of *Chlorococcum minutum* (Happey-Wood 1976b). This alga grew during summer and again in September and October following the autumn mixing of the pool. The apparent generation time in mixed water conditions after the autumn overturn was 1.3 days, less than half the apparent generation time expressed during July–August. Thus this alga may have been able to actually grow more rapidly in autumn when the cells were being circulated throughout the depth profile, being subjected to periods lacking irradiance when the cells were deeper than the extent of the euphotic zone. On the other hand, rates of actual cell division of *Chlorococcum* may have been greater than that calculated as the apparent growth rate during summer, because no allowance for losses was made. Nevertheless, the important feature was that the net rate of population increase was greater during mixed water conditions, emphasizing the importance of passive movement or transport by circulation in water turbulence for this nonmotile green alga. These effects of water circulation on the rates of population increase of *Chlorococcum* were the opposite of those previously described for *Chlamydomonas*. Thus for morphologically similar algae, motility of the chlamydomonad conferred a selective advantage under conditions of thermal stratification, but the nonmotile chlorophyte exhibited a greater net rate of population increase in turbulent water. The differences in apparent growth rates for the two algae were substantial – a twofold decrease for *Chlorococcum* and a fivefold increase for *Chlamydomonas* during stratification. This indicates that growth of the nonmotile alga was affected less by changes in water circulation than was the flagellate individual.

More subtle effects of changes in water circulation and the relationship between the mixed depth and euphotic depth have been described by Reynolds (1983) in investigations of the effects of artificial mixing on the dynamics of phytoplankton populations in large limnetic enclosures. *Sphaerocystis schroeteri,* a mucilagenous, colonial, nonmotile green alga was present in the phytoplankton from May until mid-August (Fig. 5-11). It was the dominant component in late May, but contributed a low biomass in comparison with other dominant algae, such as diatoms in the spring and blue-green algae in summer and early autumn. The population of *Sphaerocystis* increased in May from a small inoculum present in the water column. Artificial mixing was imposed in early June, resulting in an increase of the mixed depth. *Sphaerocystis* declined and was replaced for a short period by the diatoms *Fragilaria* and *Tabellaria*. Once the water column had restratified, numbers of *Sphaerocystis* increased rapidly in late June. Following artificial mixing of the water column for a second period in July, growth of the chlorophyte was arrested again. Events were qualitatively similar to those in June; diatoms increased for a short time until they were lost through sedimentation as thermal stratification became reestablished. Following the July mixing the metalimnion occupied the 1–9-m depth zone, and regrowth of *Sphaerocystis* took place. At this time the plankton was dominated by *Anabaena flos-aquae,* and as a result the 8-m artificially deepened epilimnion was extremely turbid. During this third growth phase the net exponential rate of population of *Sphaerocystis* was lower than during either the May or June growth phase; the population declined and sedimented from the water column.

Population increase of *Sphaerocystis* was prevented by artificial mixing; the green alga was unable to compete with diatoms in artificially induced mixed water conditions. Thus it is most unlikely that

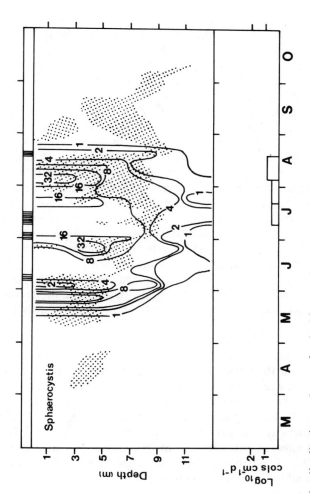

Fig. 5-11. Depth–time diagram of the distribution and rate of arrival into sediment traps of *Sphaerocystis schroeteri* in Blelham Tarn; isopleths in colonies·mL^{-1}; position of the metalimnion is shown by shading. Redrawn from Reynolds et al. (1983); used with permission.

such a nonmotile green alga will develop in mixed water conditions, when nutrients suitable for diatom growth are available. Decrease in *Sphaerocystis* occurred in August when the mixing depth in the water column exceeded the euphotic depth by a factor of more than two. Thus this alga depends on water turbulence for suspension and requires maintenance in the euphotic zone for most of the time. Reduction in the depth of the euphotic zone such as occurred here due to the presence of blue-green algae may be sufficient to prevent net growth of *Sphaerocystis* when conditions of mixed water depth and nutrients would otherwise be conducive to net growth of the alga.

Nonmotile algae such as *Sphaerocystis* have no facility or mechanism for positional recovery; they depend on water turbulence in the mixed layer to return them to the euphotic zone. Once individuals are below the thermocline they will be lost from surface waters. The density of chlorophytes is less than that of diatom cells surrounded by a heavy silica wall (Reynolds 1984a), and the particle density of *Sphaerocystis* is further reduced by the surrounding copious mucilage sheath (Walsby & Reynolds 1980). Hence, the intrinsic settling rate will be considerably lower than for diatoms and will facilitate the survival and growth of populations under conditions of stratification with a suitable light climate.

In many situations, planktonic green algae are replaced by blue-green algae, which may persist through the summer until the autumn overturn. This feature was studied in laboratory experiments by Mur, Gons, and van Liere (1978). They found that the maximum growth rate of *Scenedesmus proturbans* was 1.58 d^{-1} at incident light intensities between 30–85 $W \cdot m^{-2}$, compared with 0.86 d^{-1} for *Oscillatoria agardhii* expressed at lower light intensities of 6–25 $W \cdot m^{-2}$. Based on competition experiments carried out in continuous culture, *Oscillatoria* grew well and became the dominant alga after five days at low light (1 $W \cdot m^{-2}$) and a dilution rate (D) of 0.01 h^{-1}. At 20 $W \cdot m^{-2}$ and D of 0.03 h^{-1}, good growth of both algae was recorded for 10 days, after which *Scenedesmus* declined and *Oscillatoria* again became the dominant organism. In the experiment with the highest light intensity, 35 $W \cdot m^{-2}$ and $D = 0.03$ h^{-1}, growth of *Oscillatoria* was inhibited, but once *Scenedesmus* had reached steady-state population density, reinoculation with *Oscillatoria* resulted in decline in this green alga and increase in the blue-green alga. This resulted from the lower, more favorable light intensity for *Oscillatoria* and self-shading of *Scenedesmus*. Thus this green alga requires more time in optimum light conditions to synthesize sufficient energy to maintain itself. The inherently rapid growth rate of green algae such as *Scenedesmus* enables them to outcompete blue-greens until an ecological factor becomes limiting, in this case light. However, the strategy of rapid growth results in self-limitation due to shading, and thus results in demise of the population in naturally occurring planktonic populations when an inoculum of more slow-growing algae is present.

Replacement of green algae by planktonic blue-green algae may be related to other factors in addition to light climate. The phytoplankton of Mount Bold reservoir, South Australia, was dominated by *Dictyosphaerium pulchellum* during the early phases of thermal stratification, and this alga was subsequently replaced by two blue-green algae, *Microcystis aeruginosa* and *Anabaena spiroides* (Ganf & Oliver 1982). Exponential increase in population density of *Dictyosphaerium* occurred from late October until mid-November at a rate of 0.23 ± 0.05 d^{-1} (Fig. 5-12). Once the water column became thermally stratified in November, the growth of *Dictyosphaerium* declined. However, this was not due solely to the restricted water circulation. The water of this reservoir is turbid, typical of most Australian waters (Ganf 1976; Kirk 1979), with a shallow euphotic zone of 2.7 m. In addition, investigations by bioassay demonstrated the development of gradients of algal growth potential with minimum growth of four cultured algal species in water samples taken from the euphotic zone of the reservoir. Thus once thermal stratification had developed, growth of *Dictyosphaerium* was limited both by restricted water circulation and spatial separation between light and nutrients, due to the shallow euphotic zone and lack of available nutrients required for algal growth at depths shallower than 8 m.

To compete successfully in the mixed species assemblage of phytoplankton, nonmotile green algae appear restricted to relatively short growth periods during the year defined by a narrow range of environmental conditions. During the complete water circulation and turbulence of spring and autumn, they are unable to compete with diatoms, presumably as a result of the light climate present in the turbulent water column where the relative amount of time spent in darkness by the algal cells is increased. Nonmotile chlorophytes are able to develop rapidly during the early stages of stratification, provided the planktonic environment is optically deep and wind-induced water turbulence in the epilimnion maintains the population in the euphotic zone for a large proportion of time. If the mixed zone in the epilimnion exceeds the euphotic depth by a factor of about two, growth of these algae declines. Other factors that tend to reduce the euphotic depth, such as inherently turbid water or the growth of other algal species (particularly Cyanophyceae) that result in shading, also reduce the rate of population increase of nonmotile green algae. Because of their lack of motility, loss by sedimentation of nonmotile green algae will result

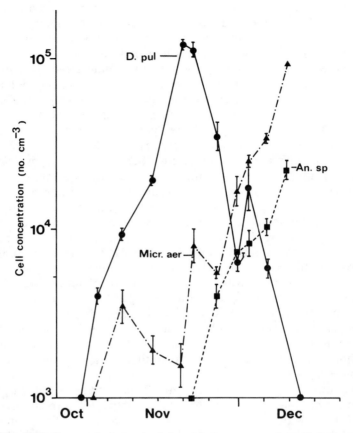

Fig. 5-12. Seasonal changes of the mean concentration of algal cells in the euphotic zone of Mount Bold Reservoir, South Australia. *Dictyosphaerium pulchellum*, D. pul.; *Anabaena spiroides*, An. sp., and *Microcystis aeruginosa* Micr. aer. Redrawn from Ganf and Oliver (1982); used with permission.

in a declining population under conditions of extremely stable stratification in a shallow epilimnion. Also, their dependence on water turbulence for transport in the surface water will reduce their ability to grow when essential nutrients become depleted. Thus the rates of population increase of nonmotile green algae depend primarily on the residence time of cells in the euphotic zone together with available nutrients in the epilimnion.

Desmids

The greatest species diversity of desmids has been described for waters of low alkalinity with pH values between 4 and 7, but these generally large unicellular algae do occur in alkaline waters (Moss 1973; Brook 1981). Desmid populations tend to be more common and diverse in species composition as "metaphyton" (Behre 1956) than in the open water of lakes and ponds. Brook (1981) stated that desmids may be carried quite frequently from benthic situations into the open water to develop as true phytoplankton, which suggests that populations of planktonic desmids are derived from survivors of actively growing populations in benthic niches. Duthie (1965), in studying desmid populations in the shallow Welsh lake Llyn Ogwen, found that a few species of desmid were truly planktonic and were found only rarely on the sediments of the lake and inflows. The majority exhibited the ability to increase in numbers both in the plankton and on the sediments. In their study of 10 lakes in northern Minnesota in the United States, Meyer and Brook (1969) found 250 species and forms of desmids, the majority of which were associated with submerged aquatic plants. Whether a benthic desmid contributes to the planktonic community appears to depend on the physical and biological characteristics of the particular lake it inhabits (Brook 1981). Desmids feature in the plankton in regions of lakes with good water circulation, and there appears to be some correlation between the morphometry and water circulation of lakes and the cell size of desmids found in their plankton. In dimictic lakes, desmid cell size ranged from 9 to 300 μm, whereas in meromictic lakes the size range was only 9 to 40 μm (Baker 1973).

Desmids generally are present in relatively low numbers in phytoplankton, although this may reflect a substantial biomass due to the relatively large size of many desmids. Although many descriptions of desmids in a wide range of habitats exist in the literature (see Brook 1981), few quantitative studies of their periodicity and population growth exist. The most detailed accounts are the studies of Canter and Lund (1966) and Lund (1971).

In Lake Windermere, Lund (1971) found that numbers of *Staurastrum lunatum* and *Cosmarium abbreviatum* began to increase in spring, whereas *Cosmarium contratum* increased later in the summer (Fig. 5-13). For most of the 25 years studied, the maximum population density was reached in September or October. The annual cycle was very regular, with one maximum in population density in autumn followed by a decline until the next summer. Cells were randomly dis-

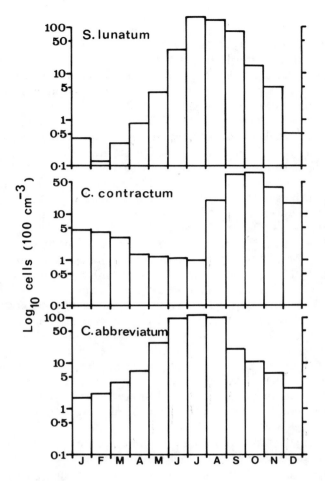

Fig. 5-13. The average monthly abundance of (a) *Staurastrum luna-tum* in Windermere, South Basin, 1945–69, (b) *Cosmarium contrac-tum,* and (c) *C. abbreviatum.* Redrawn from Lund (1971); used with permission.

tributed during isothermal periods, and greater numbers were found in the surface 10–15 m during thermal stratification, with greatest change in numbers at depths near the metalimnion. Experiments with cultures exposed at 0.5, 1, 2, and 4 m depth below the surface in the lake showed that *C. contractum* did not grow as well as the other two desmids (Fig. 5-14a). These results were not directly comparable because it was not possible to carry out all experiments on all three

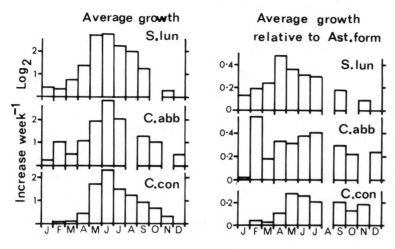

Fig. 5-14. (left panel) Average growth at 0.5, 1, 2, and 4 m depth of (a) *Staurastrum lunatum*, (b) *C. contractum,* and (c) *C. abbreviatum* for each month in which cultures were suspended in Windermere. (right panel) Average growth of (a) *Staurastrum lunatum*, (b) *Cosmarium abbreviatum,* and (c) *C. contractum,* in cultures suspended in Windermere relative to that of *Asterionella formosa* suspended at the same place and for the same length of time. Redrawn from Lund (1971); used with permission.

species at the same time each year; therefore an indirect comparison of the growth of the three desmids was made by using the growth of *Asterionella formosa* as a standard (Fig. 5-14b). This experiment confirmed that growth of *C. contractum* was considerably less than that of the other two desmids. Relative to *Asterionella,* growth of desmids was better in summer than winter. The marked exception for *C. abbreviatum* in February resulted from very little growth of the diatom. The lower growth rate of *C. contractum* in experimental cultures exposed in situ correlates with its later maximum in the plankton than the other two desmids.

The low growth rate of desmids results in their attaining population maxima later in the year than do many other green algae. Their cell size is sufficiently large that grazing is precluded as a significant factor in controlling populations. However, parasitism is a major controlling factor for the species growing during periods of warmer water with more limited circulation (Canter & Lund 1969). Of the three desmids discussed earlier, the lower growth rate of *C. contractum* dictated that the population maximum occurred after the autumn overturn, when more nutrients were available for growth, and when the cells were circulated throughout a greater volume of water and therefore were less vulnerable to parasitic infection.

Filamentous Algae

Filamentous algae are not normally considered true planktonic algae. Extremely high biomass of *Zygenema* and *Spirogyra* was found in irrigation waters associated with rice cultivation in New South Wales, Australia. These populations developed as floating mats as the last stage in algal development during the flooding period of rice growth, following populations of diatoms and flagellates (Noble & Happey-Wood 1988). It is likely that they were derived from sexual cysts and originated as benthic populations, breaking loose and developing further as planktonic algae in the nutrient-rich irrigation water in the channels and rice bays. Similarly dense growths of *Pithophora* regulated by akinete production have been intensely studied in shallow eutrophic American ponds (Lembi, Pearlmutter, & Spencer 1980).

Mention of populations of filamentous algae of significant size in descriptions of phytoplankton are rare. Sommer (1986) described plankton populations of *Mougeotia thelespora* and *Ulothrix subtilissima* in Lake Constance (Bodensee). These filamentous algae grew together with desmids in autumn, during conditions of deteriorating light climate associated with decreasing daylength and increased mixed depth of the water columm. Their survival was associated with low growth rates and with presumed nutrient supplementation by the progressive increase in the depth of mixed water. Similar to desmids, these filamentous algae were resistant to grazing due to their morphology.

STRATEGIES FOR SURVIVAL

The green algae as a whole span a wider range of cellular organization than other algal phyla. As a result the group exhibits a diverse spectrum of growth responses to critical environmental variables such as irradiance, thermal stratification, and the availability of nutrients (that is, itself, affected by the depth and stability of stratification). Biological factors such as cell size, shape, and ability to swim may also influence the potential growth response of an alga. Finally, interactions with other organisms are of importance, including allelopathy, zooplankton grazing, and, in some cases, parasitism.

Micro–green algae tend to be ubiquitous, capable of growth at any time during the year. However, maximum growth of larger green algae is most common in the natural environment during summer warm water conditions. Similarly, rapid growth has been observed under higher irradiances in comparison with other major algal groups, such as diatoms. However, the effect of higher incident irradiances on growth must be considered in combination with factors that influence light penetration, such as the depth of mixing relative to the depth of the euphotic zone, the turbidity of the water which will be affected by both organisms in suspension, such as the algae themselves, and the inherent characteristics of the water. These features represent the underwater light climate in the sense of Talling (1971). Large flagellates, frequently colonial forms, tend to be selected in situations that are optically shallow, whereas in optically-deep, stratified lakes the larger nonmotile green algae become predominant. However, manipulation of stratification to increase the depth of water mixing resulted in decline of some nonmotile green algae such as *Sphaerocystis* (Reynolds 1984b). Thus the optimal balance of light climate and depth of water mixing is extremely delicate and species specific; any change in the stratification pattern of a water column may be sufficient to result in a decrease in algal growth rate and, hence, decline in the population of chlorophytes. Population decreases of nonmotile algae may be due to greater loss of cells by sedimentation under conditions of calmer water and increased water heating that result in a shallow epilimnion. On the other hand, under conditions of increased mixing depth resulting from natural cooling of the epilimnion or artificial water mixing, a greater proportion of the algal population may be transported out of the euphotic zone such that the cells spend progressively longer periods of time in darkness or at irradiances below their compensation point.

Optimum growth rates depend on the availability of adequate nutrients. Frequently, growth of larger green flagellates is accompanied by large decreases in the inorganic nitrogen pool of the surface waters of lakes. Manipulation of nutrient levels to give increased loadings of phosphorus resulted in growth of *Sphaerocystis* at low loadings of 0.5 to 1.0 g $P \cdot m^{-2} \cdot a^{-1}$ and its replacement by *Eudorina* at heavier loadings of 1.6–2.0 g $P \cdot m^{-2} \cdot a^{-1}$ (Reynolds 1986). Thus within the same temporal niche in a single lake, nutrient loading may alter the type of green alga predominant in early summer after the diatom bloom, with growth of colonial flagellates at high phosphorus loading and the development of nonflagellate colonial greens at lower levels of phosphorus, always providing the optical climate is suitable.

Morphological features of certain types of green algae are of selective advantage under particular environmental conditions. Possession of flagella, and thus the ability to move, is advantageous under calm water conditions. This may be expressed either in a phototactic response maintaining organisms within a suitable light climate for long periods or in enabling the organism to obtain nutrients from different depth strata in the lake. Exposure to a greater volume of water for nutrient uptake is also achieved through active movement. Complex shapes and ornamentation of cells result in increased surface-area-to-volume ratio and thus a slightly decreased sedimentation rate in comparison to a sphere of similar cell volume. Possession of mucilage sheaths in the case of many of the larger colonial nonmotile chlorophytes and some desmids confers a lower specific gravity to the individuals and again a lower sinking speed in the water column thus reducing loss rates in the population.

Physiologically, rates of carbon fixation in green algae are high (Maguire & Neill 1971; Stull, Amegaza, & Goldman 1973), and the cells appear to contain a higher level of chlorophyll *a* per unit cell volume (Reynolds 1984a). However, the ecological significance of these features needs further elucidation. Uptake of a range of organic compounds has been demonstrated for 13 species of chlorococcalean algae (Pollingher & Berman 1976). Growth of *Pediastrum duplex* was enhanced by uptake of extracellular organic substrates, but only at low light (Berman, Hadas, & Kaplan 1977). This utilization of extracellular products may be of assistance in survival of populations under conditions of low irradiance, but the cellular energy input from this heterotrophic process is likely to be small in comparison to autotrophic growth.

A range of green algae have been shown to have a suppressive effect on the growth of other algae. Procter (1957) demonstrated that *Chlamydomonas* became dominant in mixtures of other algal species and this was related to the release of fatty substances. Filtrates from cultures of *Scenedesmus* and *Chlorella* inhibited the growth of *Nitzschia palea* (Jorgensen 1956). Such inhibition of growth may occur within groups of green algae. Harris (1971) found that *Pandorina morum* inhibited more colonial green algae than any of the other species tested. Thus when colonial flagellates are present in the plankton, they are likely to be limited to one species.

Grazing may have a profound effect on net rates of population increase in natural plankton populations. Particle size is critical to filtration, and thus selection by filter-feeding grazers is determined primarily by algal cell size (Reynolds 1986). Population losses to filter-feeder grazing will be far greater in the case of micro–green algae. Colonial algae such as *Sphaerocystis* may be ingested but only by larger *Daphnia,* and desmids such as *Staurastrum* are rejected. Thus the

increase in population density observed in naturally occurring planktonic micro–green algae is likely to be lower than the true growth rate,
the difference between the two determined by the grazing pressure. On
the other hand, larger algae will increase in numbers in phytoplankton
at rates approaching their true growth rate under the existing physical
and chemical conditions, because loss via grazing is minimal. Grazing
clearly has a greater effect on the population dynamics of the smaller-
sized green algae; however, this acts discontinuously because fluctuations in the community filtration rate accompany changing abundance
of suitable foods for filter feeders.

On occasion, attack by parasites may have a catastrophic effect on
the growth of a planktonic population. In the case of green algae, their
influence on the growth of desmids has been documented (Canter &
Lund 1969; Lund 1971). Infection depends on a persistent population
of algae and is more likely within conditions of restricted water circulation, such as during thermal stratification. Desmids reaching maximum standing crops before the autumn overturn were more susceptible to parasitism than were those that achieved population maxima
later in the year when circulated throughout the entire water column.

Within the biological variation of the green algae, it is possible to
identify six morphological survival strategies that have evolved within
the dynamic and complex ecosystems of freshwater lakes and ponds.
These strategies are assigned to groups or types of green algae, and it
must be emphasized that within each of these six groupings of algae
there is likely to be considerable difference between algal species. Also
there is now evidence to show that considerable physiological variation exists within single planktonic species isolated from different lakes
(Happey-Wood & Hughes 1980). Therefore, both interspecific and
intraspecific variation must be borne in mind within these strategies.

1. Micro–Green Algae (Nannoplankton and Picoplankton) in Nutrient-Poor Waters

Small-celled algae (nannoplankton and picoplankton), both motile and
nonmotile unicells, are capable of rapid growth to form a major component of the phytoplankton, persistent often throughout the year.
Small size may aid nutrient uptake in waters of low nutrient availability and result in low rates of sedimentation minimizing loss by sedimentation during thermal stratification. Motility in very small flagellates will be on a small scale and thus is likely to be of limited survival
value, particularly in turbulent waters of epilimnia of large lakes. Grazing pressure by filter feeders is minimal in lakes of low productivity.

2. Micro-Green Algae in Nutrient-Rich Waters

In eutrophic waters, micro–green algae form significant populations at any time during the year. These growth outbursts depend on the presence of a suitable inoculum, either from another algal community within the water body, from existing small planktonic populations or possibly from aerosols and are coupled with rapid growth rates enabling the small green algae to form populations of significant size between the major growth cycles of larger planktonic algae. Micro–green algae are therefore characterized by "opportunistic growth" in eutrophic waters, producing a sequence of population maxima through the year. These population maxima are likely to be greatest in late spring prior to the high grazing pressure from herbivorous zooplankton present in summer.

3. Flagellates (Larger Unicells and Colonial Volvocales)

The large, actively motile, flagellate algae are selected in conditions of stable stratification in lakes or ponds with a relatively shallow optical depth. Motility is a strategy by which individuals may optimize their environment for growth by movement to specific irradiance levels or regions of higher nutrient availability. Motility also precludes loss by sedimentation during calm water conditions. Due to colony size, grazing pressure is significant only for a short period in the life cycle immediately after release of small immature daughter colonies. Sexual reproduction appears to be important for the survival of the population as the formation of resistant zygospores, which act as an inoculum to the plankton during late spring, are induced at the end of growth cycles due to reduction in nitrogen levels in the epilimnion. Growth of these larger flagellate green algae is limited to a short period of the year and initiated under conditions of stable thermal stratification, an optically shallow epilimnion, and high levels of available nutrients. Any increase in the depth of water mixing from turbulence may result in a population decline.

4. Nonmotile Green Algae (Chlorococcales)

Chlorococcales are capable of growth in conditions of thermal stratification characterized by a relatively deep optical depth. The populations depend on water turbulence in the epilimnion for suspension. Cellular shape and the possession of mucilage will tend to reduce loss by sedimentation. Net growth or population decline is determined by alteration in the light climate of a water body. Such changes may result from different patterns of water circulation affecting the relationship between the extent of the euphotic zone and the mixed depth of the water column, which directly influences the period of time nonmotile algae are resident in the euphotic zone. Variation in the light climate may take place due to self-shading, resulting in a decrease in the rate of population increase followed by population decline. Nonmotile green algae are capable of growth only during short periods of the year when the light climate and depth of water mixing are balanced such that increases in population density outweigh losses due to sedimentation.

5. Desmids

Desmids are persistent organisms present in phytoplankton populations for much of the year, but often in very low numbers. Frequently, planktonic desmid populations are derived from benthic algal communities by wind-induced water mixing. The rates of growth are low, with maximum rates corresponding to long days and warm water of late summer. This low growth rate results in a pattern of seasonal periodicity with population maxima in late summer or autumn for the slower-growing species. A selective advantage is conferred by slow rates of population increase, since desmids growing after the autumn overturn are maintained in circulation throughout the greater water volume of the entire depth profile of a lake, and thus cells are subjected to greater nutrient availability and reduced chance of parasitic infection. However, they must have a very opportunistic light-harvesting strategy. Their large cell size results in minimal grazing pressure. Loss due to sedimentation during stratification is reduced by complex cell morphology and the possession of mucilage by some species.

6. Filamentous Algae

Filamentous algae are present in the late stage of seasonal succession either in shallow waters of high nutrient availability or in deeper lakes where they may be present with desmid species. It is likely that this late seasonal growth pattern results from long generation times and high grazing resistance due to their large size.

These six survival strategies include some of the great variety of biological types and species of green algae. The aim of this chapter has been to stimulate further discussion and particularly experimentation with green algae to increase our knowledge and understanding of this fascinating and diverse group of algae.

REFERENCES

Alexander, V. and Barsdate, R. J. (1971). Physical limnology, chemistry and plant productivity of a Taiga Lake. *Int. Rev. ges. Hydrobiol.,* 56, 825–72.

Aykulu, G. (1978). A quantitative study of the phytoplankton of the River Avon, Bristol. *Br. Phycol. J.,* 13, 91–102.

Baker, A. L. (1973). Microstratification of phytoplankton in three Minnesota lakes. Ph.D. Thesis, University of Minnesota.

Beadle, L. C. (1981). *The Inland Waters of Tropical Africa, and Introduction to Tropical Limnology.* London: Longman.

Behre, K. (1956). Die Algen besiedlung einiger Seen um Bremen und Bremerhaven. *Ver. Inst. Meersk. Bremerhaven,* 4, 221–83.

Berman, T., Hadas, O., and Kaplan, B. (1977). Uptake and respiration of organic compounds and heterotrophic growth in *Pediastrum duplex* (Meyer). *Freshwat. Biol.,* 7, 495–502.

Berman, B. and Rodhe, W. (1971). Distribution and migration of *Peridinium* in Lake Kinneret. *Mitt. Internat. Verein. Limnol.,* 19, 266–76.

Bisulputra, T. (1974). Plastids. In: *Algal Physiology and Biochemistry,* ed. W. D. P. Stewart. Berkeley: University of California Press.

Bold, H. C. (1942). The cultivation of algae. *Bot. Rev.,* 8, 69–138.

Bold, H. C., Cronquist, A., Jeffrey, C., Johnson, L. A. S., Margulis, L., Merxmiller, H., Raven, P. H., & Takhtajan, A. L. (1978). Proposal (10) to substitute the term "phylum" for "division" for groups treated as plants. *Taxon,* 27, 121–2.

Bold, H. C. and Wynne, M. J. (1985). *Introduction to the Algae,* 2d ed. Englewood Cliffs, NJ: Prentice-Hall.

Brook, A. J. (1981). *The Biology of Desmids,* Botanical Monographs, Vol. 16. Oxford: Blackwell.

Brown, R. M., Larson, D. A., and Bold, H. C. (1976). Air-borne algae. Their abundance and heterogeneity. *Science,* 143, 583–5.

Burgis, M. J., Darlington, J. P. E. C., Dunn, I. C., Ganf, G. G., Gwahaba, J. J., and McGowan, L. M. (1973). The biomass and distribution of organisms in Lake George, Uganda. *Proc. R. Soc. Lond. B.,* 184, 271–98.

Canter, H. M., and Lund, J. W. G. (1966). The periodicity of planktonic desmids in Windermere, England. *Verh. Internat. Verein. Limnol.,* 16, 163–72.

(1969). The parasitism of planktonic desmids by fungi. *Öst. bot. Z.,* 116, 351–77.

Caron, D. A., Pick, F. R., and Lean, D. R. S. (1985). Chroococcoid cyanobacteria in Lake Ontario, vertical and seasonal distribution during 1982. *J. Phycol.,* 21, 171–5.

Cassie, V. (1978). Seasonal changes in phytoplankton densities of four North Island Lakes. *N. Z. J. Mar. Freshwat. Res.,* 12, 153–66.

Cheng, D. M. H., and Tyler, P. A. (1973). Lakes Sorell and Crescent – Tasmanian paradox. *Int. Rev. ges. Hydrobiol.,* 58, 307–43.

Craig, S. R. (1984). Productivity of algal picoplankton in a small meromictic lake. *Verh. Internat. Verein. Limnol.,* 22, 351–4.

(1986). Size definition of picoplankton. In: *Photosynthetic Phytoplankton,* eds. T. Platt and W. K. W. Li. *Can. Bull. Fish. Aquat. Sci.,* 214.

Dodge, J. D. (1973). *The Fine Structure of Algal Cells.* London: Academic Press.

Duthie, H. C. (1965). Some observations on the ecology of desmids. *J. Ecol.,* 53, 695–703.

Ehrenberg, C. G. (1844). Bericht über die zur Bekanntmachung geigneten Verhandlunger der Konigl. *Preufs. Akad. Wiss. Berlin,* 9, 194–7.

Ganf, G. (1976). Primary Production. Proceedings of Australian Water Resources Council Symposium on Eutrophication, pp. 23–35. Australian Government Publications Service, Canberra.

Ganf, G., & Oliver, R. L. (1982). Vertical separation of light and available nutrients as a factor causing replacement of green algae by blue-green algae in the plankton of a stratified lake. *J. Ecol.,* 70, 829–44.

George, D. G., and Edwards, R. W. (1976). The effect of wind on the distribution of chlorophyll and crustacean zooplankton in a shallow eutrophic reservoir. *J. Appl. Ecol.,* 13, 667–90.

George, D. G., and Heaney, S. I. (1978). Factors influencing the spatial distribution of phytoplankton in a small productive lake. *J. Ecol.,* 66, 133–55.

Goldman, C. R., Mason, D. T., and Wood, B. J. B. (1972). Comparative study of the limnology of two small lakes on Ross Island, Antarctica. In: *Antarctic Terrestrial Biology,* ed. G. A. Llano, pp. 1–50. Washington American Geophysical: Union.

(1974). Carotenoids and biloproteins. In: *Algal Physiology and Biochemistry,* ed. W. D. P. Stewart. Berkeley: Univ. California Press.

Goodwin, T. W. (1980). *The Biochemistry of the Carotenoids.* Vol. 1. *Plants.* 2d ed. London: Chapman and Hall.

Happey, C. M. (1970). The estimation of cell numbers of flagellate and coccoid Chlorophyta in natural populations. *Br. Phycol. J.,* 5, 71–8.

(1976a). Vertical migration patterns in phytoplankton of mixed species composition. *Br. Phycol. J.,* 11, 355–69.

(1976b). The influence of stratification on the growth of planktonic Chlorophyceae in a small body of water. *Br. Phycol. J.,* 11, 371–81.

(1978). The application of culture methods in studies of the ecology of small green algae. *Mitt. Internat. Verein. Limnol.,* 21, 385–97.

Happey-Wood, C. M., and Hughes, D. I. (1980). Morphological and physiological variations in clones of *Asterionella formosa. New Phytol.,* 86, 441–53.

Happey-Wood, C. M., Kennaway, G. M., Chittenden, A., and Edwards, G. (1988). *J.W.G. Lund Festschrift,* Biopress.

Harris, D. O. (1971). Growth inhibitors produced by the green algae (Volvocales). *Arch. Mikrobiol.,* 76, 47–50.

Heaney, S. I., and Talling, J. F. (1980). Dynamic aspects of dinoflagellate distribution patterns in a small productive lake. *J. Ecol.,* 68, 75–94.

Heckey, R. E., Kling, H. J., and Brunskill, G. J. (1986). Seasonality of phytoplankton in relation to silicon cycling and interstitial water circulation in large, shallow lakes of central Canada. *Hydrobiologia,* 138, 117–26.

Hensen, V. (1887). Über die Bestimmung des Planktons oder des in Meere treibenden Materials an Pflantzen und Tieren. *Bericht des Deutschen wissenschaftlichen Kommission fur Meereforschung*, 5, 1–109.

Heywood, R. B. (1977). A limnological survey of the Ablation Point area, Alexander Island, Antarctica. *Phil. Trans. R. Soc. Lond. B*, 279, 27–38.

(1978). Maritime antarctic lakes. *Verh. Internat. Verein. Limnol.*, 20, 1210–15.

Heywood, R. B., Dartnell, H. J. G., and Priddle, J. (1980). Characteristics and classifications of the lakes of Signy Island, South Orkney Islands, Antarctica. *Freshwat. Biol.*, 10, 47–59.

Hobbie, J. E. (1973). Antarctic Limnology: a review. In: *Alaskan Arctic Tundra*, ed. M. E. Britton, pp. 127–68. Arctic Institute of North America Technical Papers, 25. Washington: Arctic Institute of North America.

Hutchinson, G. E. (1967). *A treatise of limnology*. Vol. II. *Introduction to Lake Biology and the Limnoplankton*. New York: Wiley.

Jorgensen, E. G. (1956). Growth inhibition and substances formed by algae. *Physiologica Pl.*, 9, 712–26.

Kalff, J., and Watson, S. (1986). Phytoplankton and its dynamics in two tropical lakes: a tropical and temperate zone comparison. *Hydrobiologia*, 138, 161–76.

Kennaway, G. M. A., and Happey-Wood, C. M. (1988). Physiological ecology of picoplankton in a Welsh mountain lake. *Br. Phycol. J.*, 23.

Kirk, J. T. D. (1979). Spectral distribution of photosynthetically active radiation in some south-eastern Australian waters. *Aust. J. Mar. Freshwat. Res.*, 30, 81–91.

Kochert, G. (1982). Sexual processes in the Volvocales. In: *Progress in Phycological Research*, Vol. 1., ed. F. E. Round & D. J. Chapman. pp. 235–56. Amsterdam: Elsevier Biomedical Press.

Kosswig, K. (1967). Der Sach wiensensee in den Ostalpen (Hochschwabgebiert) zur Limnologie eines dystrophen Gipsgewässers. *Int. Rev. ges. Hydrobiol.*, 52, 321–51.

Lack, T. J. (1971). Quantitative studies on the phytoplankton of the Rivers Thames and Kennet at Reading. *Freshwat. Biol.*, 1, 213–24.

Langmuir, I. (1938). Surface motion of water induced by wind. *Science*, 87, 119–23.

Lembi, C. A., Pearlmutter, N. L., and Spencer, D. F. (1980). Life cycle, ecology and management of the green filamentous alga, *Pithophora*. Tech. Rep. No. 130, Water Resources Center, West Lafayette, IN: Purdue University.

Lohmann, H. (1903). Nene Untersuchungen über den Reichtum des Meeres an Plankton und Über die Branchbarkeit der Verschiederen Fangmethoden. *Wiss. Meeresunters N.F.F.* 1–3, Tf. 1–7.

Lund, J. W. G. (1949). Studies on *Asterionella formosa* Hass. I. The origin and nature of the cells producing seasonal maxima. *J. Ecol.* 37, 389–439.

Lund, J. W. G. (1971). The seasonal periodicity of three planktonic desmids in Windermere. *Mitt. Internat. Verein. Limnol.*, 19, 3–25.

Maguire, B. Jr. and Neill, W. E. (1971). Species and individual productivity in phytoplankton communities. *Ecology,* 52, 903–7.

Meyer, R. L. and Brook, A. J. (1969). Freshwater algae from Itasca State Park, Minnesota. *Nova Hedwigia,* 16, 251–66.

Morris, I. (1980). *The Physiological Ecology of Phytoplankton.* Oxford: Blackwell Scientific.

Moss, B. (1969). Vertical heterogeneity in the water column of Abbot's Pond. II. The influence of physical and chemical conditions on the spatial and temporal distribution of the phytoplankton and a community of epipelic algae. *J. Ecol.,* 57, 397–414.

——— (1970). The algal biology of a tropical montane reservoir (Mlungusi Dam, Malawi). *Br. Phycol. J.,* 5, 19–28.

——— (1972). Studies on Gull Lake, Michigan. I. Seasonal and depth distribution of phytoplankton. *Freshwat. Biol.,* 2, 289–307.

——— (1973). The influence of environmental factors on the distribution of freshwater algae: an experimental study. I. Introduction and the influence of calcium concentration. *J. Ecol.,* 60, 917–32.

Moss, B., and Karim, A. G. A. (1969). Phytoplankton associations in two pools and their relationship with associated benthic flora. *Hydrobiologia,* 33, 587–600.

Munawar, M., and Munawar, I. F. (1986). Seasonality of phytoplankton in the North American Great Lakes, a comparative synthesis. *Hydrobiologia,* 138, 85–115.

Mur, L. C., Gons, H. J., and van Liere, L. (1978). Competition of the green alga *Scenedesmus* and the blue-green alga *Oscillatoria. Mitt. Internat. Verein. Limnol.,* 21, 473–9.

Noble, J. C., and Happey-Wood, C. M. (1988). Some aspects of the ecology of algal communities in rice fields and rice irrigation systems of New South Wales. *J. Aust. Inst. Agr. Sci.,* 53, 170–84.

Ong, M. H. (1977). The role of micro-green algae in algal communities. Ph.D. Thesis, University of Wales.

Ostrofsky, M. L., and Duthie, H. C., (1975). Primary productivity and phytoplankton of lakes on the Eastern Canadian Shield. *Verh. Internat. Verein. Limnol.,* 19, 732–8.

Pechlaner, R. (1971). Factors that control the production rate and biomass of phytoplankton in high-mountain lakes. *Mitt. Internat. Verein. Limnol.,* 19, 125–45.

Pollingher, U. (1986). Phytoplankton periodicity in a subtropical lake (Lake Kinneret, Israel). *Hydrobiologia,* 138, 127–38.

Pollingher, U., and Berman, T. (1976). Autoradiographic screening for potential heterotrophs in natural algal populations of Lake Kinneret. *Microb. Ecol.,* 2, 252–60.

Priddle, J., and Happey-Wood, C. M. (1983). Significance of small species of Chlorophyta in freshwater phytoplankton communities with special reference to five Welsh lakes. *J. Ecol.,* 71, 793–810.

Procter, V. W. (1957). Studies of algal antibiosis using *Haematococcus* and *Chlamydomonas. Limnol. Oceanogr.,* 2, 125–39.

Raven, J. A. (1986). Physiological consequences of extremely small size for autotrophic organisms in the sea. In: Photosynthetic Picoplankton, eds. T. Platt and W. K. W. Li., pp. 1–70. *Can. Bull. Fish. Aquat. Sci.,* 214.

Reynolds, C. S. (1976). Succession and vertical distribution of phytoplankton in response to thermal stratification in a lowland mere, with special reference to nutrient availability. *J. Ecol.,* 64, 529–51.

(1980). Phytoplankton assemblages and their periodicity in stratifying lake systems. *Holarctic Ecology,* 3, 141–59.

(1982). Phytoplankton periodicity; its motivation, mechanisms and manipulation. *Freshwat. Biol. Ass. Rep.,* 50, 60–75.

(1983). Growth-rate responses of *Volvox aureus* Ehrenb. (Chlorophyta, Volvocales) to variability in the environment. *Br. Phycol. J.,* 18, 433–42.

(1984a). *The Ecology of Freshwater Plankton.* Cambridge: Cambridge University Press.

(1984b). Phytoplankton periodicity: the interactions of form, function and environmental variability. *Freshwat. Biol.,* 14, 111–42.

(1986). Experimental manipulations of plankton periodicity in large limnetic enclosures in Blelham Tarn, English Lake District. *Hydrobiologia,* 138, 43–64.

Reynolds, C. S., and Butterwick, C. (1979). Algal bioassay of unfertilized and artificially fertilized lake water maintained in Lund Tubes. *Arch. Hydrobiol.,* 56, 166–83.

Reynolds, C. S., and Rogers, M. W. (1983). Cell- and colony-division in *Eudorina* (Chlorophyta, Volvocales) and some ecological implications. *Br. Phycol. J.,* 18, 111–19.

Reynolds, C. S., and Wiseman, S. W. (1982). Sinking losses of phytoplankton maintained in closed limnetic systems. *J. Plank. Res.,* 4, 561–600.

Reynolds, C. S., Wiseman, S. W., Godfrey, B. M., and Butterwick, C. (1983). Some effects of artificial mixing on the dynamics of phytoplankton populations in large limnetic enclosures. *J. Plank. Res.,* 5, 203–34.

Richards, M. C., and Happey-Wood, C. M. (1979). The application of pattern analysis to freshwater phytoplankton communities. *Limnol. Oceanogr.,* 24, 950–6.

Richerson, P., Armstrong, R., and Goldman, C. R. (1970). Contemporaneous disequilibrium, a new hypothesis to explain the "paradox of the plankton." *Proc. Natl. Acad. Sci.,* 67, 1710–14.

Richerson, P. J., Neale, P., Wurtsbaugh, W., Alfaro, R., and Vincent, W. (1986). Patterns of temporal variation in Lake Titicaca: A high altitude tropical lake. I. Background, physical and chemical processes and primary production. *Hydrobiologia,* 138, 205–20.

Rodhe, W. (1955). Can phytoplankton production proceed during winter darkness? *Verh. Internat. Verein. Limnol.,* 12, 117–22.

Ross, P. E., and Duthie, H. C. (1981). Ultraplankton biomass, productivity and efficiency in Lac Matamec, a precambrian shield lake. *J. Phycol.,* 17, 181–6.

Salonen, K., Jones, R. I., and Arvola, L. (1984). Hypolimnetic phosphorus retrieval by diel vertical migration of lake phytoplankton. *Freshwat. Biol.*, 14, 431–8.

Seaburg, K. G., Kaspar, M., and Parker, B. C. (1983). Photosynthetic quantum efficiencies of phytoplankton from perennially ice-covered Antarctic Lakes. *J. Phycol.*, 19, 446–52.

Sheath, R. G. (1986). Seasonality of phytoplankton in northern tundra ponds. *Hydrobiologia*, 138, 75–83.

Sieburth, J. M. N., Smetacek, V., and Lenz, J. (1978). Pelagic ecosystem structure: Heterotrophic compartments of the plankton and their relationships to plankton size fractions. *Limnol. Oceanogr.*, 23, 1256–63.

Simla, A. (1988). Spring development of *Chlamydomonas* sp. in a humic forest lake. Proc. S. I. L. Workshop "Flagellates in freshwater Ecosystems," *Hydrobiologia*.

Skreinivasa, M. R., and Nalewajko, C. (1975). Phytoplankton biomass and species composition in Northeastern Lake Ontario, April–November 1965. *J. Great Lakes Res.*, 1, 151–61.

Skuja, H. (1964). Grundzuge der Algenflora und Algenvegetation der Fjeldgegenden un Abiesko in Schwechisch Lappland. *Nova Acta R. Soc. Scient. Upsal.*, 18.

Smith, I. R. (1975). Turbulence in lakes and rivers. *Sci. Publ. Freshwat. Biol. Assoc.*, 29.

Sommer, U. (1986). The periodicity of phytoplankton in Lake Constance (Bodensee) in comparison with other deep lakes of central Europe. *Hydrobiologia*, 138, 1–7.

Sournia, A. (1980). *Phytoplankton manual.* Paris: U.N.E.S.C.O.

Starr, R. C. (1955). A comparative study of *Chlorococcum* Meneghini and other spherical zoospore-producing genera of the Chlorococcales. *Indiana Univ. Publs. Scr. Ser.*, 20, 1–111.

Stull, E. A., Amegaza, A., and Goldman, C. R. (1973). The contribution of individual species of algae to primary productivity of Castle Lake, California. *Verh. Internat. Verein. Limnol.*, 19, 630–4.

Swale, E. M. F. (1969). Phytoplankton in two English rivers. *J. Ecol.*, 57, 1–23.

Talling, J. F. (1971). The underwater light climate as a controlling factor in the production ecology of freshwater phytoplankton. *Mitt. Internat. Verein. Limnol.* 19, 214–43.

— (1986). The seasonality of phytoplankton in African lakes. *Hydrobiologia*, 138, 139–60.

Tilzer, M. M., Paerl, H. W., and Goldman, C. R. (1977). Sustained viability of aphotic phytoplankton in Lake Tahoe (California, Nevada). *Limnol. Oceanogr.*, 22, 84–91.

Tyler, P. (in press). Flagellate zonation in a polyhumic karst lake – biological responses. Proc. S. I. L. Workshop "Flagellates in Freshwater Ecosystems," *Hydrobiologia*.

Vincent, W. F. (1983). Phytoplankton production and winter mixing: contrasting effects in two oligotrophic lakes. *J. Ecol.*, 71, 1–20.

Walsby, A. E., and Reynolds, C. S. (1980). Singing and floating. In: *The Physiological Ecology of Phytoplankton,* ed. I. Morris, pp. 371–412. Oxford: Blackwell.

White, E., Downes, M., Gibbs, M., Kemp, L., Mackenzie, L., and Payne, G. (1980). Aspects of the physics, chemistry and phytoplankton biology of Lake Taupo. *N. Z. J. Mar. Freshwat. Res.,* 14, 139–48.

Chapter 6

GROWTH AND SURVIVAL STRATEGIES OF PLANKTONIC DIATOMS

ULRICH SOMMER

INTRODUCTION

Two approaches may be taken in attempting to explain the species composition of natural communities. The comparative descriptive approach seeks distributional patterns in data sets from as many natural ecosystems as possible. The ecophysiological approach *(sensu lato)* attempts to explain differences in the distribution of species (or groups of species) by differences in their experimentally established requirements for growth factors and from their resistance to mortality factors. If a taxon is neither becoming extinct nor overgrowing competing taxa then there must be an overall balance in the "costs" and "benefits" of its specific traits. Locally and temporarily, however, this balance may be unstable. This local and temporary instability is an explanation for spatial and temporal changes in species composition. Here, I address the question: "What are the specific costs and benefits of being a planktonic diatom?" A reasonable answer would allow us to predict the conditions under which diatoms would tend to dominate phytoplankton communities.

Diatoms comprise 10,000 to 20,000 species of diverse sizes and shapes. Planktonic species vary in size from 2 μm to 2 mm (i.e., by four orders of magnitude). This is about the same size range as exists from the smallest bryophytes to the largest canopy trees in a forest. Is it meaningful under such circumstances, then, to discuss the "ecology of plankton diatoms"? Shouldn't one restrict oneself to the ecology of single species? The consideration of the ecology of a higher taxon becomes justified if its members share a common feature that distinguishes them from other taxa and causes particular survival problems in a given habitat. There is no doubt that the siliceous cell covering – the frustule – constitutes such a feature of diatoms.

Superficially, the siliceous frustule seems counteradaptive to plank-

227

tonic survival because it makes diatoms dependent on an additional nutrient that is less readily recycled than nitrogen and phosphorus, and because it makes diatoms heavier and, therefore, more prone to sinking losses than other phytoplankton. Other phytoplankton need much less silica (or none at all) and play a far less significant part in the biogeochemical cycling of silica. Despite the apparent "costs" of the frustule for diatom growth and survival, diatoms are among the most successful taxa in freshwater and marine phytoplankton. This suggests that the "costs" are balanced by some "benefits," which may be directly or indirectly related to the siliceous frustule.

Therefore, the title of this chapter could well be "Consequences of the Siliceous Frustule for Planktonic Life." These consequences shall be discussed in the following order: nutrition and growth, sinking, recycling of Si, sex and perennation. In the last section I shall try to provide a synthesis and address the question: "Under which conditions are diatoms a successful component of phytoplankton?" A subsection on the calculation of in situ growth and loss rates is also provided as a tool by which others may analyze their own diatom data. There will be no discussion of distributional patterns or ecophysiological questions that are not diatom-specific (e.g., photosynthesis, metabolic effects of temperature, grazing). Relevant information on these topics may be found in Lewin and Guillard (1963) Werner (1977b), Reynolds (1984) and references in Chapter 1.

SILICON AS A NUTRIENT

Silicon in Natural Waters

There seems to be a general consensus that the only significant dissolved component of silicon in natural waters is orthosilicic acid $(Si (OH)_4)$; a few reports on the occurrence of other dissolved species in natural waters are generally doubted (Aston 1983). At pH <9, orthosilicic acid is present only in the undissociated state. The concentrations of orthosilicic acid in natural waters are generally measured by the molybdenum-blue method (Strickland & Parsons 1969), which gives a limit of detectability of about 0.03 $\mu mol \cdot L^{-1}$ Si (and 0.3 $\mu mol \cdot L^{-1}$ if measured in an autoanalyzer). Dissolved Si concentrations in the sea range from nondetectable to 200 μmol Si$\cdot L^{-1}$, and in lakes they range from nondetectable to about 400 μmol Si$\cdot L^{-1}$. Surface waters are always undersaturated with regard to silicon, the solubility being 1000 μmol Si$\cdot L^{-1}$ at 0°C and 2464 μmol Si$\cdot L^{-1}$ at 25°C in seawater (Spencer 1983). The differences in solubility between freshwater and seawater are minor. The concentrations of orthosilicic acid often show marked seasonal periodicity closely related to the seasonality of diatoms (e.g., Sommer & Stabel 1983).

Particulate silicon in natural waters has two main components: silicate minerals and biogenic diatom silica. The latter is often considered to be opal, but Kamitani (1971) states that recent diatom shells resemble silica gel, and fossil ones are more like opaline silica. The greater solubility of diatom silica in comparison to quartz and silicate minerals can be used for selective determination of diatom Si (Tessenow 1966).

The Silicon Content of Diatoms

Since the classic study of Einsele and Grim (1938), the silicon content of freshwater planktonic diatoms has been studied on several occasions (Table 6-1). There is a tendency for values from cultures to be lower than those from natural samples. For example, reported Si contents of cultured *Asterionella formosa* cells range from 11 to 52 pg· cell^{-1} (see Table 7.4 in Paasche 1980), and values from plankton samples are 70 to 200. Werner (1977a) suspected that the plankton data are overestimates, but it might be possible that the discrepancy arises from the diminution of cell size during prolonged cultivation. Certainly, the highest values reported by Einsele and Grim (1938) seem somewhat incredible, because a Si content of 0.56 pg Si/μm^3 cell volume *(Cyclotella melosiroides)* would require that 57% of the cell volume be occupied by silica (if a density of 2.1 is assumed for diatom silica).

By employing complicated fractionation methods, Werner (1966, 1977a) separated three fractions of diatom silica: cell-wall silica (which composed 96–98% of total silica), plasmatic silicic acid, and a fraction that could be extracted by 100°C water and polymerized by 0.1 N HCl. Due to the conservative behavior of wall silica, the total Si content changed only by a factor of 1.6 between Si-rich and Si-starved cells. The content of plasmatic silicic acid decreased by a factor of 10 under prolonged Si starvation, and acid-polymerizable silicic acid was completely exhausted.

Silicon in Diatom Cell Metabolism

The historical view that silicon is utilized only as a structural component of the diatom cell wall has been revised mainly through the studies of Werner (1966, 1977a). Using Si-starvation experiments and by inhibiting orthosilicic acid uptake by germanic acid, he demonstrated the close coupling between silicate metabolism and other metabolic

Table 6-1. *Si-content of natural diatom populations expressed per cell or per unit cell volume.*

Species	pg Si·cell^{-1}	pg Si·μm^3	Source
Asterionella	45–200	0.14–0.19	Einsele & Grim (1938)
formosa	70	0.16	Bailey-Watts (1976a)
	71–80	0.11	Happey (1970)
	100–190	0.15–0.27	Sommer & Stabel (1983)
Fragilaria	70–200	0.01–0.11	Einsele & Grim (1938)
crotonensis	100–140	0.09–0.14	Sommer & Stabel (1983)
Tabellaria	170–350	0.14–0.16	Einsele & Grim (1938)
fenestrata			
Synedra acus	450–1200	0.14–0.19	Einsele & Grim (1938)
	120	0.36	Sommer & Stabel (1983)
Melosira	60	0.14	Einsele & Grim (1938)
granulata	90–110	0.09–0.11	Sommer & Stabel (1983)
Melosira	120–180	0.19	Einsele & Grim (1938)
italica			
Cyclotella	100	0.26–0.28	Einsele & Grim (1938)
glomerata			
Cyclotella	350	0.51–0.56	Einsele & Grim (1938)
melosiroides			
Cyclotella	900	0.46	Einsele & Grim (1938)
comta			
Stephanodiscus	4000	0.16	Einsele & Grim (1938)
astraea			
Stephanodiscus	22–26	0.11–0.13	Sommer & Stabel (1983)
binderanus			
Stephanodiscus	5–8.5	0.10–0.17	Sommer & Stabel (1983)
hantzschii			

pathways. Exponentially growing *Cyclotella cryptica* cells (generation time 5 h) stopped cell division less then 1 h after transfer into a Si-free medium. Inhibition of DNA synthesis and protein synthesis occurred after 4 h, and inhibition of chlorophyll synthesis after 5 h. After 12 h, apparent photosynthesis was decreased by 80%, and fatty acid synthesis was strongly increased. Consequently, Si-starved diatom cells are characterized by increased lipid content of cells and decreased protein and carbohydrate contents. Moed, Hoogveld, and Apeldoorn (1976) used the accumulation of oil droplets in diatom cells as indicators of Si limitation in situ. Silicon starvation can be distinguished from N-

and P-starvation, which are characterized by increased carbohydrate contents and not lipid contents. Protein content decreases as in the case of Si starvation, but starvation symptoms usually do not occur until a few cell divisions have taken place after transfer into a P- or N-free medium (Werner 1966).

Not all diatom species studied so far stop cell division immediately following deprivation of external silicon supply, as does *Cyclotella cryptica. Skeletonema costatum, Thalassiosira pseudonana,* and *Asterionella formosa* undergo one more cell division by building less silicified new frustules (Lund 1950; Moed 1973; Paasche 1973a; Sommer & Stabel 1983). Analysis of field data (Lund 1950; Sommer & Stabel 1983) and experiments (Moed 1973) have shown that *Asterionella formosa* has to "pay" for this further cell division by increased cell mortality (Fig. 6-1). The analysis of Sommer and Stabel has also shown that *Fragilaria crotonensis, Melosira granulata,* and *Stephanodiscus binderanus,* species that stop cell division under Si stress, did not similarly experience increased mortality.

Silicon Limitation

Three types of saturation models describe the kinetics of nutrient limitation: uptake versus external concentration models (e.g., Dugdale 1967), reproductive rate versus intracellular concentration (cell quota) models (e.g., Droop 1973), and reproductive rate versus external concentration models (Monod 1950). For a more detailed review of the physiology of nutrient limitation see Turpin (Chapter 8, this volume). The Monod model is the most popular one among field ecologists because it relates the ecologically most relevant response μ, the reproductive rate, directly to an environmental parameter S, the concentration of the limiting nutrient:

$$\mu = \frac{\mu_{max} S}{S + k_s} \qquad (1)$$

However, because this model integrates several steps (i.e., concentration-dependent uptake, cell-quota-dependent feedback to uptake, cell-quota-dependent reproduction), it is only applicable under certain conditions. If, as realized in chemostat culture, nutrient uptake and population growth rate are in equilibrium, then the three types of limitation models are mathematically equivalent (Burmaster 1979). If the nutrient regime is variable, as often happens in nature and in batch cultures, the carryover of nonequilibrated cell quotas from previous periods of different nutrient availability will lead to departures from

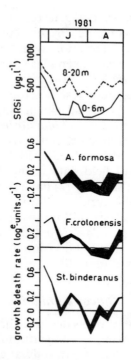

Fig. 6-1. Response of the reproductive rate (μ) and the rate of cell mortality (δ) of three diatoms to summer depletion of soluble reactive silicon in Lake Constance. Population dynamic rates are calculated according to equations presented in text. The top limit of black area characterizes μ; the thickness of the black area, δ. The theoretically impossible case of μ < 0 is either caused by sampling error (plankton patchiness) or by neglect of grazing (see text). Note that the mortality rate of *Asterionella formosa* responds far more strongly to Si depletion than does that of *Fragilaria* and *Stephanodiscus*. Redrawn after Sommer (1987).

reproductive rates predicted by the Monod equation. However, because of the absence of significant intracellular storage, this problem is negligible in the case of Si-limited diatom growth, as compared to N- and P-limited growth. In several cases, however, the addition of threshold concentrations to the Monod equation has been found necessary (Tilman and Kilham 1976). The Monod parameters for Si-limited growth of several diatoms are compared in Table 6-2.

The applicability of the Monod relationship to Si-limited growth of diatoms, even under non-steady-state conditions, has important consequences for recent arguments about nutrient limitation in situ (Goldman, McCarthy, & Peavey 1979). These authors argued that seston elemental composition (C:N:P ratios) suggested absence of nutrient limitation. Rapid buildup of high cell quotas at locally enriched microsites and subsequent nonlimited growth at the expense of these cell quotas were suggested as an explanation. The smallness of the plasmatic silicate pool relative to the amount of silicate needed for the construction of new thecae makes this mechanism impossible in the case of Si-limited diatom populations.

The in situ reproductive rate of diatoms can be quantified either directly by use of the mitotic index technique (McDuff & Chisholm 1981) or by quantification of the losses (Sommer 1984a and later discussion). The latter approach has been used to analyze the growth response of several diatoms in Lake Constance to regularly occurring summer depression of ambient Si concentrations (Fig. 6-1). There is no doubt about the coincidence between the depression of Si(OH)$_4$ and the depression of growth rate (μ).

Diatoms as Nutrient Competitors

An increasing number of phytoplankton competition studies during recent years (e.g., Holm & Armstrong 1981; Sommer 1983; Tilman & Kiesling 1984) have shown that the disadvantage of being silica dependent is counterbalanced by the fact that diatoms are good competitors for other nutrients, especially phosphorus. The equilibrium result of nutrient competition can be predicted from Monod kinetics (for a detailed account of resource-based competition theory see Tilman 1982). Under steady-state conditions, growth rate equals loss rate (λ), and the renewal rate of nutrients equals the consumption rate. Population density and residual nutrient (R^*) concentration are constant,

$$\text{if } \mu = \lambda, \text{ then}$$
$$R^* = \frac{k_s\lambda}{\mu_{max} - \lambda} \tag{2}$$

Competitive superiority of species 1 over species 2 is defined by $R^*_1 < R^*_2$, because the result of species 1 lowering the nutrient concentration of R^*_1 is that species 2 will have a $\mu_2 < \lambda_2$, which leads to its displacement. At high loss rates relative to μ_{max}, competitive ability depends more on μ_{max} ("velocity selection"); at low loss rates it depends more on the initial slope of the Monod curve (μ_{max}/K_s; "affinity selection"). If nutrient supply rates are chosen in proportions such

Table 6-2. *Monod-type growth parameters μ_{max} (ln units·d^{-1}), k_{Si} ($\mu mol \cdot L^{-1}$), and threshold concentration (Si_0), if appropriate, for Si-limited growth of diatoms.*

Freshwater

Species	Temp. (°C)	k_{Si} ($\mu mol \cdot L^{-1}$)[a] or $k_{Si} + Si_0$	$\mu_{max}(d^{-1})$	Source
Synedra filiformis	20	19.7	1.1	Tilman (1977)
Synedra ulna	8	4.9 + 0.2	0.16	Tilman, Mattson & Langer (1981)
	13	3.8 + 0.3	0.71	
	20	4.0 + 0.6	0.65	
	24	4.4 + 0.3	0.78	
Asterionella formosa	4	1.3 + 0.6	0.35	Tilman et al. (1981)
	8	1.6 + 0.2	0.52	
	13	2.5 + 0.2	0.79	
	20	3.7 + 0.1		
Fragilaria crotonensis	20	1.5	0.62	Tilman (1981)
Tabellaria focculosa	20	19.0	0.74	Tilman (1981)

Diatoma elongatum	20	0.85	1.51	Tilman (1981)
Cyclotella meneghiniana	20	1.44	0.92	Tilman & Kilham (1976)
Stephanodiscus minutus	10	0.31	0.71	Kilham (1984)
	20	0.88	0.15	
Synedra sp.	10	6.07	0.71	Kilham (1984)
	15	4.84	0.79	
	20	3.14	0.82	
Marine				
Thalassiosira pseudonana estuarine clone	20	0.48 + 0.7	2.44	Paasche (1973a,b)
	20	1.24 + 0.22	2.78	Paasche (1973a,b)
	20	0.98	2.51	Guillard, Killam, & Jackson (1973)
Sargasso clone	20	0.19	1.77	Guillard et al. (1973)
Thalassiosira nordenskiöldii	3	0.088	0.61	Paasche (1975)
	10	0.022	0.87	Paasche (1975)

[a] Or k_{Si} + Si_0 if there is a threshold concentration.

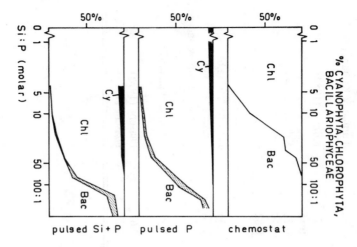

Fig. 6-2. The different effects of perfect steady state ("chemostat"), a nutrient regime with weekly pulses of P ("pulsed P"), and weekly pulses of Si and P ("pulsed Si + P") on the outcome of diatom–green algae–blue-green algae competition expressed in percentage contribution to total biomass. *Black,* Cyanophyta; *white,* Chlorophyta (minimum during weekly oscillations); *Dark shading,* Bacillariophyceae (minimum during weekly oscillations), *light shading* (range of oscillations). Si:P ratios are supply ratios (based on total loading per weekly cycle). Note that diatoms are displaced more at increasingly higher Si:P ratios as more environmental variability is included. Redrawn after Sommer (1985a).

that different species are limited by different nutrients, as many species can coexist as there are limiting nutrients.

In P-limited steady-state competition experiments conducted thus far, only Si limitation could prevent diatoms from outcompeting Cyanobacteria (Holm & Armstrong 1981; Tilman & Kiesling 1984) or green algae (Sommer 1983; Tilman and Kiesling 1984). If Si was supplied in sufficient amounts to prevent Si limitation, the diatoms became dominant. In all P-competition experiments with natural freshwater multispecies inoculae, *Synedra* spp. were the final winners (Sommer 1983; Smith, & Kalff 1983; Tilman & Kiesling 1984). Sommer (1985a) has also studied non-steady-state competition by adding one (P) or two (P and Si) nutrients at discrete weekly intervals into otherwise continuous cultures (Fig. 6-2). Two results were as follows: (1) More species could coexist than under strict steady-state conditions. (2) The domain of diatom dominance was displaced along the

Table 6-3. *Initial slope [μ_{max}/k_s] of Si- and P-limited growth kinetics and optimum atomic Si:P ratios for diatoms.*[a]

	$\mu_{max}/k_{Si}(d^{-1}/\mu M)$	μ_{max}/k_P	Optimum Si:P ratio
Cyclotella meneghiniana	0.9	3.2	6
Stephanodiscus minutus	0.85	5.2	-n.d.-
Fragilaria crotonensis	0.41	73	20
Asterionella formosa	0.35	98	87
Synedra filiformis	0.056	217	527
Tabellaria flocculosa	0.038	45	33

[a]Data of Tilman (1977, 1981) and Kilham (1984).

Si:P ratio gradient in comparison to steady-state competition. The periodic incidence of nonlimiting P concentrations caused a relative advantage for small, coccoid green algae (with higher μ_{max} but lower P affinity than diatoms), and P-storing Cyanobacteria. The additional introduction of a pulsed Si regime further increased the relative disadvantage for diatoms because of their inability to utilize the pulses for storage and the absence of continuous supply. However, at the highest Si:P ratios *Synedra acus* still was the dominant species.

Tilman (1981) found an inverse rank order among competitive abilities for Si and P. The data for *Stephanodiscus minutus* (Kilham 1984) and *Cyclotella meneghiniana* (Tilman 1977) fall on the same continuum. In Table 6-3 the diatom species are ranked according to the initial slope (μ_{max}/k_s) of their Si-limited Monod growth curve. Only *Tabellaria flocculosa* does not conform to the compensating tradeoff for kinetic constants between Si and P characteristics, being the worst competitor for Si and only a moderate competitor for P. Again, a *Synedra* species is the best competitor for P.

There are two hypotheses to explain the competitive superiority of diatoms for nonsiliceous nutrients (i.e., P). Munk and Riley (1952) suggested that diatoms, by sinking faster than other algae with similar geometrical properties, are less susceptible to the development of depleted microzones around the cells. Werner (1977b), on the other hand, suggested that diatom silica might act as an effective adsorbing agent for dissolved substances at low concentrations. He supported his hypothesis by an analogy in purification technology and by the use of silicates as catalysts. Further experimental research along this line would be desirable.

DIATOMS AND SINKING LOSES

Mechanical Theory of Sinking

Sedimentation is the terminal fate of any particle suspended in lakes that is heavier than water. Under the simplifying assumption of laminar flow, the terminal sinking velocity (V_T) of a spherical particle depends on the radius r, the density of the medium ρ, the density ρ' of the particle, and the viscosity η of the medium, as described by Stoke's law:

$$V_T = \frac{2}{9} gr^2 (\rho' - \rho)\eta^{-1} \tag{3}$$

If particles are not spherical, a correction factor for form resistance (ϕ^{-1}) has been used. For details on form resistance of different geometrical bodies see Walsby and Reynolds (1980).

Because phytoplankton are only slightly more dense than water, small changes in their density lead to large changes in sinking velocity. The density of diatom silica is about 2.1 g·mL^{-1}. In Table 6-1 Si contents of freshwater diatoms range from 0.07 to 0.56 pg·Si μm^{-3} cell volume, the majority being between 0.10 and 0.20 pg·μm^{-3}. These translate into SiO$_2$ weights of 0.10 to 1.1 pg SiO$_2$·μm^{-3} for the extremes and 0.21 to 0.43 pg SiO$_2$·μm^{-3} for the majority of cells. If the remainder of the cell material is assumed to have a density of 1 g·mL^{-1} (a conservative estimate), cell densities range of 1.079 to 1.63 g·mL^{-1} for the extremes and 1.11 to 1.225 g·mL^{-1} for the midrange. If we assume that nonsiliceous algae have a density of about 1.04 g·mL^{-1}, then a diatom in freshwater (density 1 g·mL^{-1}) will sink at least twice as fast as a nonsiliceous alga of similar geometrical properties. The majority of species will sink about 3 to 5.5 times faster and the heaviest ones 16 times faster.

Measured densities of diatom cells are within the foregoing range of calculated values. Wiseman, Jaworski, and Reynolds (1983) obtained densities for *Asterionella formosa* of 1.15 to 1.26 mg·mL^{-1} by isopycnic density gradient centrifugation in Ficoll. Density gradient centrifugation of natural phytoplankton from Lake Constance in Percoll

(Sommer unpubl. data) yielded mean densities of 1.08 g·mL^{-1} for *Stephanodiscus binderanus,* 1.13 g·mL^{-1} for *Asterionella formosa, Fragilaria crotonensis,* and *Melosira granulata,* and 1.163 for *Diatoma elongatum.* Nonsiliceous algae ranged from 1.019 *(Pandorina morum)* to 1.067 (*Cosmarium* sp.), with the exception of buoyant Cyanophyta (< 1 g·mL^{-1}).

Consideration of density therefore makes it evident that sinking must be of greater evolutionary importance to diatoms than to other planktonic algae. They must have either evolved increasing form resistance (e.g., spines, setae, rostra; see Barber & Haworth 1981) or adapted their life cycle to periodical sinking. However, it must be remembered that sinking is not disadvantageous under all circumstances. As mentioned earlier, it might facilitate nutrient uptake by reducing nutrient-depleted microzones (Munk & Riley 1952). Also, Lund (1950) suggested that sedimentation may act as an escape mechanism for *Asterionella formosa* if Si deficiency in the euphotic zone would lead to increased mortality.

The velocity at which cells leave the surface strata is not identical with V_T. Turbulence will resuspend the cells, and the apparent sinking velocity will be the difference between the terminal sinking velocity and resuspension. If turbulence is perfect (an infinite number of mixing events), the loss of particles from a mixed water layer can be calculated according to Reynolds (1984):

$$N_t = N_0 e^{V_T t / z_m} \qquad (4)$$

where N_0 is the starting concentration of particles, N_t is the concentration at time t (d), V_T is the terminal sinking velocity (m·d^{-1}), and z_m is the thickness of the mixed layer. If mixing is imperfect and only one mixing event per day is assumed, then

$$N_t = N_0 \left(1 - \frac{V_T}{z_m} \right)^t \qquad (5)$$

Equation (5) is probably more realistic because diurnal heating generates a temperature gradient in the epilimnion, and nocturnal cooling leads to homothermy within the epilimnion and subsequent mixing. In both equations, V_T and z_m have a compensatory effect on sedimentary losses. The higher the value of V_T, the more a species depends on deep mixing to minimize sinking losses. It follows that diatoms will require greater mixing depths for persistence within the near-surface layers than will other algae of similar size and form resistance.

In Situ Sedimentation Studies

Sedimentation in situ can be studied by use of sediment traps. They should be constructed so that they provide realistic estimates of the sedimentary particle flux in the water column. Walsby and Reynolds (1980) doubted that the number of particles passing through the aperture of any type of sediment traps was identical to the real particle flux. However, several authors (Hargrave & Burns 1978; Blomqvist & Håkanson 1981) have shown that cylindrical traps provide realistic estimates, whereas conical traps either underestimate (if the aperture is larger than the bottom area) or overestimate (if the bottom is larger than the aperture) the particle flux. In cylindrical traps it is important to avoid the loss of material already settled to the bottom of the traps as a result of resuspension by turbulence. This can best be avoided by high, narrow cylinders. Bloesch and Burns (1980) recommended a height–diameter ratio of 10:1 for turbulent lakes and 5:1 for calm lakes. Sinking velocity (V) can be estimated according to Hargrave and Burns (1979) by dividing the sedimentary flux (F, $N \cdot m^{-2} \cdot d^{-1}$) by the concentration of settling particles above the trap (C, $N \cdot m^{-3}$):

$$V = FC^{-1} \tag{6}$$

Equation (6) assumes a homogenous distribution of particles above the trap, both vertically and horizontally, which is certainly an oversimplification of natural conditions.

The annual average sinking velocities of selected phytoplankton species in Lake Constance are shown in Fig. 6-3 (Sommer 1984b). With the exception of small *Stephanodiscus hantzschii* (the only nannoplanktonic alga found in countable amounts in the traps), diatoms sink faster than nonsiliceous algae. Hypolimnetic sinking velocities in most case are one order of magnitude higher than epilimnetic ones. Survival during the sinking process was analyzed by comparing the total annual catch of intact cells in the 20-m traps (lower limit of euphotic zone) with the 120-m trap (20-m bottom distance). The quicker a species sank, the more intact cells reached the lowermost trap ($p < 0.001$). A "half-life" of 18 ± 4.5 d (95% confidence limits) for sinking cells could be derived, irrespective of taxonomic differences.

Sinking velocities vary not only interspecifically over several orders of magnitude but also within a single species temporally (Fig. 6-4). If sinking velocities were only hydrographically controlled (mixing depth, intensity of turbulence), the curves of the different species would have been parallel to each other. With the exception of *Asteri-*

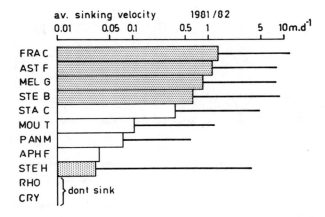

Fig. 6-3. Average sinking velocities, V_s, of principal phytoplankton species of Lake Constance (data from Sommer 1984b). Histogram bars : V_s from 0–20 m; Extended lines: V_s from 20–120 m depth. Diatoms (shaded): FRA C, *Fragilaria crotonensis;* AST F, *Asterionella formosa;* MEL G, *Melosira granulata;* STE B, *Stephanodiscus binderanus;* STE H, *Stephanodiscus hantzschii;* Nonsiliceous algae (white): STA C, *Staurastrum cingulum;* MOU T, *Mougeotia thylespora;* PAN M, *Pandorina morum;* APH F, *Aphanizomenon flos-aquae;* RHO, *Rhodomonas* spp.; CRY, *Cryptomonas* spp. The latter two are very common in Lake Constance but were never found in traps.

onella and *Fragilaria* in 1981 this was apparently not the case. However, clear minima of the sinking velocity occurred during periods of exponential increase, whereas stationary phases and decline phases were characterized by maxima of the sinking velocity. Those periods were also the periods of maximum incidence of dead cells.

The Lake Constance sedimentation study (Sommer 1984b) revealed three arguments supporting the assumption that cells with a reduced viability or even moribund cells are primarily destined for high sedimentary losses: (1) The calculated half-life of 18 days for sinking phytoplankton is short in relation to the well-known dark survival abilities of diatoms (see later discussion); (2) exponential growth phases coincide with minima of the sinking velocity; (3) the exponential loss rate due to sinking is temporally correlated with the loss rate due to cell death (see later section). The particular causes of mortality – silicon deficiency and fungal parasitism – seemed to make no difference. The literature contains both studies that support the assumption of a physiological dependence of sinking velocities (field studies: Eppley, Holmes, & Strickland 1967; Smayda 1970; Reynolds 1976; experimen-

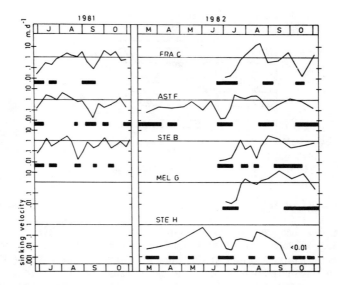

Fig. 6-4. Temporal variability of sinking velocities of Lake Constance diatoms (20-m trap). Abbreviations as in Fig. 6-3. Velocities are on a log scale; the horizontal line marks $1 \ m \cdot d^{-1}$. Black horizontal bars indicate periods of exponential population growth. Note the several order of magnitude varability of V_s and the coincidence of V_s minima with periods of exponential growth. Redrawn from Sommer (1984b).

tal studies: Jaworski, Talling, & Heaney 1981; Wiseman et al. 1983) and examples of the elimination of viable diatoms population by sudden decrease of the mixing depths, especially in small, sheltered lakes or artificial enclosures (Lund 1971; Knoechel & Kalff 1975; Reynolds, Wiseman, & Clarke 1984).

Physiological Control of Sinking?

The dependence of sinking velocities on the nutritional status of cells is difficult to interpret in terms of Stoke's law. In particular, the increase of sinking velocities as a consequence of silicon deficiency seems paradoxical. Silicon deficiency should lead to less silicified frustules and consequently to reduced density. In a detailed experimental study, Wiseman et al. (1983) were not able to explain the two- to sixfold difference in sinking rate between exponentially growing and senescent cells of *Asterionella formosa* either by changes in density or by changes in cell geometry.

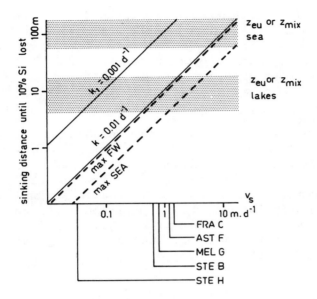

Fig. 6-5. Distance (*y*-axis) that a dead diatom cell can sink at a given sinking velocity (*x*-axis) until 10% of frustule silica is lost as predicted by the dissolution constant k_1 (diagonal lines). The broken lines characterize the maximum k_1 for freshwater diatoms (recalculated according to Bailey-Watts 1976b) and for marine diatoms (Kamitani 1982). Average epilimnetic sinking velocities of Lake Constance diatoms are shown on the *x*-axis. Abbreviations as in Fig. 6-3. Shaded areas characterize typical extension of euphotic zones or mixing zones in the sea (around 100 m) and in stratified lakes (around 10 m).

Observed apparent sinking velocities of diatom cells at times exceed calculated V_T, although they should theoretically be less. If we assume a density of water of $1 \text{ g} \cdot \text{mL}^{-1}$, a density of *Asterionella* of $1.2 \text{ g} \cdot \text{mL}^{-1}$, and a viscosity of water of $1 \times 10^{-3} \text{ kg} \cdot \text{m}^{-1} \cdot \text{s}^{-1}$, the terminal sinking velocity of a sphere of 11 μm diameter (the spherical equivalent of an *Asterionella* cell of 700 μm³) should be $1.13 \text{ m} \cdot \text{d}^{-1}$. Figures 6-4 and 6-5 include periods of time when higher values for the apparent sinking velocity were observed. At present, only speculations are possible. The only parameter of Stoke's formula that can conceivably be altered in a way to increase V_T is particle size. If cells or colonies stick together, they form larger particles. Aggregates of a few millimeters can often be seen in fresh hypolimnetic samples. Unfortunately, when samples are mixed for enumeration of cells, they tend to break apart, especially after fixation with Lugol's iodine. So it is difficult to determine the importance of such aggregates in nature. Flocculation of algal cells may be brought about when physiological changes lead to a breakdown of the electrostatic surface charge of cells (zeta potential, Grünberg & Soeder 1967) or by excretion of organic compounds, but these causes still remain to be adequately investigated.

RECYCLING OF SILICON

The solubility of diatom frustules in water and the undersaturation of silicic acid in natural waters facilitate recycling of diatom Si. However, the potential existence of a short Si cycle entirely within the productive zone, as has been described for C, N, and P, depends on the velocity of dissolution, on the sinking velocity, and on the mixing depth. Kamitani and Riley (1979) found a constant logarithmic decline of particulate diatom Si during the first phase of dissolution (described by the constant k_1), and a slower logarithmic decay during a second phase (described by k_2):

$$\frac{dC}{dt} = - k_1(C_0 - C) \qquad (7)$$

or

$$k_1 = -\ln\frac{[(C_0 - C)/C_0]}{t} \qquad (8)$$

where C_0 is the initial concentration of particulate diatom Si. The constant k_1 for a variety of marine diatoms ranges from 0.0018 to 0.031 day^{-1} (Kamitani and Riley 1979; Kamitani 1982). Data for freshwater diatoms can only be obtained by recalculation of literature data under the assumption that the first phase constant (k_1) remains valid during the entire experiment. According to the graphs in Kamitani and Riley (1979) and Kamitani (1982), this assumption seems justified if less than 50% of the initially present Si has been lost. Recalculation of the data of Bailey-Watts (1976b) gives values ranging from undetectable dissolution ($k_1 = 0$) to $k_1 = 0.012$ d^{-1} for freshwater planktonic diatoms (mixed samples). The data of Parker, Conway, and Yaguchi (1975) permit the calculation of k_1 values of 0.0039 to 0.0047 d^{-1}, which lie well within the range of Bailey-Watts' data.

A k_1 of 0.331 d^{-1} (maximum of Kamitani) infers a dissolution rate of 10% of Si after 3.5 days, or 50% after 23 days. A k_1 of 0.012 d^{-1} (maximum of Bailey-Watts) infers 10% dissolution after 8.8 days and 50% after 58 days. Figure 6-5 projects the implications of frustule solubility, sinking velocity (abscissa), and depth of the productive zone (or of the mixed surface layer). The diagonal lines for a given k_1 show how far a cell will have sunk (ordinate) before 10% of frustule silica are lost by dissolution. The stippled areas indicate typical depth ranges of the euphotic zone or the mixing depth in the sea (100 meters or more) and in lakes during stratified periods (a few meters to 20 meters). The annual average epilimnetic sinking velocities of dead diatom cells and colonies from Lake Constance (Sommer 1984b) are indicated at the abscissa. The figure shows that the frustules of the larger diatoms *(Stephanodiscus binderanus, Melosira granulata, Asterionella formosa, Fragilaria crotonensis)* will have lost only about 10% of their silicon after sinking a distance corresponding to the thickness of the productive layer typical for lakes, even if we assume the maximum k_1 found by Bailey-Watts (1976b) for freshwater diatoms. Only small *Stephanodiscus hantzschii* cells would lose about 97.5% of their Si during 10 m of sinking (probably less, because k_2 is smaller than k_1).

It follows that there will be a significant short cycle of Si if small nannoplankton diatoms are the main consumers of Si, but epilimnetic recycling will be negligible if large net plankton diatoms are the main consumers. In the second case, Si will be recycled either in the hypolimnion (Parker et al. 1975 for Lake Michigan) or in the sediment (Tessenow 1966, several north German lakes). Although Tessenow (1966) found that the mechanical breakdown of frustules by zooplankton grazers can increase the velocity of dissolution two- to threefold, Reynolds et al. (1982) demonstrated that large diatoms tend to be less edible for filter feeders than are nannoplankton ones. Moreover, Ferrante and Parker (1977) found that formation of fecal pellets by zooplankton retards dissolution of the frustule fragments and increases sinking velocities. In the sea the order of magnitude difference in mixing depths to fresh waters is sufficient by itself to permit the existence of a short cycle of Si dissolution (Nelson & Goering 1977; Kamitani 1979).

The epilimnetic short cycle of Si that occurs in the sea and among nannoplankton diatoms in lakes presumably runs much more slowly than the short cycles of phosphorus and nitrogen (Dugdale & Goering 1967), although I am unaware of comparative measurements having been made. This difference has to be kept in mind if chemostat competition experiments are compared with lake data. The Si:P or Si:N ratio (see Table 6-3 and associated text) calculated from ambient concentrations is *n*-fold higher than the ecologically relevant ratio of supply rates (Tilman 1982), where *n* is the proportion of the cycling velocities of the respective nutrients. In a mixed plankton, where diatoms sink faster than other algae, the loss of Si from the productive zone will be proportionately greater than the loss of other nutrients. As a consequence, the Si:P and the Si:N ratio decline during the course of a diatom bloom, causing increasing competitive disadvantage for the diatoms. Regeneration of orthosilicic acid in the euphotic zone can mainly be brought about by vertical transport from deeper layers (i.e., by increased vertical mixing). It is interesting from an evolutionary point of view that both the demographic consequences of sedimentation and the slow recycling of Si lead to similar consequences for the hydrographic requirements for mixing of diatoms.

SEXUALITY, SPORE FORMATION, AND PERENNATION

Sexuality of diatoms has been studied for a long time (for reviews see Lewin & Guillard 1963; Geitler 1973; Drebes 1977). It was discovered that the formation of the zygote (auxospore) is a mechanism to restore the initial cell size after continued cell divisions have led to a progressive size decrease (McDonald–Pfitzer rule). Environmental stimuli can only induce sexuality within a defined size range of cells (von Stosch 1965).

Diatom sexuality is very costly from a demographic point of view. The two meiotic divisions give rise to only one or two gametes per diploid mother cell. After fusion of the gametes only one or two zygotes will result. Then it takes two further mitoses to produce one initial cell from one auxospore because diatom wall formation is only possible following a nuclear division (Geitler 1963). Thus, four consecutive division events produce only one or two cells from the two mother cells, whereas four consecutive vegetative divisions would have produced 32 cells. It should therefore be expected that sexuality is restricted to periods when resource limitation prevents further population growth. However, this is not demographically possible, because only in rapidly dividing populations may many cells reach the size of sex inducibility (Lewin & Guillard 1963). Therefore, the diatom zygote, in contrast to that of many other algal groups, is not suitable as a resting stage, and auxospores continue to develop directly into new vegetative cells without an interruption of growth by prolonged dormant periods.

The high demographic costs and the unsuitability of the zygote as a resting stage make it no surprise that many planktonic diatoms have circumvented size decrease and sexuality, probably by some elasticity of the girdle bands. Especially among members of the family Fragilariaceae, the most important family of freshwater plankton diatoms, sexuality is absent in the pelagic species (Geitler 1973).

Recently Lewis (1984) has drastically revised the traditional way of interpreting the interdependence of size reduction and sexuality. Since the example of at least some species show that the McDonald–Pfitzer rule of continuous diminution can be circumvented, it cannot longer be argued that sexuality is an expensive but inevitable consequence of the structure of the diatom frustule. However, if we assume that the gains (recombination) and losses (demographic costs) of sexuality are balanced in a way that makes the occurrence of sexuality at long, supra-annual intervals most advantageous, then the continuous size decrease of cells is an excellent internal timer for sexuality working over numerous generations.

Asexual resting stages have been mainly reported for marine planktonic diatoms. Resting spores (Hargraves & French 1983) differ in valve structure from vegetative cells, whereas resting cells (Anderson 1975, 1976) differ only in plasmatic structure. Germination from resting stages is usually triggered by addition of nutrient plus light, and resting stage formation is triggered by nutrient (especially N) deficiency (Hargraves & French 1983). Among freshwater planktonic diatoms, only the few *Rhizosolenia* spp. (a predominantly marine genus) have resting spores.

Many diatom species lack specialized resting stages, and this is especially true for the majority of freshwater planktonic forms. The only prominent example of distinctive resting cells in freshwater plankton diatoms is *Melosira italica* (Lund 1954), for which the resting stage plays an important role in the seasonal cycle, being able to survive about three years under dark, anaerobic conditions. Most freshwater species, however, pass unfavorable periods by maintaining a low, slowly declining stock of vegetative cells either suspended in the water or lying on the sediment surface. The relative importance of pelagic survivors certainly is variable from lake to lake, depending mainly on morphometric and hydrographic features. During stratified periods, only littoral sediment survivors are able to recolonize the lake. It follows that pelagic survival will be more important in deep lakes with steep slopes, due to reduced vertical mixing zones, whereas in shallow lakes or in lakes with at least important shallow portions, survival on the sediment surface may be more important.

It seems that diatoms without specialized resting stages do not tolerate anaerobic conditions (Lund 1954). However, prolonged dark survival under aerobic, cold-water conditions has been proved for a variety of marine diatoms (Smayda & Mitchell-Innes 1974; Antia 1976), and this also seems probable for freshwater diatoms. In deep, monomictic lakes where the average light intensity of the mixed water layer is too low to support algal growth during winter circulation, diatoms tend to stop growth later in the year and to decline in population size less rapidly than many other algae (e.g., Lake Constance, Sommer 1981).

CONCLUDING DISCUSSION

We have considered aspects of mineral nutrition of diatoms, the relationship between size reduction and sexuality, the problem of perennation, the role of sedimentary losses, and the problem of silica recycling. These aspects of diatom ecology have been selected because they are directly or indirectly connected with the siliceous frustule. Before we attempt to define the environmental conditions under which diatoms may be successful, we must emphasize that, as is the case for most other phytoplankton species, diatoms are photosynthetic organisms. They thus depend on light, their metabolic rates respond to temperature, and they are susceptible to a variety of nonsedimentary loss processes (e.g., grazing, parasitism, hydrological washout).

In seasonal environments such as temperate lakes, the first step in answering the question: "Under what conditions are diatoms successful?" is to consider the question: "When are diatoms successful?" The next step is the analysis of the concomitant processes that determine the balance of population increase and decrease.

A Method for Distinguishing Growth and Loss Processes in Natural Diatom Populations

The seasonal fluctuations of algal populations are the combined result of growth and loss. A moderate increase in population density may as easily be the result of moderate reproduction and negligible losses or the result of vigorous reproduction and moderate losses. In order to evaluate the relative importance of growth-controlling factors (light, temperature, nutrients) and of loss factors (grazing, sinking, physiological death), it is necessary to separate growth (= reproduction) rates and loss rate. This is easier for diatom species than for many other taxa, because the silica frustule remains visible and countable after cell death.

The observed rate of increase k can easily be calculated:

$$N_2 = N_1 e^{k(t_2 - t_1)} \quad \text{or} \quad k = \frac{\ln N_2 - \ln N_1}{t_2 - t_1}$$

The observed rate of increase is composed of the growth (= reproduction) rate μ and the loss rate λ:

$$k = \mu - \lambda \tag{9}$$

In organisms reproducing by binary fission, μ can be calculated from the fraction of cells undergoing mitosis (p_D) per unit time, usually per day:

$$\mu = \ln(1 + p_D) \tag{10}$$

This method is particularly applicable for organisms with synchronous or phased cell divisions, which is typical for many eukaryotic phytoplankton. It requires close interval sampling over a 24-h cycle and thorough knowledge of the duration of the morphologically distinct division stage chosen for analysis of p_D (see McDuff & Chisholm 1981). Alternatively, μ can be calculated from Si consumption if a diatom bloom is monospecific and import of Si into the photic zone can be neglected (Reynolds et al. 1982). In absence of direct analyses, μ can also be estimated by complete analysis of losses:

$$\lambda = \gamma + \sigma + \pi + \delta + \tau \cdots \tag{11}$$

where γ is the grazing rate, σ is the rate of sinking losses, π is the loss rate due to parasitism, δ is the loss rate due to physiological death, and τ is the loss rate due to hydrological washout. In some cases there may be good reason to consider one or more negligible, which facilitates analysis.

The grazing rate can be calculated according to Reynolds et al. (1982) from the proportional volume of water that is filtered by the zooplankton per day (G) and the probability of a cell contained in the filtered volume being ingested and killed (Φ, grazing probability index according to Vanderploeg & Scavia 1979):

$$\gamma = \Phi G \tag{12}$$

In fact, both Φ and G are difficult to determine (see Thompson, Ferguson, & Reynolds 1982; Ferguson, Thompson, & Reynolds 1982). Even if monospecific zooplankton of uniform size are used in grazing experiments (Knisely & Geller 1986), the confidence limits of Φ are uncomfortably broad. However, in general for filter feeders in Lake Constance (*Daphnia hyalina, D. galeata,* and *Eudiaptomus gracilis*) Φ tends to be high (> 0.5) for small *Stephanodiscus hantzschii,* moderate (around 0.25) for filamentous diatoms like *Melosira grannulata, Stephanodiscus binderanus* as well as for *Asterionella formosa,* and low (< 0.1) for *Fragilaria crotonensis.* For raptorial feeders *(Cyclops),* Φ trends are reversed, with preference for the larger particles.

Sedimentary losses can either be estimated from trap catches (N_s) or from terminal sinking velocities (V_T) and mixing depths (z_m), provided that the productive zone does not extend beyond the pycnocline. N_s is the integral of instantaneous sedimentary losses during the exposure period t_1 to t_2 (Sommer 1984a,b):

$$N_s = \int_{t_1}^{t_2} \sigma N_1 \, e^{k(t_2 - t_1)} \, dt \tag{13}$$

or

$$\sigma = \frac{N_s k}{N_1(e^{k(t_2-t_1)} - 1} \tag{14}$$

If the epilimnion is perfectly turbulent, σ can also be calculated as (Reynolds 1984)

$$\sigma = \frac{V_T}{z_m} \tag{15}$$

or if only one mixing event per unit time is assumed (Reynolds & Wiseman 1982), then

$$\sigma = \frac{\ln(1 - V_T/z_m)}{t} \tag{16}$$

The death rate δ can be calculated as the increase of dead cells accumulating in sediment traps plus the increase in the water column as follows (Sommer 1984a,b):

$$\delta = \frac{N_D k}{N_1(e^{k(t_2-t_1)} - 1)} \tag{17}$$

According to analyses by autoradiography (Knoechel & Kalff 1978), cells in which all chloroplasts are disintegrated can be counted as "dead." If frustules are excreted intact, herbivorous zooplankton σ also covers some part of γ.

Parasitism by aquatic fungi is a well-known cause for diatom mortality (Van Donk 1983). However, as long as the time interval between infection and death is unknown, it is impossible to calculate π from the proportion of cells showing signs of parasitism. At present, it is more convenient to treat π as a fraction of δ. Since fungal parasitism does not destroy the frustule, it is often difficult to decide whether an empty frustule is a victim of parasitism or some other mortality-induc-

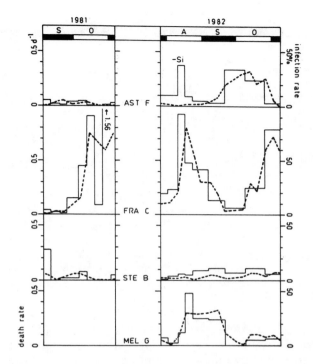

Fig. 6-6. Death rate (δ) of diatoms (bar diagrams) in response to percentage of populations infected by chytrids ("infection rate," broken line). Abbreviations as in Fig. 6-3. Note that only one mortality maximum (*Asterionella,* August 1982) does not coincide with an infection maximum. This mortality maximum is ascribed to Si deficiency. Redrawn after Sommer (1987).

ing factor. Fig. 6-6 shows how important parasitism can be in diatom mortality. With the exception of Si-deficiency-induced mortality of *Asterionella formosa* in August 1982, all mortality maxima in Lake Constance coincide with maximal proportions of parasitized cells.

The loss rate due to washout, τ, can be calculated by dividing the discharge per unit time of the outlet by the volume of the mixed layer. In large lakes it will be usually negligible.

The equations used in this section assume horizontal homogeneity in the distribution of organisms (i.e., point samples are representative for the respective water layer). Most planktologists have realized that the homogeneity assumption is unrealistic; however, logarithmic transformation of population density data dampens the numerical

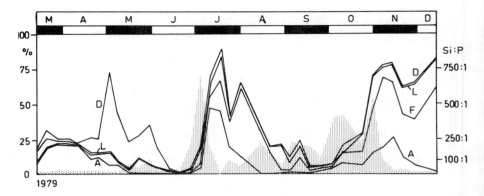

Fig. 6-7. Time-delayed responses of diatoms (percentage of total phytoplankton biomass) to Si:P ratios (shaded area) in Lake Constance euphotic zone. A, *Asterionella formosa;* F, Fragilariaceae; L, Large diatoms (netplankton); D, all diatoms. Note that the only diatom maximum not preceded by a Si:P maximum (early May) is composed of nannoplanktic centric species (mainly *Stephanodiscus hantzschii*), which have low Si:P optima (see Table 6-3).

error in rate calculations. At the commonly used sampling interval of a week, the difference between a 10-fold and a 20-fold increase in cell density is reflected by only a difference of $k = 0.33$ d^{-1} and $k = 0.43$ d^{-1}, respectively. Trap catches are less sensitive to patchiness in plankton distribution because the exposure of a trap over a period of several days in combination with water movements above the trap will integrate over most of the patchiness.

Under What Conditions Are Diatoms Successful?

In an environment with temporally variable species composition, dominance over competitors is achieved by a better growth-minus-loss balance. Usually, maximum dominance will follow the occurrence of the most favorable combination of growth and loss factors with a certain delay because it takes some time until an "inoculum" becomes a population peak. Among the growth factors, high Si:P ratios have been shown to favor diatoms over other algae. However, due to the slower recycling of silicon, diatom growth will quickly reduce the Si:P ratio (Fig. 6-7) and change the environmental conditions in favor of other taxa unless deep mixing increases the import of Si from deeper water layers.

Not only the necessity for silica replenishment but also the avoidance of sedimentary losses make diatoms depend on vertical turbulent

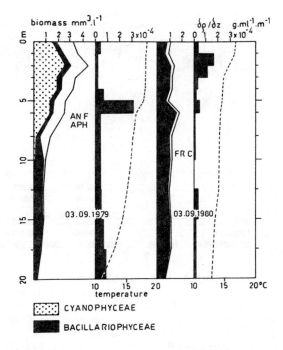

Fig. 6-8. Differences in composition and vertical distribution of late summer phytoplankton in relation to thermal stability in 1979 (left-hand panels) and 1980 (right-hand panels). Biomass of phytoplankton as cell volume: *stippled,* Cyanophyta (mainly *Aphanizomenon* and *Anabaena*); *shaded,* Bacillariophyceae (1979 mainly *Melosira granulata* and *Stephanodiscus binderanus,* 1980 mainly *Fragilaria crotonensis*); *white,* others (1979 mainly *Dinobryon*). Broken line: water temperature. Black bars: vertical density gradient. Note the pycnocline from 5 to 6 m in 1979. Redrawn after Sommer (1987).

exchange of water. The experiments of Reynolds et al. (1984) have clearly demonstrated that artificial mixing in large enclosures ("Lund tubes") could either prolong diatom dominance or induce new increases of diatoms. The same can be shown by analysis of interannual differences in the otherwise regular succession of phytoplankton in Lake Constance (Sommer 1981, 1985b). During thermally stable years the July peak of diatoms is ended by Si depletion and followed first by a maximum of *Ceratium hirundinella* and then by cyanobacteria together with *Dinobryon* spp. During that period a low biomass of diatoms persists in the lower half of the euphotic zone. During less stable summers there will be no near-surface stratum of *Cyanophyta,* and diatoms dominate throughout the entire euphotic zone (Fig. 6-8).

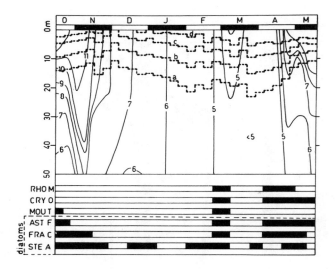

Fig. 6-9. Maximum depth for growth of various autumn and winter algae (thick broken lines) shown together with depth–time diagram of temperature). Maximum depths for growth have been calculated weekly based on experimentally determined minimum light requirements, vertical extinction coefficients and average weekly surface irradiance. The lines represent from bottom to top: (a) *Fragilaria crotonensis* and *Stephanodiscus astraea* (2.2 J·cm^{-2}·d^{-1}), (b) *Asterionella formosa* (13 J·cm^{-2}·d^{-1}), (c) *Mougeotia thylespora* and *Cryptomonas ovata* (53 and 63 J·cm^{-2}·c^{-1}), (d) *Rhodomonas minuta* (138 J·cm^{-2}·d^{-1}).

Because of the absence of a pronounced thermocline in Lake Constance, temperature profiles usually fail to show clear year to year differences in stratification, but conversion of temperature data into density shows either a pronounced density gradient from 5–6-m depth during 1979 (cyanobacteria dominant) or the absence of such a gradient in 1980 (diatoms dominant).

Deep mixing is not only advantageous for sinking avoidance and the nutrition of diatoms, but it also leads to exposure to lower average light intensities than would permanent residence in near-surface water entrained by strong thermal stratification. It can therefore be postulated that diatoms should either be (1) restricted to waters with low light attenuation, such as oligotrophic waters, and well mixed seasons in more eutrophic waters, or (2) have lower minimum light requirements than other algae. The latter can be tested by simple exposure experiments with autumn and winter phytoplankton from Lake Con-

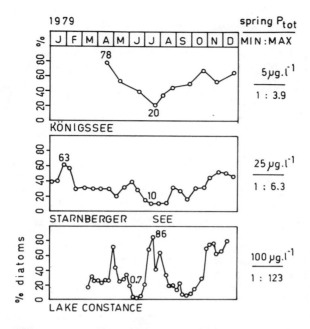

Fig. 6-10. Seasonal variability of diatom contribution to total phytoplankton biomass in three deep South German lakes of different trophic status, as indicated by total phosphorus during spring overturn. Minimum to maximum ratio refers to percentage of total biomass contributed by diatoms; minima and maxima are also shown numerically in graphs (data for Königsee: Siebeck 1982; data for Starnberger See: Lenhart & Steinberg 1982).

stance (Fig. 6-9). Natural phytoplankton samples were diluted with hypolimnetic water in order to avoid nutrient limitation, and the phytoplankton suspensions were exposed in bottles at different depth. The diatoms consistently demonstrated growth at greater depths than did other algae. The experimental findings were consistent with a later cessation of autumn growth and a slower decline during winter, the two main causes for the predominance of diatoms in winter plankton.

The other proposed postulate becomes an argument centering upon water transparency in regulating diatom abundance. High water transparency not only compensates optically for increased mixing depth, but it also permits a longer residence within the euphotic zone for a sinking alga under stable conditions. Because both water transparency and Si:P or Si:N ratios tend to decrease with cultural eutrophication, it is tempting to predict a decreasing importance of diatoms. However, water transparency and nutrient supply are highly seasonal in eutrophic lakes, including transient periods of conditions favorable for diatoms. I therefore only postulate that the relative importance of diatoms should be more seasonal in more eutrophic lakes. Figure 6-10 demonstrates this to be true for three deep central European lakes, each with a different trophic status. Both proposals for explaining diatom seasonal distribution in lake plankton would thus appear to be supported by available data.

REFERENCES

Anderson, O. R. (1975). The ultrastructure and cytochemistry of resting cell formation in *Amphora coffaeformis* (Bacillariophyceae). *J. Phycol.*, 11, 272–81.

    (1976). Respiration and photosynthesis during resting cell formation in *Amphora coffaeformis* (Ag.) Kutz. *Limnol. Oceanogr.*, 21, 452–6.

Antia, N. J. (1976). Effects of temperature on the darkness survival of marine microplanktonic algae. *Microb. Ecol.*, 3, 41–54.

Aston, S. R. (1983). Natural water and atmospheric chemistry of silicon. In: *Silicon Geochemistry and Biogeochemistry*, ed. S. R. Aston, pp. 77–100. London: Academic Press.

Bailey-Watts, A. E. (1976a). Planktonic diatoms and some diatom-silica interactions in a shallow eutrophic Scottish loch. *Freshwat. Biol.*, 6, 69–80.

    (1976b). Planktonic diatoms and silica in Loch Leven, Kinross, Scotland: a one month silica budget. *Freshwat. Biol.*, 6, 203–13.

Barber, H. G. and Haworth, E. Y. (1981). A guide to the morphology of the diatom frustule. *Freshwat. Biol. Assoc. Scient. Publ.*, 41.

Bloesch, J. and Burns, N. M. (1980). A critical review of sedimentation trap technique. *Schweiz. Z. Hydrol.*, 42, 15–55.

Blomqvist, S. and Håkanson, L. (1981). A review on sediment traps in aquatic environments. *Arch. Hydrobiol.*, 92, 101–32.

Burmaster, D. E. (1979). The continuous culture of phytoplankton: mathematical equivalence among three steady-state models. *Am. Nat.*, 113, 123–34.

Van Donk, E. (1983). The effect of fungal parasitism on the succession of diatoms in Lake Marseeveen I. (The Netherlands). *Freshwat. Biol.*, 13, 241–51.

Drebes, G. (1977). Sexuality. In: *The Biology of Diatoms*, ed. D. Werner, pp. 250–83. Oxford: Blackwell Scientific.

Droop, M. R. (1973). Some thoughts on nutrient limitation in algae. *J. Phycol.*, 9, 264–72.

Dugdale, R. C. (1967). Nutrient limitation in the sea: dynamics, identification and significance. *Limnol. Oceanogr.*, 12, 685–95.

Dugdale, R. C. and Goering, J. J. (1967). Uptake of new and regenerated forms of nitrogen in primary productivity. *Limnol. Oceanogr.*, 12, 196–206.

Einsele, W. and Grim, J. (1938). Über den Kieselsäuregehalt planktischer Diatomeen und dessen Bedeutung für einige Fragen ihrer Ökologie. *Z. Botanik*, 23, 545–90.

Eppley, R. W., Holmes, R. W., and Strickland, J. D. H. (1967). Sinking rate of marine phytoplankton measured with fluorometer. *J. Exp. Mar. Biol. Ecol.*, 1, 191–208.

Ferguson, A. J. D., Thompson, J. M., and Reynolds, C. S. (1982). Structure and dynamics of zooplankton communities maintained in closed systems, with special reference to algal food supply. *J. Plank. Res.*, 4, 523–43.

Ferrante, J. G. and Parker, J. I. (1977). Transport of diatom frustules by copepod fecal pellets to the sediments of Lake Michigan. *Limnol. Oceanogr.*, 22, 92–8.

Geitler, L. (1963). Alle Schalenbildungen der Diatomeen treten als Folge von Zell – oder Kernteilungen auf. *Ber. dt. bot. Ges.*, 75, 393–6.

—— (1973). Auxosporenbildung und Systematik bei pennaten Diatomeen und die Cytologie von *Cocconeis*-Sippen. *Öst. Bot. Z.*, 122, 299–321.

Goldman, J. C., McCarthy, J. J., and Peavey, D. G. (1979). Growth rate influence on the chemical composition of phytoplankton in oceanic waters. *Nature*, 279, 210–15.

Grünberg, U. and Soeder, C. J. (1967). Zeta-Potential Synchron kultivierter Chlorellen. *Naturwiss.*, 54, 205–6.

Guillard, R. R. L., Kilham, P., and Jackson, T. A. (1973). Kinetics of silicon-limited growth in the marine diatom *Thalassiosira predona* Hassle and Heimdal. (*Cyclotella nana* Hustedt). *J. Phycol.*, 9, 233–7.

Happey, C. M. (1970). The effects of stratification on phytoplanktonic diatoms in a small body of water. *J. Ecol.*, 58, 635–51.

Hargrave, B. T. and Burns, N. M. (1979) Assessment of sediment trap collection efficiency. *Limnol. Oeanogr.*, 24, 1124-36.

Hargraves, P. E. and French, F. W. (1983). Diatom resting spores: significance and strategies. In: *Survival Strategies of the Algae*, ed. G. A. Fryxell, pp. 49–69. Cambridge: Cambridge University Press.

Holm, N. P. and Armstrong, D.E. (1981). Role of nutrient limitation and competition in controlling the populations of *Asterionella formosa* and *Microcystis aeruginosa* in semicontinuous culture. *Limnol. Oceanogr.*, 26, 622–34.

Jaworski, G. H. M., Talling, J. F., and Heaney, S. I. (1981). The influence of carbon dioxide depletion on growth and sinking rate of two planktonic diatoms in culture. *Br. Phycol. J.*, 16, 395–410.

Kamitani, A. (1971). Physical and chemical characteristics of biogenic silica. *Mar. Biol.*, 8:89–95.

—— (1982). Dissolution rates of silica from diatoms decomposing at various temperatures. *Mar. Biol.*, 68, 91–6.

Kamitani, A. and Riley, J. P. (1979). Rate of dissolution of diatom silica walls in seawater. *Mar. Biol.*, 55, 29–35.

Kilham, S. S. (1985). Silicon and phosphorus growth kinetics and competitive interactions between *Stephanodiscus minutus* and *Synedra* sp. *Verh. Internat. Verein. Limnol.*, 22, 435–9.

Knisely, K. and Geller, W. (1986). Selective feeding of four zooplankton species on natural lake phytoplankton. *Oecologia,* 69, 86–94.

Knoechel, R. and Kalff, J. (1975). Algal sedimentation the cause of diatom blue-green succession. *Verh. Internat. Verein. Limnol.,* 19, 745–54.

———(1978). An in situ study of the productivity and population dynamics of five freshwater plankton diatom species. *Limnol. Oceanogr.,* 23, 195–218.

Lenhart, B. and Steinberg, C. (1982). Zur Limnologie des Starnberger Sees. *Bayer. Landesamt f. Wasserwirtschaft, München,* 1–284.

Lewin, J. C. and Guillard, R. R. L. (1963). Diatoms. *Ann. Rev. Microbiol.,* 17, 373–408.

Lewis, W. M. (1984). The diatom sex clock and its evolutionary significance. *Am. Nat.,* 123, 73–80.

Lund, J. W. G. (1950). Studies on *Asterionella formosa* Hass. II. Nutrient depletion and the spring maximum. *J. Ecol.,* 38, 1–35.

———(1954). The seasonal cycle of the plankton diatom *Melosira italica* (Ehr.) Kutz. subsp. *subarctica.* Müll. *J. Ecol.,* 42, 151–79.

———(1971). An artificial alteration of the seasonal cycle of the plankton diatom *Melosira italica* subsp. *subarctica* in an English lake. *J. Ecol.,* 59, 521–33.

McDuff, R. and Chisholm, S. W. (1981). The calculation of in situ growth rates of phytoplankton populations from fractions of cells undergoing mitosis: a clarification. *Limnol. Oceanogr.,* 27, 783–8.

Moed, J. R. (1973). Effect of combined action of light and silicon depletion on *Asterionella formosa* Hass. *Verh. Internat. Verein. Limnol.,* 18, 1367–74.

Moed, J. R., Hoogveld, H. L., and Apeldoorn, W. (1976). Dominant diatoms in Tjeukemeer (The Netherlands). II. Silicon depletion. *Freshwat. Biol.,* 6, 355–62.

Monod, J. (1950). La technique de la culture continue: theorie et applications. *Ann. Inst. Pasteur Lille,* 79, 390–410.

Munk, W. H. and Riley, G. A. (1952). Adsorption of nutrients by aquatic plants. *J. Mar. Res.,* 11, 215–40.

Nelson, D. M. and Goering, J. J. (1977). Near surface silica dissolution in the upwelling region of northwest Africa. *Deep Sea Res.,* 6, 351–67.

Paasche, E. (1973a). Silicon and the ecology of marine plankton diatoms. I. *Thalassiosira pseudonana* growth in a chemostat with silicate as limiting nutrient. *Mar. Biol.,* 19, 117–26.

———(1973b). Silicon and the ecology of marine plankton diatoms. II. Silicate uptake kinetics in five diatom species. *Mar. Biol.,* 19, 262–9.

———(1975). Growth of the plankton diatom *Thalassiosira nordenskiöldii* Cleve at low silicate concentrations. *J. Exp. Mar. Biol. Ecol.,* 75, 173–83.

———(1980). Silicon. In: *The Physiological Ecology of Phytoplankton,* ed. I. Morris, pp. 259–84. Oxford: Blackwell Scientific.

Parker, J. I., Conway, H. L., and Yaguchi, E. M. (1975). Dissolution of diatom frustules and recycling of amorphous silicon in Lake Michigan. *J. Fish. Res. Board Can.,* 34, 551–4.

Reynolds, C. S. (1976). Sinking movements of phytoplankton indicated by a simple trapping method. II. Vertical activity ranges in a stratified lake. *Br. Phycol. J.,* 11, 293–303.

(1984). *The Ecology of Freshwater Phytoplankton.* Cambridge: Cambridge University Press.

Reynolds, C. S., Thompson, J. M., Ferguson, A. J. D., and Wiseman, S. W. (1982). Loss processes in the population dynamics of phytoplankton maintained in closed systems. *J. Plank. Res.,* 4, 561–600.

Reynolds, C. S. and Wiseman, S. W. (1982). Sinking losses of phytoplankton in closed limnetic systems. *J. Plank. Res.,* 4, 489–522.

Reynolds, C. S., Wiseman, S. W., and Clarke, M. J. O. (1984). Growth and loss-rate responses of phytoplankton to intermittant artificial mixing and their potential application to the control of planktonic algal biomass. *J. Appl. Ecol.,* 21, 11–39.

Siebeck, O. (1982). Der Königsee, eine limnologische Projektstudie. *Abt. Limnologie d. Zoolog. Inst. Univ. München,* 131 pp.

Smayda, T. J. (1970). The suspension and sinking of phytoplankton in the sea. *Oceanogr. Mar. Biol. Ann. Rev.,* 8, 353–414.

Smayda, T. J. and Mitchell-Innes, B. (1974). Dark survival of autotrophic, planktonic marine diatoms. *Mar. Biol.,* 25, 195–202.

Smith, R. E. and Kalff, J. (1983). Competition for phosphorus among co-occurring freshwater phytoplankton. *Limnol. Oceanogr.,* 28, 448–64.

Sommer, U. (1981). The role of r- and K-selection in the succession of phytoplankton in Lake Constance. *Acta Oecologia/Oecol. Gener.,* 2, 327–42.

(1983). Nutrient competition between phytoplankton species in multispecies chemostat experiments. *Arch. Hydrobiol.,* 96, 399–416.

(1984a). Population dynamics of three planktonic diatoms in Lake Constance. *Holarct. Ecol.,* 7, 257–61.

(1984b). Sedimentation of principal phytoplankton species in Lake Constance. *J. Plank. Res.,* 6, 1–14.

(1985a). Comparison between steady state and non-steady state competition: Experiments with natural phytoplankton. *Limnol. Oceanogr.,* 30, 337–48.

(1985b). Seasonal succession of phytoplankton in Lake Constance. *BioScience,* 35, 351–7.

(1987). Factors controlling the seasonal variation in phytoplankton species composition – a case study for a deep, nutrient-rich lake. *Prog. Phycol. Res.,* 5, 110–73.

Sommer, U. and Stabel, H. H. (1983). Silicon consumption and population density changes of dominant planktonic diatoms in Lake Constance. *J. Ecol.,* 71, 119–30.

Spencer, S. R. (1983). Marine biogeochemistry of silicon. In: *Silicon Geochemistry and Biogeochemistry,* ed. S. R. Aston, pp. 101–42. London: Academic Press.

von Stosch, H. A. (1965). Manipulierung der Zellgröße von Diatomeen im Experiment. *Phycologia,* 5, 21–44.

Strickland, J. D. H. and Parsons, T. R. (1968). A practical handbook of seawater analysis. *Bull. Fish. Res. Board Can.,* 169, 1–311.

Tessenow, U. (1966). Untersuchungen über den Kieselsäuregehalt der Binnengewässer. *Arch. Hydrobiol. Suppl.,* 32, 1–136.

Thompson, J. M., Ferguson, A. J. D., and Reynolds, C. S. (1982). Natural filtration rates of zooplankton in a closed system: The derivation of a community grazing index. *J. Plank. Res.,* 4, 549–59.

Tilman, D. (1977). Resource competition between planktonic algae: An experimental and theoretical approach. *Ecology,* 58, 338–48.

(1981). Test of resource competition theory using four species of Lake Michigan algae. *Ecology,* 62, 802–15.

(1982). *Resource Competiton and Community Structure.* Princeton: Princeton University Press.

Tilman, D. and Kiesling, R. L. (1984). Freshwater algal ecology. Taxonomic trade-offs in the temperature dependence of nutrient competitive abilities. In: *Current Perspectives in Microbial Ecology,* eds. M. J. Klug and C. A. Reddy, pp. 314–9. Amer. Soc. Microbiol.

Tilman, D. and Kilham, S. S. (1976). Phosphate and silicate growth and uptake kinetics of the diatoms *Asterionella formosa* and *Cyclotella meneghiniana* in batch and semicontinuous culture. *J. Phycol.,* 12, 375–83.

Tilman, D., Mattson, M., and Langer, S. (1981). Competition and nutrient kinetics along a temperature gradient: An experimental test to a mechanistic approach to niche theory. *Limnol. Oceanogr.,* 26, 1020–53.

Vanderploeg, H. A. and Scavia, D. (1979). Calculation and use of selectivity of feeding: zooplankton grazing. *Ecol. Model.,* 7, 135–49.

Walsby, A. E. and Reynolds, C. S. (1980). Sinking and floating. In: *The Physiological Ecology of Phytoplankton,* ed. I. Morris, pp. 371–411. Oxford: Blackwell Scientific.

Werner, D. (1966). Die Kieselsäure im Stoffwechsel von *Cyclotella cryptica,* Reimann, Lewin und Guillard. *Arch. Mikrobiol.,* 55, 278–308.

(1977a). Silicate metabolism. In: *The Biology of Diatoms,* ed. D. Werner, pp. 110–49. Oxford: Blackwell Scientific.

(1977b). *The Biology of Diatoms.* Oxford: Blackwell Scientific.

Wiseman, S. W., Jaworski, G. H. M., and Reynolds, C. S. (1983). Variability in sinking rate of the freshwater diatom *Asterionella formosa* Hass: The influence of the excess density of colonies. *Br. Phycol. J.,* 18, 425–32.

Chapter 7

GROWTH AND REPRODUCTIVE STRATEGIES OF FRESHWATER BLUE-GREEN ALGAE (CYANOBACTERIA)

HANS W. PAERL

INTRODUCTION

Of the major freshwater phytoplankton groups, blue-green algae (cyanobacteria) are highly individualistic from physiological, taxonomic, and ecological perspectives. Perhaps the most clear-cut distinction between cyanobacteria and other phytoplankton taxa rests in the fact that the former are prokaryotes, exhibiting well-defined morphological and physiological similarities to the eubacteria (Stanier 1977). Among prokaryotes, however, cyanobacteria reveal a set of unique physiological and morphological characteristics, the most outstanding of which are O_2-evolving photosynthesis (photosystems I and II), associated biliprotein accessory pigments, the apparent lack of flagellar motility, and, from an ecological perspective, the tendency to periodically accumulate as high-density "bloom" populations in surface and near-surface mesotrophic and eutrophic waters (Fogg 1969; Fogg et al. 1973; Carr & Whitton 1982).

In examining growth and reproductive strategies, our focus will be on physiological, morphological, and ecological mechanisms respon-

I would like to thank B. Bright and J. Garner for their assistance with manuscript preparation and V. Page and H. Page for illustrations. I am grateful to R. Fulton for discussion and information regarding trophic impacts of cyanobacteria, and to C. S. Reynolds for critically reviewing this chapter. Research activities described in this chapter were supported by National Science Foundation grants BSR 8314702, OCE 8500 740, North Carolina Sea Grant Project R/MER-1, and the University of North Carolina Water Resources Research Institute Grant A-122 NC.

sible for promoting coexistence and dominance by cyanobacteria in planktonic and benthic freshwater communities. Some of these mechanisms are unique to cyanobacteria, and others are shared with other phytoplankton groups. Particular attention will be paid to ecophysiological adaptations, which have been responsible for the remarkable success of these prokaryotic photoautotrophs during the evolution of phytoplankton communities.

MORPHOLOGICAL DIVERSITY OF PLANKTONIC CYANOBACTERIA

Intracellular prokaryotic characteristics, such as the lack of defined nuclei, chloroplasts, and organelles, are commonly shared among planktonic cyanobacteria. However, cellular morphological features, including structurally (and functionally) differentiated cells, cell shapes, dimensions, and solitary versus colonial habits, display a wide spectrum of diversification among members of this phytoplankton group. Depending on physiological requirements and environmental constraints during growth phases, many cyanobacterial genera are able to regulate magnitudes of colonial and filamentous development as well as cell differentiation. Although some limited alterations in cell sizes can be observed among diverse species exposed to varying degrees of nutrient availability, typically the freshwater cyanobacteria can be segregated according to size categories, shapes, and aggregative (colonial/filamentous) behavior. Four basic morphological groups can be characterized using those criteria: (1) unicellular picoplankton (0.2–3 μm in diameter), (2) colonial coccoid nanno- and microplankton (2–10 μm in diameter), (3) solitary filamentous microplankton, and (4) colonial filamentous microplankton (C. S. Reynolds, personal communication).

Among the smallest known planktonic cyanobacteria are the unicellular coccoid, ovoid, and rod-shaped picoplanktonic species that compare with the eubacteria in both shapes and sizes. Picoplankton, ranging in linear dimension from 0.2 to 3 μm, generally consist of solitary cells. Careful microscopic analyses reveal the presence of picoplankton throughout euphotic and aphotic zones of marine (Waterbury et al. 1979; Fogg 1982) and freshwater (Bailey-Watts, Bindloss, & Belcher 1968; Paerl 1977; Reynolds 1984) habitats exhibiting diverse trophic states. Cyanobacterial picoplankton can at times account for a significant fraction of phytoplankton productivity and biomass in oligotrophic lakes and seas having extensive euphotic zones (Paerl 1977; Paerl & Mackenzie 1977). Genera such as *Synechococcus* and *Cyano-*

dictyon characteristically inhabit (and can occasionally dominate) both near-surface epilimnetic and deep-euphotic hypolimnetic waters, forming distinct layers (chlorophyll maxima) in large oligotrophic lakes such as Lake Tahoe (Richerson, Lopez, & Coon 1978; Paerl unpublished), Lakes Superior and Michigan (Munawar & Munawar 1975), and Lake Taupo, New Zealand (Paerl & Mackenzie 1977; W. F. Vincent, personal communication), among others. Although it is believed that cyanobacterial picoplankton are well adapted (in terms of optimal surface-to-volume ratios) to nutrient-depleted oligotrophic waters, these genera also thrive (but seldom dominate) in nutrient enriched meso- and eutrophic waters.

A second morphological group includes colonial coccoid, ovoid, and rod-shaped genera. Individual cell sizes are distinctly larger than the solitary picoplankton; generally, sizes may range from 2–3 to 10 μm in diameter. Representative genera include the common bloom formers *Gomphosphaeria* and *Microcystis* as well as the subdominant genus *Chroococcus,* all of which typically inhabit mesotrophic and eutrophic lakes, ponds, and rivers. *Microcystis,* in particular, can form thick surface scums consisting of large aggregated colonies, each containing at least several hundred cells, during calm, thermally stratified periods. Cells comprising colonies are all vegetative; no cellular differentiation into heterocysts and/or akinetes is apparent in this group. Vegetative cells can, however, revert to inactive resting spores during periods unfavorable for active growth. Resting spores often reside in sediments during winter months when planktonic growth is either absent or greatly restricted.

The filamentous cyanobacteria can be divided into two groups according to aggregative or colonial behavior. Solitary (nonaggregated) filamentous genera such as *Oscillatoria* and *Lyngbya* consist of cylindrical cells forming long straight filaments. These genera often inhabit distinct subsurface (metalimnetic) layers and can occasionally rapidly alter their vertical position in the water column by buoyancy regulation to seek light and nutrient regimes suitable for optimal growth. *Oscillatoria* or *Lyngbya* may at times become dominant and form subsurface blooms. However, more often these genera appear as subdominants in mesotrophic and eutrophic waters. Vegetative cells in filaments can differentiate to form akinetes, a type of resting spore (discussed later).

Lastly, numerous filamentous genera can form aggregates of filaments resulting in large, buoyant, suspended or benthic colonies that are at times discernible with the naked eye. A majority of aggregated filamentous genera are capable of N_2 fixation following morphological and physiological differentiation of vegetative cells into heterocysts during nitrogen-deficient (and usually phosphorus-sufficient) periods. Dominant genera in this group include the N_2 fixers *Anabaena, Aphanizomenon, Gloeotrichia,* and *Nostoc.* These genera are capable of forming akinetes that are particularly evident during unfavorable growth periods.

The common planktonic (and occasionally benthic) freshwater cyanobacterial genera have been summarized in Table 7-1 in accordance with both individual morphological and functional characteristics. Upon examining Table 7-1, we see that the evolution of cyanobacterial phytoplankton has witnessed a vast amount of structural and physiological diversity. The ecological bases and rationales for such diversification will form the framework of this chapter.

GROWTH AND LIMITING FACTORS

Physical and Chemical Tolerances: The Cyanobacterial Paradox

Despite their initial success, diversification, and proliferation during the creation and evolution of earth's biosphere (Knoll 1977; Schopf & Walter 1982), cyanobacteria are in many ways paradoxical microalgae when considering their environmental requirements and tolerances. Well known as "pioneer organisms," cyanobacteria characteristically are initial colonizers and dominant phytoplankton in such inhospitable habitats as recently filled volcanic craters, geothermal pools, alpine and boreal ponds and lakes, highly polluted (either with organic and/or inorganic wastes) lacustrine and riverine systems (Table 7-1; Lund 1965; Brock 1967; Fogg et al. 1973; Reynolds & Walsby 1975; Gibson & Smith 1982). Numerous cyanobacterial genera are able to survive extensive desiccation (Potts & Whitton 1979), and it is not uncommon to find both N_2- and non-N_2-fixing genera coexisting as soil and aquatic algae in geographically diverse regions (Geitler 1932, 1960). Cyanobacteria are often sole inhabitants of extremely nitrogen-deficient waters of varying trophic states because of the ability of certain genera to biologically fix atmospheric N_2 (Pearsall 1932; Fogg 1969; Gibson & Smith 1982). Cyanobacteria can also thrive under extremely low ambient concentrations of free metals (Murphy, Lean, & Nalewajko 1976). Similarly, because of their ability to store phosphorus internally as polymeric polyphosphate bodies during periods when ambient orthophosphate supplies exceed growth requirements, cyanobacteria can survive and maintain growth during subsequent periods of external phosphorus deficiency (Healey 1973, 1982; Grillo & Gibson 1979).

Table 7-1. *Common planktonic cyanobacterial genera: their morphologies, growth characteristics, and chief habitats.*

Filamentous genera	Morphology	Habitats
N₂-fixers (heterocystous)		
Anabaena	Filaments with intercalary heterocysts can grow as large aggregates of filaments, mucilage associated with filaments but no sheaths	Chiefly planktonic subdominant in oligotrophic to eutrophic lakes, rivers and ponds, can form nuisance blooms in eutrophic waters, especially those enriched with phosphorus and organic matter, prefers stratified stagnant waters during calm warm periods for maximum bloom development. Certain species known to be toxic to invertebrate and vertebrate consumers
Anabaenopsis	Short filaments having terminal heterocysts, usually found as solitary filaments, lacking sheaths	Planktonic, largely subdominant in mesotrophic to eutrophic waters. Seldom forms extensive blooms
Aphanizomenon	Filaments having intercalary heterocysts, often found in bundles surrounded by confluent mucous sheath	Planktonic, subdominant in oligotrophic waters but can be dominant during late spring through summer months in mesotrophic to eutrophic lakes and ponds. Can produce massive bright green surface blooms. Often dominant under phosphorus-enriched conditions

Table 7-1. *(cont.)*

Filamentous genera	Morphology	Habitats
Calothrix	Filaments usually having basal but sometimes intercalary heterocysts, filaments either solitary or in fascicles	Mainly epiphytic and epilithic but sometimes planktonic. Common in streams varying from oligotrophic to eutrophic, also found on submerged structures (wood, stone) in lakes, ponds and rivers of varying trophic states
Gloeotrichia	Aggregated filaments having basal heterocysts, usually found in common gelatinous matrix	Occurs both planktonically and attached. Found as subdominant in oligotrophic to eutrophic ponds, lakes, and streams. Often attached to submerged macrophytes and wooden structures. Seldom dominant as blooms
Nodularia	Straight to slightly curved filaments having both intercalary and terminal heterocysts. Heterocysts appear depressed. Filaments solitary or aggregated in mucous colonies having thin sheaths	Planktonic, often forming extensive blooms in P-enriched lakes, rivers, ponds, and shallow pools; prefers standing waters
Nostoc	Filaments having intercalary heterocysts, found both solitary and as colonies (planktonic) or colonial in an attached mucoid (often spherical) semi-transparent mass, having the appearance of a small green grape	Occurs both in lakes and streams of varying trophic states, often found attached to rocks and wood in oligotrophic alpine lakes and streams. Common in plankton, but seldom forms extensive blooms

Rivularia	Colonial, filamentous with basal heterocysts, filaments radiating from center in spherical to hemispherical attached colonies. Filaments are tapered with cells closest to heterocysts having largest diameters	Almost exclusively found on wet rocks, decaying wood, and submersed vegetation. Also found in geothermal and mineral springs/streams as mat components
Scytonema	Filaments having false branches and thick sheaths, rarely gelatinous. False branches originating at or near intercalary heterocysts	Occurs chiefly as fuzzy tangled mats attached to submersed rocks, wood, or plants. Occasionally found as small irregular colonies in plankton. Also common in geothermal springs, and has been found in a range of trophic states
Tolypothrix	Filaments having firm sheaths, of varying thickness, lateral branches arising below the heterocysts. Usually found as paired filaments. Has basal heterocysts	Attached or free-floating in standing water, also found attached to rocks in slowly flowing streams. Found in diverse trophic states. Common in slow-flowing mountain streams, alpine lakes, and extreme environments (arctic and antarctic streams, lakes, and rivers)
N₂-fixing genera (nonheterocystous)		
Lyngbya	Solitary or clumped filaments often mixed with other algae. Filaments composed of uniseriate nonbranched trichomes within nongelatinous membranelike sheaths. Cells in filaments usually shorter than wide. N₂-fixing capabilities only present in some species	Common both planktonically and epiphytically, preferred habitats include mesotrophic to eutrophic standing waters, ponds, and slow-moving streams. Commonly associated with other planktonic and benthic algae. Rarely forms nuisance blooms, and if so, blooms are usually dispersed through the water column. Can be found in extremely nutrient-rich habitats

Table 7-1. *(cont.)*

Filamentous genera	Morphology	Habitats
Microcoleus	Filaments either solitary or aggregated into twisted rope-like bundles. Many filaments glide or creep readily. Often extensively epiphytized by bacteria. N_2-fixation has not been conclusively shown in axenic filaments	Almost exclusively associated with submersed surfaces, including sand, gravel, wood, and decaying organic matter. Can inhabit a range of trophic state systems. Also common as a mat constituent in geothermal springs, dried lake beds, salt flats, and mud flats
Oscillatoria (many non-N_2-fixing species present)	Filaments solitary or interwoven to form mats. Nonbranching filaments without sheaths. Cells cylindrical, shorter than broad in long filaments, longer than broad in short filaments. Many smooth gliding forms present	Common in mesotrophic to eutrophic (including hypereutrophic) lakes, ponds, and stagnant waters. Often subdominant during *Anabaena* and *Aphanizomenon* blooms, but can also form extensive dominant blooms in nitrogen and phosphorus enriched waters. Many species known to be epiphytic or epilithic. This genus is common in extreme environments, including thermal springs, desert lakes, salt ponds, dry lake beds, arctic tundra ponds, and alpine lakes
Plectonema	Filaments branched, solitary, or forming mats. False branching singular or paired filaments, having firm sheaths	Almost always found attached as feltlike colonies and mats on submersed rocks, vegetation and wood. Quite common in flowing waters of diverse trophic states

Genus	Morphology	Ecology and distribution
Phormidium (many species do not appear to fix N_2)	Solitary or aggregated thin filaments usually combined with other algal species in colonies, mats, and flocs. Thin watery mucous sheaths present, but difficult to detect in many species	Seldom planktonic, more often living in consortia with other benthic mat microorganisms. Most commonly found on muds and stones of shallow streams and ponds of varying trophic states. Also present in mountain streams

Non-N_2-fixing filamentous genera

Genus	Morphology	Ecology and distribution
Arthrospira	Filaments solitary, cylindrical, forming loose spirals, cross walls distinct, sheaths absent	Planktonic. Common, but rarely dominant in mesotrophic to eutrophic waters, including shallow ponds, streams, and lakes
Spirulina	Filaments usually twisted, unicellular, cylindrical. Cross walls indistinct.	Mostly planktonic, although some species can inhabit benthic algal mats. Usually subdominant with other bloom-forming cyanobacteria in mesotrophic to eutrophic waters, but occasionally can produce blooms

Non-filamentous, colonial, non-N_2-fixing genera

Genus	Morphology	Ecology and distribution
Chroococcus (*Coccochloris*)	Coccoid to oval cells either single-celled or arranged in microscopic to macroscopic colonies, often in a thick gelatinous matrix	Very common subdominant planktonic genus in mesotrophic to hypereutrophic waters, often associated with nuisance bloom genera, but seldom forming blooms by itself
Gomphosphaeria	Cells ovoid to ovoid-cylindric in diameter, arranged near the surface of colonies, in a gelatinous matrix, either firm or semiviscous. Pseudovacuoles present	Common in plankton of mesotrophic to eutrophic lakes, ponds, and slow-moving streams. May form blooms during summer months

Table 7-1. (cont.)

Filamentous genera	Morphology	Habitats
Microcystis	Globose cells aggregated in microscopic to macroscopic colonies, having a gelatinous matrix. Cells often possess gas vacuoles, leading to highly buoyant colonies	Common plankton of mesotrophic to hypereutrophic lakes, rivers and ponds. Can dominate as nuisance blooms in slow-flowing or stratified waters. Prefers nitrogen and phosphorus enriched waters. Certain strains known to be toxic to invertebrate and vertebrate consumers
Non-filamentous colonial N₂-fixer		
Gloeocapsa	Coccoid cells arranged in sets of cells ranging from 2 to 10 encapsulated in a spherical mucilagenous matrix. Sets are often found aggregated in larger colonies that vary from microscopic to macroscopic size	Usually subdominant planktonic organisms in mesotrophic to eutrophic lakes and streams. Can form dark slimy masses on rocks, submersed wood in streams and shallow ponds
Non-filamentous single celled		
Synechococcus (N₂ fixation charachteristics uncertain)	Single truncate ovoid or truncate-subcylindrical often becoming ovoid-cylindrical, up to three times as long as broad. These olive-yellowish cells generally range from <2–15 µm in diameter	Extremely common but seldom dominant in plankton of oligotrophic to eutrophic lakes and ponds. Can form appreciable portion of phytoplankton biomass in deep, hypolimnetic waters where it thrives under low light levels. This genus is often overlooked in microscopic analyses of freshwater phytoplankton due to small cell sizes

A variety of bloom-forming genera can withstand extensive (daily to weekly) periods of exposure to high (supersaturated) photosynthetically active radiation (PAR) flux as well as accompanying ultraviolet radiation striking the surface waters they inhabit. Coexisting eukaryotic phytoplankton often reveal immediate and severe photooxidative damage in response to such irradiance conditions (Eloff, Steinitz, & Shilo 1976; Paerl et al. 1985).

Despite the acclaim as microorganisms for all seasons and climates, cyanobacteria do exhibit rather profound sensitivity toward rapid but minor shifts in environmental conditions. For example, it is rather ironic that persistent blooms of either the filamentous N_2-fixing *Anabaena* and *Aphanizomenon* or the colonial non-N_2-fixing *Microcystis* can thrive under excessively warm 35°–40°C) and high PAR (>1500 $\mu E \cdot m^{-2} \cdot s^{-1}$) conditions in surface waters (Fogg 1969; Reynolds & Walsby 1975; Kellar & Paerl 1980; Robarts 1984; Paerl et al. 1985), whereas a rapid weather-related drop in temperature of only 5°C maintained for several days can suddenly terminate blooms (Fogg 1969; Paerl 1985). Furthermore, sudden shifts in wind speed and/or direction can result in similar negative impacts on bloom occurrence and intensity (Ganf & Viner 1973; Reynolds & Walsby 1975; Paerl 1985). There are instances where a 12–24-h windy period following calm days led to the termination of surface blooms and concurrent loss of dominance by cyanobacteria in the planktonic community (Wirth & Dunst 1967; Lorenzen & Mitchell 1975; Reynolds & Walsby 1975).

Another area of susceptibility to environmental perturbation involves sudden shifts in ionic properties of ambient waters (Carpelan 1964; Mohleji & Verhoff 1980). Even though many cyanobacterial genera reveal a high degree of tolerance or a favorable response to excessive nutrient and/or pollutant loading, significant accompanying shifts in total ionic strength or salinity of affected waters can at times cause dramatic alterations in cyanobacterial dominance (MacLeod 1971; Melamed-Harel & Tel-or 1981). In nutrient-enriched North Carolina estuarine waters, a relatively minor shift in salinity (0.2–0.5 ppt total salts) can cause dramatic shifts in phytoplankton community composition. At less than 0.2 ppt salinity, massive blooms of either *Microcystis aeruginosa* (in N-sufficient waters) or individual blooms of *Anabaena flos-aquae/A. circinalis/A. spiroides* and *Aphanizomenon flosaquae* (in N-depleted waters) can be commonplace. At salinities exceeding 0.5 ppt potential dominance by these nuisance genera is virtually eliminated, although some Chrococcoid-type cyanobacteria can remain as subdominants under these conditions (Paerl et al. 1984). Cyanobacteria are also known to exhibit extreme sensitivity to specific sources of terrigenous (allochthonous) runoff. In particular, a high degree of humic- and fulvic-acid loading may alter the potential for both cyanobacterial dominance and bloom potential (Prakash 1971).

A large number of cyanobacterial genera show a generally negative growth response to acidic (pH < 6.0) conditions (Fogg et al. 1973). It has been well documented that, as a group, cyanobacteria have a distinct preference for neutral to alkaline waters (Griffiths 1939; King 1970; Fogg et al. 1973; Shapiro 1973) while being rapidly replaced by eukaryotic phytoplankton (particularly chrysophytes and chlorophytes) under acidic conditions (Shapiro 1973). Undoubtedly, diverse environmental factors come into play when explaining this general observation because pH shifts affect the solubility, chemical speciation (availability), and hydration characteristics of a wide variety of essential phytoplankton nutrients (Stumm & Morgan 1981); membrane transport characteristics (Walsby 1982); extra- and intracellular enzyme reactions (Stewart 1974); photosynthetic electron transport; and osmotic potential of cytoplasm (Stewart 1974; Walsby 1982).

In summary, one is left with a paradoxical conclusion. Given the extensive evolutionary history (including the transition from anoxic to fully oxygenated conditions) that cyanobacteria have endured (Margulis 1975; Margulis, Walker, & Rambler 1976), these microorganisms exhibit noticeable intolerance to relatively small shifts among a variety of basic physical–chemical variables. Significantly, it is these variables that regulate hydrological and environmental conditions of freshwater ecosystems.

The Physical Environment: Its Growth-Regulating Role

In exploring this apparent paradox of mixed tolerance levels, we must consider temporal and spatial scales within which environmental fluctuations occur in aquatic ecosystems. The rapidity, endurance, and seasonal stability with which temporal environmental changes occur are critical to all members of the phytoplankton community (Harris 1980). However, degrees of environmental stability over time are particularly relevant with respect to cyanobacterial success and dominance (Gibson et al. 1971; Ganf & Horne 1975; Reynolds & Walsby 1975; Reynolds 1980; Paerl 1986). Physical–chemical and biotic heterogeneity can lead to gradients and barriers among a wide range of environmental variables (Hutchinson 1961; Richerson, Armstrong, & Goldman 1970). What may be construed as a distinct environmental fluctuation on a large scale actually represents a compositional or averaged picture of countless individual small-scale regimes having gradients or boundary regions between them along vertical and horizontal transects of the water column.

When we specifically consider the physical environment, it becomes evident that cyanobacteria, although often exhibiting favorable growth and/or reproductive characteristics under conditions that we might perceive as "extreme," strongly prefer the maintenance of environmental stability. A clear-cut case is the progression of nuisance blooms in relation to physical stratification of the water column. Generally, nuisance bloom genera, including *Microcystis, Anabaena,* and *Aphanizomenon,* prefer a thermally and sometimes chemically stable, stratified water column as opposed to thoroughly mixed, near-surface waters (Ganf & Horne 1975; Reynolds 1980). Among these genera, vertical position in the water column can be rapidly adjusted by buoyancy alteration (Walsby 1972).

Buoyancy is regulated by varying intracellular gas vacuole formation and by adjusting cellular ballast. Both mechanisms are regulated through the synthesis and breakdown of specific products of photosynthesis (Dinsdale & Walsby 1972; Grant & Walsby 1977; Konopka 1984), and both mechanisms are closely coupled to photosynthetic potentials and histories (Van Liere & Walsby 1982). At suboptimal photosynthetic rates, cellular production and accumulation of recently fixed photosynthetic products is low. Depressed photosynthate levels limit ballast production because carbohydrates are the foremost of a variety of dense, ballast-providing condensates (Konopka 1984). Furthermore, dissolved photosynthetic products (i.e., sugars) constitute a pool of osmotically active compounds. Hence, depressed photosynthate production tends also to lower the cell's osmotic potential (Van Liere & Walsby 1982). When osmotic pressure is relatively low, gas vacuole synthesis can proceed at relatively high rates. Conversely, if cell osmotic potential is high due to accumulation of photosynthate, then gas vacuole formation is reduced. In effect, cell turgor pressure (i.e., wall pressure) provides a mechanical control on gas vacuole formation by either "squeezing out" or collapsing existing gas vacuoles and minimizing new gas vacuole synthesis when turgor pressures are high (Van Liere & Walsby 1982). Accordingly, cells having relatively low rates of photosynthesis tend to increase buoyancy because gas vacuole formation can proceed in a relatively unhindered fashion and cell ballast accumulation would be minimal. In contrast, cells having recently maintained optimal photosynthetic rates contain relatively large amounts of ballast and few gas vacuoles. Thus, cells having experienced poor photosynthetic histories often tend to exhibit maximum buoyancy whereas optimal photosynthetic histories lead to negatively buoyant cells.

Under stratified conditions, PAR, CO_2, and inorganic nutrient supplies often prove limiting along well-defined vertical gradients (Walsby

& Booker 1980; Klemer, Feuillade, & Feuillade 1982; Paerl & Ustach 1982). Alteration of cell orientation along such gradients is a desirable, if not essential, capability. Under such environmental constraints, buoyancy-mediated vertical migration provides an effective means of rapid cell reorientation. Both field and laboratory measurements of vertical migration rates among major bloom-forming cyanobacteria have yielded some impressive values. Reynolds (1972) found naturally occurring colonies of *Anabaena circinalis* to migrate upward at speeds in excess of 20 cm·h^{-1}, and clumps of *Microcystis aeruginosa* sampled from Lake George, Uganda (Ganf 1975) migrated with vertical speeds in excess of 3 m·h^{-1}. Our buoyancy studies on the Chowan River, North Carolina (Paerl & Ustach 1982; Paerl 1982b), yielded vertical migration speeds ranging from 40 cm to 2.75 m h^{-1} for *Aphanizomenon flos-aquae* in a thermally stratified water column. Even faster buoyancy rates (in excess of 10 m·h^{-1}) have been observed by Imberger and Humphries (unpubl.). Such migration rates greatly exceed eukaryotic (dinoflagellates, chlorophyceans) flagellar swimming speeds (Loeblich 1966; Throndsen 1973). Clearly, in strongly stratified water columns having stable gradients of PAR, CO_2, and nutrient concentration, cyanobacteria able to rapidly compensate buoyancy and ballast have a distinct advantage over eukaryotic taxa lacking an efficient means of vertical migration. On a short-term basis (minutes to hours), cyanobacterial bloom species are able to rapidly alter their vertical orientation to follow shifting gradients of growth-related environmental factors. Persistent positive buoyancy leads to the accumulation of surface-scum populations in stabilized, near-surface water layers (Fogg 1969; Paerl 1983a; Robarts 1984). Such populations can effectively trap more of the incoming PAR and atmospheric sources of CO_2 as well as inorganic nutrients while also shading subsurface populations of nonbuoyant phytoplankton (Van Liere & Walsby 1982). Given the fact that naturally occurring cyanobacterial bloom species can possess effective ways of protecting their photosynthetic apparatus and other labile cellular constituents against photooxidation in UV-rich surface waters (Eloff et al. 1976; Paerl & Kellar 1979; Paerl, Tucker, & Bland 1983; Paerl et al. 1985), surface-scum formation could be viewed as a highly adaptive mechanism directed at optimal utilization of various environmental resources while contemporaneously restricting PAR transmittance (and perhaps CO_2 diffusion) to potentially competitive subsurface phytoplankton populations (Fig. 7-1).

The foregoing discussion provides a rationale for the link between physical stability (i.e., vertical stability) and the establishment and proliferation of migratory cyanobacterial bloom genera. Buoyancy regu-

Daytime

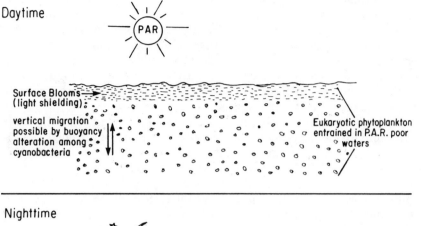

Surface Blooms
(light shielding)

vertical migration
possible by buoyancy
alteration among
cyanobacteria

Eukaryotic phytoplankton
entrained in P.A.R. poor
waters

Nighttime

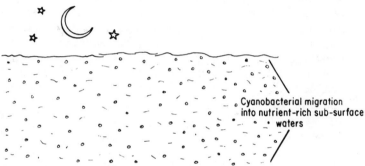

Cyanobacterial migration
into nutrient-rich sub-surface
waters

Fig. 7-1. *Top:* Illustration depicting daytime cyanobacterial surface
blooms effectively outcompeting subsurface eukaryotic phytoplankton for photosynthetically active radiation under physically stratified
conditions. In addition to forming the thick surface scums shown
here, cyanobacterial bloom species are capable of vertical migration
in order to avoid midday photoinhibitory periods and to periodically
dwell in subsurface nutrient-enriched waters.
Bottom: Nighttime conditions where, by virtue of bouyancy compensation, cyanobacteria are capable of migrating throughout the water
column, including nutrient-rich hypolimnetic waters.

lation in stratified water columns also provides an important mechanism for the maintenance of metalimnetic cyanobacterial genera,
including populations of *Oscillatoria* (Klemer et al. 1982; Konopka
1982), *Synechoccoccus, Chrococcus,* and *Phormidium* (Pavoni 1963;
Vollenweider, Munawar & Stadelman 1974; Priscu & Goldman 1983).

Among these diverse genera, requirements for physical stability such as nonmixed hydrological conditions, constant temperature, and irradiance regimes are of paramount importance for ensuring successful habitation and potential dominance. Granted migratory means of ensuring physical environmental stability, specific cyanobacterial genera can occupy habitats characterized by temperature, irradiance, and nutrient extremes including hot springs, volcanic lakes, and hypereutrophic lakes, rivers, and reservoirs.

As previously stated, disruption of physical stability generally leads to a rapid loss of dominance by both nuisance and nonnuisance cyanobacterial genera in phytoplankton assemblages (Wirth & Dunst 1967; Gibson et al. 1971; Lorenzen & Mitchell 1975; Reynolds 1975). Under nonstratified (mixed) conditions, buoyancy or ballast regulation is often ineffective at maintaining cyanobacteria in preferred vertical locations. As a consequence, turbulent mixing freely circulates cyanobacteria along with other phytoplankton groups. This change in physical "ground rules" forces cyanobacteria to compete more directly for thermal, irradiance, and nutrient regimes with phytoplankton groups known to thrive under well-mixed conditions, including green algae, diatoms, chrysophytes, and some microflagellate genera (Harris, Haffner, & Piccinin 1980; Talling 1982). Shifts in phytoplankton dominance in response to the periodicity and magnitude of vertical mixing are often most profound in eutrophic impoundments that can exhibit massive cyanobacterial blooms when stagnant (Fogg 1969). Summer periods (days to weeks) of stratification often usher in such blooms in these systems, whereas persistent periodic short-term (hourly to daily) destratification can lead to periods of noncyanobacterial dominance. Limnologists have utilized artificial physical control mechanisms in moderating and eradicating bloom potentials among mesotrophic and eutrophic impoundments used for recreation, sport fishing, or sources of drinking water. Artificial destratification using pumps, bubblers, and enhancement of flushing rates (reduction in water retention time) is a current approach to destratifying small impoundments that are periodically plagued with cyanobacterial blooms during summer stagnancy (Yount 1966; Wirth & Dunst 1967; Lorenzen & Mitchell 1975). Physical means of controlling cyanobacterial blooms are often impractical, however, in large lakes, reservoirs, and river systems.

Cyanobacterial reliance on physical stability appears fully compatible with evolutionary considerations. Within geological time frames, it is well known that cyanobacteria have had to endure hundred- to million-year periods of atmospheric cooling and warming (ice ages), desiccation, volcanic eruptions, and resultant shifts in the geochemical composition of lakes and oceans, as well as altered concentrations of

CO_2, O_2, and S, and also contrasting PAR and UV transmittance characteristics of the atmosphere (Cloud 1976; Schopf & Walter 1982). Within and between each set of catastrophic events, long-term (at least yearly) stability prevailed (Schopf & Walter 1982). Within such time frames it would appear likely that genetic variability and resulting physiological adaptations among cyanobacteria could easily have accounted for survival and diversification among genera known to have existed in aquatic habitats since the Precambrian period. Furthermore, cyanobacteria have developed a myriad of symbiotic associations with eukaryotic microalgae (e.g., *Rhizosolenia-Richelia*), fungi, lichens, higher plants (e.g., *Azolla-Anabaena, Gunnera-Nostoc;* Stewart, Rowell, & Rai 1980), and diverse animals that include protochordates (Ascidians-*Prochloron;* Lewin 1976), sponges (*Ulosa*-Chroococcaceae; Rützler 1981), bryozoa, shrimps, and even polar bears (Lewin 1981). In each case the cyanobacterial endosymbiont is exposed to microenvironmental conditions quite different from the ambient environment. Endosymbiosis itself may have been the product of adaptations to long-term environmental alterations during the evolution of eukaryotic forms of life on earth. Lastly, the evolution and enrichment of atmospheric molecular oxygen can largely be attributed to the advent and proliferation of cyanobacterial oxygenic photosynthesis (Cloud 1976; Margulis et al. 1976). Resultant dissolved oxygen levels in freshwater and marine habitats have steadily increased during the nearly 1 billion years since the rise of oxygenic photosynthesis. As instigators of atmospheric O_2 enrichment, cyanobacteria have had to develop diverse ways of dealing with this potentially toxic (to diverse biochemical reactions) end product of photosynthesis (Stewart 1974). Given steady enrichment with atmospheric O_2, cyanobacteria have accommodated genetically and physiologically to such environments (see Fogg et al. 1973; Carr & Whitton 1982).

Considering the restricted sets of environmental tolerances (including physical stability) exhibited by cyanobacterial planktonic species, there is reason to believe that specific geological transition periods (most notably the ice ages) may have signaled long-term absences of certain genera in particular geographic locales (Barghoorn & Schopf 1965; Cloud 1976; Schopf & Walter 1982). However, given high degrees of mobility, multiple potential habitats (soils, water, as endosymbionts), and ease of dispersal (water and air transport) of spores (Brown, Larson, & Bold 1964), it is tempting to speculate that unfavorable transient growth periods were ineffective in causing extinction of a significant number of terrestrial and aquatic cyanobacterial species. More likely, events such as ice ages served to select for cold-tolerant and desiccation-resistant genera that can today be found in tundra soils and ponds, polar seas and lakes, and embedded in glacial and sea ice (see Drouet & Daily 1956; Holm-Hansen 1964).

Nutrition and Growth Relationships

Role of Organic Matter. Enrichment of seas and lakes with biogeochemically formed organic matter most likely provided necessary energy and growth resources for the variety of heterotrophic prokaryotes thought to have been pioneer microbial inhabitants during Archean–Precambrian periods of the earth's history (Cloud 1976; Holland 1978). Prior to the advent of anoxic photosynthesis (by pigmented sulfur bacteria) and limited chemolithotrophic production, preformed organic matter provided the chief source of reduced carbon compounds available for biosynthetic and catabolic processes. Fossil records strongly suggest that cyanobacterial evolution and diversification were initiated in the organic matter-rich "primordial soups" that Archean oceans and lakes represented (Margulis 1982; Schopf & Walter 1982). Modern freshwater cyanobacteria have retained numerous means of utilizing organic matter to satisfy energy, growth, and metabolite requirements (Stewart 1974). Even though photoautotrophy represents the main route by which growth and energy requirements are met among cyanobacterial genera, heterotrophic utilization of dissolved organic matter represents an ancillary means of obtaining energy-rich carbon compounds and exogenous growth factors (Khoja & Whitton 1971; Droop 1974; Stanier & Cohen-Bazire 1977). Through the application of tracer techniques (kinetic uptake assays and microautoradiography), it can be demonstrated that diverse cyanobacteria readily assimilate a variety of organic compounds at naturally occurring trace quantities. Such assimilation occurs during the photosynthetically active daytime and at night (Saunders 1972; Paerl 1985). In the laboratory, some cyanobacterial strains (both N_2-fixers and non-N_2-fixers) can be shown to subsist entirely on organic compounds under aphotic conditions (Hoare, Hoare, & Moore 1967; Droop 1974). Hence, it would appear that to a limited extent, cyanobacteria have retained the ability to assimilate exogenous organic matter.

Several early systematic and ecological studies aimed at relating the promotion of cyanobacterial growth and dominance to specific environmental variables have hinted at a relationship between organic matter enrichment and the presence of cyanobacteria in phytoplankton communities (see Brook 1959; Vance 1965; Fogg 1969). In a study of numerous lakes in the English Lake District, Pearsall (1932) concluded that the tendency of certain lakes to periodically host blooms of well-known genera such as *Anabaena, Aphanizomenon,* and *Micro-*

cystis was related to organic matter content, which at that time was assayed as "albuminoid" dissolved nitrogen content.

More recent examinations of eutrophic and hypereutrophic lakes rich in dissolved organic matter generally substantiate early observational data that such waters are particularly susceptible to cyanobacterial blooms (Fogg 1969; Reynolds & Walsby 1975). Microautoradiographic examinations of isotopically labeled organic matter assimilation by phytoplankton in such waters reveal that both bloom cyanobacteria and their bacterial epiphytes are particularly active in assimilating a range of organic compounds (Paerl & Kellar unpublished data). *Anabaena oscillarioides* in eutrophic Lake Rotongaio, New Zealand, revealed assimilation of the sugars glucose, maltose, and galactose at trace (natural) concentrations (Fig. 7-2a,b). This cyanobacterium also assimilated a wide range of amino acids despite the fact that it simultaneously exhibited N_2 fixation. Likewise, *Anabaena circinalis* examined during blooms in the Chowan River, North Carolina, was capable of assimilating a range of sugars and amino acids during both N_2-fixing and non-N_2-fixing periods (Fig. 7-3). *Microcystis aeruginosa,* a non-N_2-fixing species, dominant in the highly eutrophic Neuse River, North Carolina, was particularly adept at assimilating a variety of amino acids, but was somewhat less active in assimilating the sugars glucose and fructose (Fig. 7-4). The 6-carbon sugar alcohol mannitol is also widely assimilated by both N_2-fixing and non-N_2-fixing naturally occurring and cultured cyanobacterial genera (Fig. 7-5a,b).

Bacteria associated with these cyanobacteria also assimilate a wide range of organic substrates, including those already mentioned (Fig. 7-2). Without the aid of microautoradiography it is therefore difficult to distinguish true cyanobacterial "uptake" of organics from uptake associated with microbial epiphytes. Despite this interpretative problem, investigations of diverse lakes and rivers suggest that cyanobacterial utilization of organics is commonplace (Saunders 1972; Paerl 1985).

Assimilated organic compounds could serve several useful purposes. Foremost, the utilization of such organics via the oxidative pentose phosphate cycle provides energy and growth resources (Droop 1974; Stewart 1974). This form of carbon utilization might play a particularly important role either in turbid or highly colored waters where PAR transmittance is restricted or in deep aphotic waters where heterotrophic metabolism could prove necessary in maintaining growth and viability. Second, utilization (including metabolism and respiration) of organic matter by cyanobacteria and associated bacteria leads to localized regions of enhanced O_2 consumption (Paerl 1984b; 1985). Localized O_2 consumption near oxygen-sensitive heterocysts or N_2-fix-

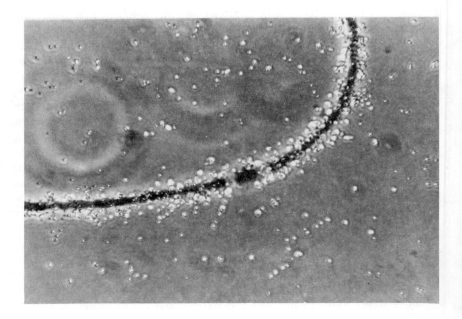

Fig. 7-2. (a) Microautoradiograph of ^{3}H galactose (uniformly labeled) assimilation by the cyanobacterium *Anabaena oscillarioides*, sampled from Lake Rotongaio, New Zealand. Silver grains exposed (reduced) by ^{3}H decay are shown as white dots superimposed over a dark *A. oscillarioides* filament in this phase contrast micrograph. Due to heavy labeling of filaments some elevated "background" exposure (reduced silver grains not associated with filaments) is evident in this microautoradiograph.

ing loci in nonheterocystous cyanobacteria can help protect nitrogenase from oxygen inactivation, consequently promoting optimal rates of N_2-fixation in nitrogen-depleted, oxygenated waters (Paerl 1978; Paerl & Kellar 1978, 1979). Paerl (1985) has shown that enrichment with several hexose sugars and mannitol enhanced N_2-fixation rates in *Anabaena oscillarioides* independently of organic nutrient additions (P, Fe, Mo, and other trace metals, Fig. 7-6).

Diverse organic compounds can act as chelators of essential trace metals (Stumm & Morgan 1981). Chelation of metals such as iron, manganese, copper, and cobalt retains such metals in the water column, albeit in a chemically bound manner, thereby minimizing losses by coprecipitation and sedimentation in chemically insoluble forms.

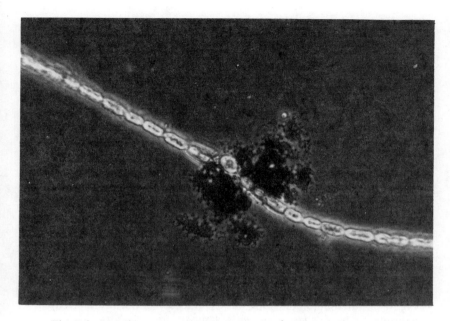

Fig. 7-2. (b) Microautoradiograph showing ^{3}H amino acid (uniformly labeled amino acid mixture) assimilation by bacteria specifically associated with heterocysts of *Anabaena oscillarioides*. Labeling is shown as dense clusters of black (exposed) silver grains surrounding the heterocyst. Several bacteria broke loose from the heterocyst during microautoradiograph preparation; these bacteria (rod-shaped) also appear heavily coated by dense clusters of silver grains. Note that the *A. oscillarioides* filament reveals no significant assimilation of the ^{3}H amino acid mixture.

Cyanobacteria are known to excrete powerful hydroxamate chelators (siderochromes) that are effective in specifically sequestering essential trace metals at exceedingly low ambient concentrations (Murphy et al. 1976; Simpson & Neilands 1976). It may be possible that such chelators can also be effective in harvesting essential metals from weaker natural organic chelators. Natural organic chelators may also play a role in immobilizing potentially toxic metals such as copper, cadmium, and mercury to which cyanobacteria can exhibit a high degree of sensitivity (Johnston 1964; Sunda & Guillard 1976; McKnight & Morel 1979). In this manner, the presence of dissolved organic matter may make specific freshwater habitats more amenable to cyanobacterial growth and dominance.

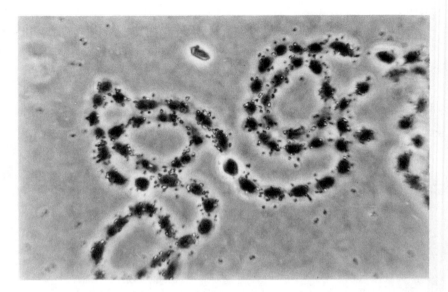

Fig. 7-3. Microautoradiograph of ^{3}H glucose (uniformly labeled) assimilation by *Anabaena circinalis* filaments sampled from the Chowan River, North Carolina. Exposed silver grains appear as small white, grey, and black dots superimposed over dark *A. circinalis* filaments in this phase contrast micrograph. Labeling of these filaments was substantially less dense than that shown for *A. oscillarioides* (see Fig. 7-2). Accordingly, "background" exposure also appears reduced.

If we consider some of the potential benefits of organic matter to cyanobacterial growth, it is not surprising that organic-matter-enriched waters are particularly favorable habitats for cyanobacterial growth and proliferation. Indeed, some of the most notorious cyanobacterial nuisance sites include organic-matter-enriched agricultural ponds, sewage treatment ponds, and waters receiving industrial processing wastes rich in oxidizable organics.

Role of Inorganic Matter. Among the essential inorganic nutrients required by freshwater cyanobacteria, phosphorus and nitrogen are often cited as being in shortest supply (Gerloff & Skoog 1957; Schindler 1975, 1977; Healey 1982). Phosphorus, carbon, and sulfur are largely derived from mineral (rock, soil, volcanic origin) sources, whereas nitrogen is most commonly a product of the breakdown of organic matter derived from atmospheric ammonia and nitrate

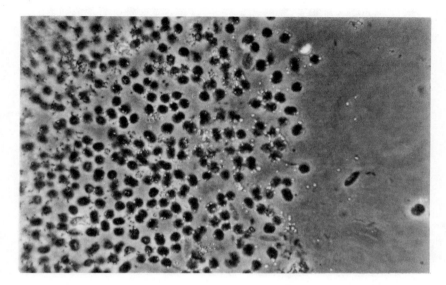

Fig. 7-4. Representative microautoradiograph of ^{3}H amino acid (uniformly labeled amino acid mixture) assimilation by the colonial non-N$_2$-fixing cyanobacterium *Microcystis aeruginosa* sampled from the eutrophic Neuse River, North Carolina. This phase contrast micrograph illustrates the labeling pattern within a portion of a large colony. ^{3}H labeling is shown as grouped white silver grains superimposed over dark aggregated *M. aeruginosa* cells.

sources or the biological conversion of atmospheric dinitrogen gas (N$_2$).

Phosphorus, as orthophosphate, is perhaps the most often-cited growth-limiting inorganic nutrient in freshwater (Likens 1972; Schindler 1977). In most freshwater habitats physically amenable to cyanobacterial growth, phosphorus availability often dictates magnitudes of dominance and bloom formation (Healey 1982). Cyanobacteria as a group do not appear to possess cellular phosphorus requirements or uptake characteristics (affinities) dramatically different from eukaryotic phytoplankton (Healey 1982; Rhee 1982). As with many eukaryotic taxa, cyanobacteria are capable of intracellular storage of assimilated phosphorus during periods of supply in excess of growth requirements (Healey 1982). For storage, assimilated orthophosphate is enzymatically polymerized as polyphosphates concentrated in microscopically visible intracellular polyphosphate bodies (Jensen

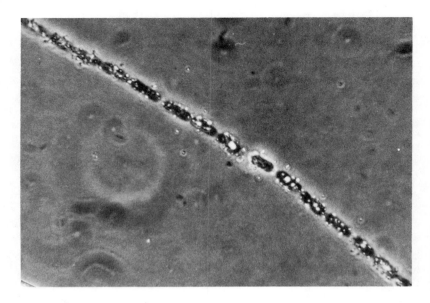

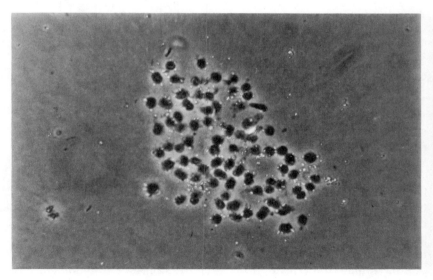

Fig. 7-5. Microautoradiographs of ^{3}H mannitol (uniformly labeled) assimilation among both N_2-fixing *(Anabaena oscillarioides)* and non-N_2-fixing *(Microcystis aeruginosa)* cyanobacterial genera. Figure 7-5a illustrates ^{3}H labeling of an *A. oscillarioides* filament (obtained from culture), and Fig. 7-5b illustrates labeling of a small *M. aeruginosa* colony (obtained from the Neuse River, North Carolina).

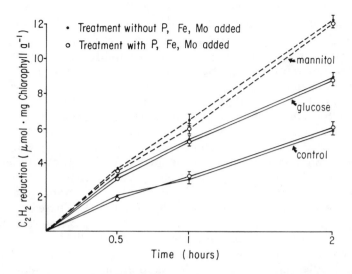

Fig. 7-6. Impact of the hexose sugar glucose (1 mM) and sugar–alcohol mannitol (1 mM) both singly and in combination with phosphorus (P, 50 μM), iron (Fe, 10 μM), and molybdenum (Mo, 5 μM) on acetylene reduction (N_2-fixation) in axenic populations of *Anabaena oscillarioides*. Results of 2-h long acetylene reduction incubations following 24 h previous nutrient enrichment are shown. Control conditions indicate that neither glucose or mannitol were added. Phosphorus, iron, and molybdenum were supplied to one set of controls, but these inorganic nutrients revealed no significant stimulation above basic culture conditions (Chu-10 medium used). Results are reported per unit of chlorophyll *a*, thereby normalizing them for *A. oscillarioides* biomass. Variability among triplicate samples for each treatment is shown.

1968; Stewart, Pemble, & Al-Ugaily 1978). Such reserves are vital sources of phosphorus during transient periods of phosphate depletion. Cyanobacteria capable of rapid vertical movement by buoyancy alteration have the additional advantage of short-term migrations into phosphate-rich aphotic waters bordering surface sediments during physically stratified periods when phosphate depletion may be a feature of epilimnetic and surface waters. In this manner, bloom species might "load up" on phosphate during periods clearly not favoring surface growth (nighttime or times of extreme phosphate depletion), accumulate intracellular phosphate as polyphosphate reserves, and subsequently return to surface waters to pursue intense photoautotrophic growth even in ambient waters exhibiting more-or-less continuous phosphate depletion (Schindler et al. 1981).

Certainly, among N_2-fixing genera long-term phosphate availability is a factor strongly regulating N_2-fixing capabilities and bloom formation in nitrogen-depleted waters (Schindler 1977). An often-used approach in managing systems favoring dominance by N_2-fixing genera is to restrict external phosphorus inputs (Beeton 1965; Edmonson 1969; Schindler 1971, 1975; Cronberg, Gelin, & Larsson 1975; Vollenweider 1976; Baalsrud 1982). During the development of lake management and restoration techniques, it was quickly recognized that relatively high phosphorus inputs tended to favor cyanobacterial dominance and bloom formation (Beeton 1965; Schindler 1977). Such dominance can be even more ensured under conditions of relatively low nitrogen availability (Niemi 1979; Smith 1983). Therefore, in relatively large and physically complex aquatic systems not amenable to periodic hydrological flushing or artificial destratification, constraints on phosphorus inputs have, by and large, been the most effective and economically viable techniques for suppressing cyanobacterial bloom formation (Beeton 1965; Edmondson 1969; Vollenweider 1976).

Nitrogen is an essential element in proteins (amino acids), nucleic acids, and a variety of key molecules involved in biosynthesis and metabolism (chlorophylls, electron carriers, enzymes, chelators, vitamins, etc.). As such, demands for exogenous nitrogen required for growth and reproduction are consistently high. Prokaryotic heterotrophs, widely believed to be the initial microbial "producers" in ancient seas and lakes (Margulis 1982), presumably utilized dissolved ammonia and geochemically formed organic and inorganic nitrogen compounds as biologically available sources of nitrogen (Cloud 1976; Holland 1978). Atmospheric ammonia enrichment may have been an additional source of biologically utilizable nitrogen during the Archean–Precambrian period (Holland 1978). Eventually, the chief sources of "combined" nitrogen (NH_3, NO_2^-, NO_3^-, organic nitrogen) became depleted as much of this nitrogen source was incorporated by developing and proliferating microbial communities. Once this reservoir was depleted, the sole remaining source of "new" atmospheric nitrogen was N_2. At this point in the evolution of microorganisms strong selective pressure no doubt arose for microbes to directly utilize N_2 as a nitrogen source. However, in order to utilize N_2 biologically, the highly stable triple bond within the dinitrogen molecule must be cleaved (Stiefel 1977). The reduced Archean–Precambrian atmospheric conditions were favorable for reduction of N_2 by means of cleaving the triple bond and producing biologically available NH_3. Such requirements set the stage for biological nitrogen fixation where through the employment of an enzyme complex capable of reducing N_2 to NH_3 (nitrogenase) atmospheric N_2 became available for microbial growth (Yates 1977; Postgate 1978).

Nitrogen Fixation. Nitrogen fixation is of widespread occurrence among obligate anaerobic heterotrophic and chemolithotrophic bacteria (Postgate 1978), photosynthetic bacteria (also obligate anaerobes) and cyanobacteria (Stewart 1975). It has yet to be found in eukaryotic microorganisms, higher plants, or animals. As stated earlier, a unique characteristic of cyanobacteria is that they are oxygenic, photosynthetic prokaryotes. Equally unique, and from an ecological perspective at least as important, is the fact that numerous members of this group are capable of fixing atmospheric N_2 under ambient oxic conditions (Fogg et al. 1973; Carr & Whitton 1982). The very fact that oxygenic photosynthesis and highly reductive N_2-fixation can accompany each other in cyanobacteria certainly seems paradoxical from a biochemical perspective. Nitrogen fixation is rapidly inactivated (irreversibly in cell-free extracts but reversible under most in vivo conditions) in the presence of molecular oxygen (Yates 1977). Accordingly, the development of compatability between oxygenic photosynthesis and anoxic N_2-fixation must be regarded as a major achievement during the evolution and diversification of aquatic cyanobacteria.

Various physiological and morphological adaptations allow certain cyanobacterial genera to fix atmospheric N_2 as their dominant nitrogen source in aerobic nitrogen-deficient waters. The development of metabolically specialized thick-walled cells termed *heterocysts* among several well-known filamentous genera represents a classic example of parallel physiological and morphological adaptations facilitating N_2-fixation under aerobic conditions (Wolk 1982). Heterocysts are morphologically and physiologically differentiated vegetative cells that possess characteristics necessary for N_2-fixation (Land & Fay 1971; Wolk 1979). During the initial stages of differentiation (which is promoted under nitrogen-deficient conditions) precursors to heterocysts (proheterocysts) reveal distinct changes in pigmentation, generally changing from the bright green granular appearance of vegetative cells to the yellow-hue transparent appearance of mature heterocysts. This color change signals changes accompanying the modification of the light-trapping apparatus (photosystems I and II) and the reaction centers of photosynthesis. The "yellowing" of heterocysts can specifically be attributed to a loss of biliproteins during differentiation (most distinct of which is blue-colored phycocyanin) while chlorophyll *a* and carotenoids remain present (Wolk 1982). Photosystem I activity is retained while photosystem II activity ceases to function (Donze,

Haveman, & Schiereck 1972). As a result, heterocysts remain active in trapping PAR through a functional photosystem I and transferring high-energy-electron to intermediary electron acceptors such as NAD and ferrodoxin. These electrons are subsequently transferred to the nitrogenase enzyme complex situated in heterocysts (Apte, Rowell, & Stewart 1978; Wolk 1982). Moreover, the elimination of photosystem II activity also negates photolysis of H_2O, which normally leads to molecular O_2 production (Fay et al. 1968). Potential O_2 inactivation of heterocyst nitrogenase activity is thereby eliminated. However, the elimination of H_2 as a photosystem II electron donor in photosynthesis necessitates an alternative source of electrons for photosystem I. This source appears to be the neighboring vegetative cells that supply heterocysts with reduced carbon compounds (Wolk 1968, 1982).

Heterocysts undergo high rates of endogenous respiration, ATP synthesis, and hydrogenase activity (Yates 1977; Bothe, Distler, & Eisbrenner 1978; Bothe, Yates, & Cannon 1982). Such enhanced metabolic activities appear to be directed at producing energy required for nitrogenase-mediated N_2-fixation as well as the scavenging and reduction of molecular O_2, which could potentially inhibit nitrogenase activity. Despite the protection afforded by thickened cell envelopes and polar plugs, these morphological adaptations are not absolutely effective in shielding O_2. This can be demonstrated in the filamentous genus *Anabaena* when exposed to O_2-supersaturated conditions, which often depress and inhibit N_2-fixation rates (Stewart & Pearson 1970; Paerl & Kellar 1978). Alleviation of O_2 supersaturation by flushing with an inert gas such as N_2 or Ar leads to immediate enhancement of N_2-fixation rates (Fig. 7-7). O_2 supersaturation is a common feature of surface waters supporting N_2-fixing *Anabaena* blooms.

Several nonheterocystous cyanobacterial genera are also capable of N_2-fixation. However, these organisms require oxygen-free conditions near the site of N_2-fixation. The filamentous genera *Oscillatoria* and *Microcoleus* appear to compensate for the absence of heterocysts either temporally by confining most of their N_2-fixing efforts to nonphotosynthetic periods (nighttime, Bautista & Paerl 1985; Stal & Krumbein 1985), or spatially by growing in colonies which are physiologically differentiated (Carpenter & Price 1976; Bryceson & Fay 1981) or are closely associated with bacteria that are capable of locally removing O_2 around filaments via respiration (Paerl & Kellar 1978). Numerous filamentous and single-celled cyanobacteria can also reside in dimly lit hypoliminia, in surface sediments, or in periphyton communities where their metabolic and energetic needs for reduced carbon compounds may be met either by extremely low rates of photosynthesis (accompanied by low O_2 evolution and growth rates) or by heterotro-

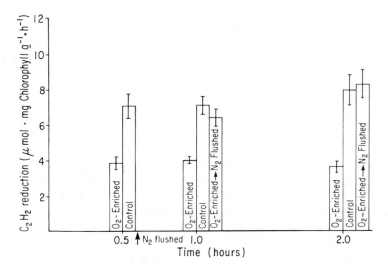

Fig. 7-7. Effects of aqueous O_2 supersaturation (180% O_2 saturation) on acetylene reduction rates (normalized for biomass as chlorophyll *a*) in *Anabaena oscillarioides*. Oxygen supersaturated conditions were created by substituting the headspace with O_2-enriched air in sealed 60-mL serum bottles containing *A. oscillarioides*. Control conditions revealed 95% O_2 saturation. Oxygen saturation determinations were made with a YSI model 54 ARC dissolved O_2 meter equipped with a model 5739 O_2 temperature probe at the initiation of the experiment and during subsequent acetylene reduction assays. After ½ h of O_2 supersaturation, several sets of serum bottles were flushed with N_2 gas, thereby lowering O_2 saturation to approximately 80–90%. Nitrogen flushed bottles were then assayed in parallel with control and O_2-supersaturated bottles. Nitrogen flushing revealed that initial inhibition of acetylene reduction by O_2 supersaturation was reversible and could be rapidly relieved in *A. oscillarioides*. Variability among triplicated treatments is shown.

phic utilization of ambient soluble organic compounds (Reuter, Loeb & Goldman 1983). In this manner, low but uninhibited rates of N_2-fixation could be maintained.

Among colonial nonheterocystous cyanobacteria, the best-documented case for physiological differentiation between internal and external portions of colonies is in the genera *Oscillatoria* and *Microcoleus*. Bundlelike colonies of the filamentous marine *Oscillatoria* (formerly called *Trichodesmium*) reveal partitioning of photosynthetic activities between high activity (in terms of CO_2 assimilation) peripheral filaments and low activity internal filaments. Carpenter & Price (1976) linked the formation of reduced (in terms of O_2 evolution) internal regions to N_2-fixing potentials of *Trichodesmium*. Subsequent determinations of internal redox potentials using the tetrazolium salt 2,3,5-triphenyl-2-tetrazolium chloride (TTC) have similarly indicated that relatively low photosynthetic O_2 evolution rates in internal regions prevailed in comparison to external regions (Bryceson & Fay 1981; Paerl & Bland 1982). The reduction of TTC in internal regions was closely coupled to N_2-fixing potentials among colonial *Trichodesmium*. In fact, if *Microcoleus* bundles were supplied with TTC prior to conducting N_2-fixation (acetylene reduction) assays, nitrogenase activity was effectively blocked (Paerl & Bland 1982). These results strongly suggest that TTC reduction and nitrogenase activity share common sources of reducing power in internal regions of colonies. Additional evidence that internal O_2-reduced microenvironments are important in supporting N_2-fixation can be obtained by comparing N_2-fixation potentials among *Microcoleus* populations composed of single filament versus colonial (aggregated) forms. Aggregated forms consistently demonstrate optimal rates of N_2-fixation per unit chlorophyll *a* (Fig. 7-8). Lastly, when aggregated *Microcoleus* populations are mildly sonicated to disrupt colonies, N_2-fixation rates are immediately suppressed (Fig. 7-8).

It can be concluded that both permanent structural modifications such as heterocyst formation and periodic aggregation are effective means by which potentially harmful O_2 can be spatially separated from the N_2-fixing nitrogenase complex in a variety of cyanobacteria. Such morphological adaptations complement temporal phasing of photosynthesis and N_2-fixation (Gallon, Kurz & LaRue 1975; Stal & Krumbein 1985).

Physiological O_2 removal near the site of N_2-fixation offers additional protection from potential inhibition of nitrogenase activity. A variety of mechanisms play a protective role: (a) activity of superoxide dismutase and catalase enzymes, by which the potentially harmful free radical form of O_2 (superoxide) is enzymatically converted to H_2O_2 and eventually to H_2O (Henry, Gogotov, & Hall 1978); (b) the oxyhydrogen or "knallgas" reaction by which an uptake hydrogenase coupled to an ATP-yielding electron transport system reduces O_2 to H_2O with H_2 acting as an electron donor (Bothe, Distler, & Eisbrenner 1978), the reduction providing the energy requirements for ATP synthesis; (c) photorespiration by vegetative cells through which O_2 is utilized to produce the highly oxidized dicarboxylic acid, glycolate (Tolbert 1974); (d) carotenoid synthesis, which requires O_2 to produce the various carotenes and xanthophylls common to cyanobacteria (Goodwin 1980); (e) extracellular envelope (mucilage) production, which is

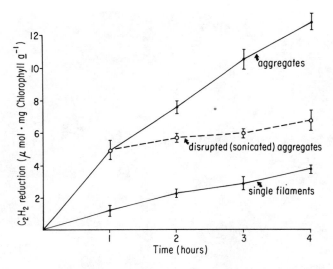

Fig. 7-8. Acetylene reduction rates in *Microcoleus chthonoplastes* populations either consisting of single or aggregated filaments. After a 1-h acetylene reduction assay period, a set of aggregated *Microcoleus* samples was mildly sonicated to disrupt the bundles (leading to a population of single filaments). The dashed line shows subsequent rates of acetylene reduction among disrupted aggregates. Results from all samples were normalized for chlorophyll *a*. Variability among triplicate samples for each treatment is shown.

known to retard inward diffusion of O_2 (Wolk 1982); and (f) specific associations with heterotrophic bacteria leading to localized O_2 consumption (Paerl 1982a).

Cyanobacterial–Bacterial Associations. On the whole, bacterial associations with diverse filamentous, colonial, and single-celled cyanobacterial genera are commonplace in nature (Paerl 1978; Caldwell & Caldwell 1978; Lupton & Marshall 1981). Such associations can exhibit symbiotic characteristics (Paerl 1982a). Prior to recent ecological and physiological studies of such associations, it was popularly believed that bacterial associations with cyanobacteria represented various stages of senescence or breakdown of host cyanobacterial populations. However, it has been shown repeatedly that both groups of microorganisms benefit from each other's presence (Paerl 1982a).

A particularly fortuitous association is that of heterotrophic bacteria and polar regions of heterocysts (Paerl 1982a). The polar regions are known to be sites of organic matter losses from host cyanobacteria

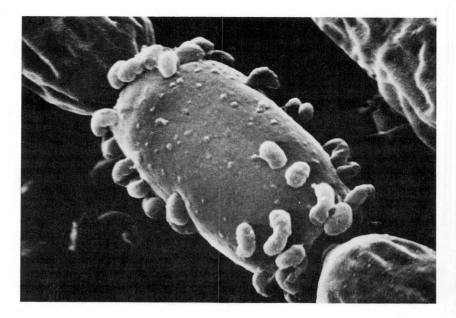

Fig. 7-9. (a) Scanning electron micrograph showing a specific bacterial association with the heterocyst of *Aphanizomenon flos-aquae,* sampled from Clear Lake, California (sampling courtesy of Dr. A. J. Horne). Note the particular preference bacteria exhibit for the polar regions joining heterocysts to adjacent vegetative cells.

(Fogg et al. 1973; Paerl 1984a). Associated bacteria are known to be chemotactically attracted to a variety of organic carbon (sugars, organic acids) and nitrogen (amino acids) compounds (Gallucci & Paerl 1983; Paerl & Gallucci 1985). When observed microscopically, these bacteria demonstrate random encounters and repulsion when coming into contact with vegetative cells. However, following encounters with heterocysts, especially polar regions, bacteria tenaciously remain oriented in such regions, followed by attachment and rapid growth (Fig. 7-9a,b). As metabolite transfer points, polar regions are likely to be more highly susceptible to potential inward diffusion of O_2 from ambient waters because the thick envelopes and cell walls along the sides of heterocysts generally offer stable structural protection against O_2 diffusion. By exhibiting a certain degree of "leakiness," polar regions of heterocysts attract heterotrophic bacteria, which in turn locally consume O_2 during the metabolic utilization and breakdown of excretion products (Paerl 1985). Associated bacteria can in

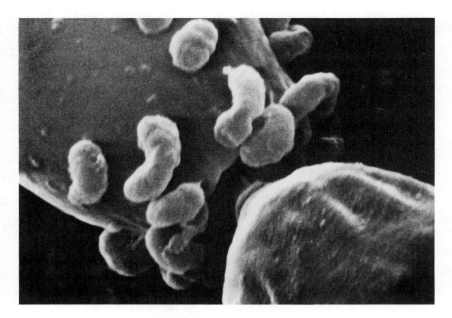

Fig. 7-9. (b) A magnified view of attached bacteria in the polar region. Critical point drying steps during sample preparation have led to some shrinkage, exposing the polar plug connecting adjacent cells.

this manner be considered as a metabolic barrier to potential O_2 inactivation of nitrogenase activity.

Experimentally, it can be shown among filamentous N_2-fixing cyanobacteria that optimal N_2-fixation rates are often achieved in the presence of bacteria forming specific associations with host cyanobacteria. As an example, axenically (bacteria-free) grown *Anabaena oscillarioides* populations reveal from 10 to 50% lower rates of N_2-fixation (per amoung of chlorophyll *a*) than do populations grown in the presence of associated bacteria (Fig. 7-10a). Similar results are obtained when examining *A. oscillarioides* growth rates (Fig. 7-10b).

In the laboratory, numerous N_2-fixing and non-N_2-fixing cyanobacterial species are incapable of growth and propagation without the presence of bacteria (Fogg 1982; Ohki & Fujita 1982); clearly, this serves as testimony to the interdependence, if not symbiotic behavior, embodied in bacterial–cyanobacterial associations. In early studies of several such associations, Lange (1967) and Kuentzel (1969), among others, noted that cyanobacteria associated with bacteria often

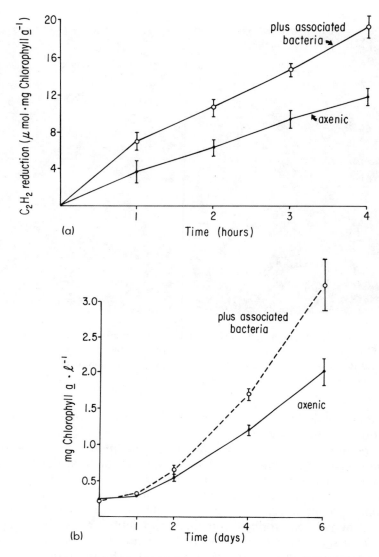

Fig. 7-10. (a) Acetylene reduction (normalized for biomass as chlorophyll *a*) in axenic versus bacteria-associated populations of *Anabaena oscillarioides*. Hourly subsamples were analyzed from triplicated samples representing each treatment. Variability among triplicates is shown at each sampling interval. (b) Comparative growth curves (biomass estimated as chlorophyll *a*) for axenic versus bacteria-associated *Anabaena oscillarioides*. Variability among triplicated samples is shown during a six-day growth period.

attained higher growth rates when the association was supplied with organic carbon compounds (simple sugars). In searching for a mechanism for this response, it was observed that photosynthetic depletion of ambient CO_2 (and to some extent HCO_3^-) occurred among the dense cyanobacterial populations used, thus raising the pH of surrounding poorly buffered water to values in excess of 10. Under such conditions the presence of associated bacteria was particularly effective in optimizing host cyanobacterial growth. It was subsequently concluded that bacterial mineralization of organic matter either released by cyanobacteria or added to these waters produced CO_2 that was immediately utilized for photosynthesis by CO_2-starved cyanobacteria. In this manner associated bacteria optimized cyanobacterial growth. More recent studies by Burris, Wedge, and Lane (1981), Pearl (1982a), Schiefer and Caldwell (1982), and others have confirmed that localized CO_2 depletion during bloom conditions in scums and mats can commonly occur, especially under nutrient (N, P)-enriched conditions. Bacterial associations with cyanobacteria, in part, help alleviate such environmental limitations. Other benefits thought to be derived from associations with diverse bacteria include (a) phycosphere recycling of phosphorus (as PO_4^{3-}) and nitrogen (as NH_3) through bacterial metabolism; (b) supplementation of growth factors, including vitamins; (c) production of metal-specific chelators and surface active compounds by bacterial epiphytes that might play roles in regulating trace metal availability and colonial aggregation characteristics.

From an evolutionary perspective, physically and metabolically coupled associations between freshwater cyanobacteria and a variety of eubacteria have aided in overcoming a host of potential environmental limitations. The ability to locally remove O_2 during the development of an O_2-rich biosphere seems advantageous to any organisms possessing O_2-labile enzyme systems such as nitrogenase. Similarly, phycosphere recycling of essential inorganic nutrients would be highly advantageous under the increasingly common conditions of nutrient limitation accompanying the development and proliferation of freshwater phytoplankton communities. Lastly, the associations with bacteria capable of producing and supplying specific essential growth factors and vitamins would seem to be a viable alternative to endogenous synthesis (and accordingly an increased need for biochemical, physiological, and structural specialization) of such substances.

The "cost" for such associations was relatively simple and affordable from a metabolic standpoint. The fact that passive and active excretions of products of photosynthesis and N_2-fixation appear commonplace among many cyanobacterial genera (Walsby 1974; Nalewajko 1978) may be viewed as a fortuitous set of events. By being "leaky," cyanobacteria can potentially attract a wide range of hetero-trophic bacteria. In exchange for the nutritional needs such excretion products provide, bacterial recycling or synthesis of specialized com-pounds provide a benefit for host cyanobacteria. In this regard, it is fascinating that among filamentous N_2-fixing cyanobacteria, the het-erocyst–vegetative cell junction is a "weak link" where filament break-age is often initiated. It would seem paradoxical that such a highly spe-cialized region of filaments would prove to be highly susceptible to physical disruption. However, when we consider the fact that mutually beneficial bacterial associations are often promoted in this leaky region, there may exist a possible ecological explanation for such a structural "weakness."

A fine line appears to exist between mutualistic and antagonistic bacterial associations. Often such different interactive phases may be separated temporally over a few days or spatially over a few centime-ters. Indeed, one of the more spectacular and ill-understood sequential phases of blooms is the rapidity with which a cyanobacterial popula-tion can "crash" as a rotting mass of decomposing organic matter (Fogg 1969). Under such conditions, bacteria "associated" with cyano-bacteria act largely as highly opportunistic heterotrophic decomposers, invading the remains of previously dominant cyanobacterial popula-tions. Why and how does this change from mutalistic to antagonistic behavior develop within the bacterial community?

Careful observations of cyanobacterial blooms in relation to fluc-tuations in essential and sometimes limiting environmental factors reveal that crashes of blooms are often initially linked to an abrupt change in environmental conditions or a cessation in physical–chem-ical stability of ambient waters rather than to specific antagonistic infestations by associated microorganisms (Ganf & Viner 1973; Rey-nolds & Walsby 1975; Paerl 1983b). For example, a rapid depletion of essential nutrients through a change in water mass characteristics brought about by flushing, rainfall, or sudden nutrient depletion by a massive bloom can lead to cyanobacteria becoming nutrient stressed (Paerl 1983b). As with most other aquatic organisms, environmental stress leads to altered physiological characteristics ("stress response") and a greater degree of susceptibility to infections, disease, and death. Likewise, sudden changes in water column temperature, irradiance, and stability represent environmental changes that may occur faster than an organism's ability to adjust to such changes. Again, physiolog-ical stress is the likely outcome of such a scenario, and increased sus-ceptibility to infections and microbial attack (Daft & Stewart 1971) could be expected.

Environmentally induced crash conditions are commonplace in water bodies faced with periodic cyanobacterial blooms (Fogg 1969).

During late summer and early fall such events are particularly evident, often resulting from rapid weather changes (cooling), nutrient depletion, and changes in water column stability (increased vertical mixing or "turnover"). In this context, bacterial antagonism is viewed by this author as a *result* rather than a *cause* of physiological stress imposed on cyanobacterial bloom populations. Therefore, in view of unforeseeable environmental changes, the physiological characteristics of cyanobacteria and their microbial cohorts are radically altered. In tropical regions such crash events might last a few days to weeks, after which reestablished physical–chemical stability can again usher in cyanobacterial populations capable of eliciting a mutually beneficial association with heterotrophic microorganisms (Ganf & Viner 1973; Ganf & Horne 1975). However, in temperate regions the cyanobacterial community may need to "overwinter" physically unstable conditions as cysts or akinetes (Nichols & Adams 1982) until the following growth season when such stability is reestablished long enough to support and promote blooms.

COMPETITION AND TROPHIC INTERACTIONS

Diverse genera of nuisance cyanobacteria produce toxins (Collins 1978, Gorham & Carmichael 1980). This trait is only known to be shared with eukaryotic dinoflagellates, but chemical compositions of the toxins are quite dissimilar (Schantz 1971; Collins 1978). Toxin production by freshwater strains of *Microcystis* and *Anabaena* has been most thoroughly examined. *Microcystis* toxin, or microcystin, is composed of several alkaloids, peptides, and cyclic polypeptides. It exhibits hepatic and neurotoxic properties when administered to laboratory mice (Collins 1978). The toxin is potent enough to cause death in such animals within several minutes. Accordingly, microcystin has been alternatively named the *fast death factor* (FDF). Strains of *Anabaena flos-aquae* are known to produce several highly potent toxins collectively termed *very fast death factors* (VFDF) (Gorham & Carmichael 1980). As with *Microcystis,* this toxin is composed of both alkaloids and peptides. However, VFDF toxins are readily water soluble, whereas *Microcystis* FDF generally behaves like an endotoxin. Due to its high solubility and low molecular weight, neuro- and hepatotoxic impacts of VFDF can affect aquatic organisms that have not necessarily ingested intact *Anabaena* cells (Gorham et al. 1964).

Although it is tempting to characterize toxicity as a potentially adaptive and competitive feature within plankton communities, the ecological and physiological rationales as well as environmental impacts of cyanobacterial toxin production remain unclear and puzzling. General ecological explanations for toxicity among nuisance genera include (a) exotoxin (extracellular toxin) production, which negatively impacts potentially competitive phytoplankton popula- tions, resulting in a reduction of species diversity and potential dom- inance by nuisance cyanobacterial taxa, and (b) endotoxins (intracel- lular toxins), which discourage grazing and consumption by herbivores, thereby initiating an avoidance response toward cyanobac- teria along with increased grazing pressure on nontoxic eukaryotic phytoplankton taxa. Both explanations clearly emphasize the linkage of toxin production to the competitive success of cyanobacteria. In natural waters, however, such "ideal" rationales for toxin production often lack confirmation. In fact, paradoxical, if not contradictory, impacts of cyanobacterial toxin production are often observed. There are numerous cases where toxins produced by nuisance cyanobacteria (i.e., *Anabaena, Microcystis, Aphanizomenon*) have shown no detect- able effect on other eukaryotic members of the phytoplankton com- munity (see Gorham & Carmichael 1980; Fulton & Paerl in prep.). Such toxins do, however, cause sickness and death among terrestrial animals that are not common grazers on cyanobacteria (Gorham & Carmichael 1980). As a result, specific cyanobacterial toxins that can kill a mouse within 1 min or that lead to cattle deaths over a period of several days often fail to adversely impact noncyanobacterial members of the phytoplankton community, dominant zooplankton, other inver- tebrates, or fish species. Cyanobacterial dominance under such cir- cumstances cannot be directly attributed to toxicity. More often, resul- tant dominance and bloom phenomena can be explained through alternative competitive interactions such as surface-scum formation and shading of underlying eukaryotic phytoplankton populations, N_2- fixing capabilities during nitrogen-depleted periods, mutualistic and symbiotic interactions specific to cyanobacterial–bacterial consortia, and rapid vertical mobility among specific cyanobacterial genera that promote ready access to both rich hypolimnetic nutrient supplies and near-surface PAR sources.

To varying degrees, cyanobacterial endotoxins and exotoxins can adversely affect a range of freshwater zooplankton known to be dom- inant grazers in both eutrophic and mesotrophic waters (Gentile 1971; Gorham & Carmichael 1980). Depending on the specific toxicities of cyanobacterial strains as well as diverse sensitivities of resident zoo- plankton species, a wide range of toxic and negative trophic effects (poor palatability) have been noted by workers. The physiological and/ or mechanistic impacts of cyanobacterial toxins on diverse members of the zooplankton community are poorly understood. Virtually all controlled laboratory observations of such impacts have been confined to the cladoceran *Daphnia*. A general review of observational and experimental work reveals that feeding behavior among cladocera as a group is often negatively affected by filamentous or colonial cyanobacteria (Burns & Rigler 1967; Arnold 1971; Porter 1977; Webster & Peters 1978; Porter & Orcutt 1980). From a mechanical perspective, the cladocerans that preferentially graze on a range of small cells (bacteria to single-celled algae: 1–20 μm) filter and subsequently assimilate filamentous and colonial cyanobacteria inefficiently. The presence of multicellular cyanobacteria seems to negatively affect some cladocerans more severely than others; this differential effect may be related to constrasting body sizes among cladocerans. In addition to mechanical constraints, poor tastes of cyanobacteria also lead to restricted ingestion and assimilation among cladocerans. Clark (1978), Porter and Orcutt (1980), and Lampert (1981) demonstrated that unicellular *Microcystis* or small fragmented *Anabaena* filaments that represented no mechanical constraints were nevertheless avoided by cladocerans. Similar cell densities of identical size-range chlorophyceans were readily grazed and assimilated by the same grazers. These experiments clearly demonstrate poor palatability ("taste"), possibly due to toxicity associated with these cyanobacteria.

If neither mechanical nor chemical constraints are present, many cladocerans will ingest cyanobacteria. Subsequent assimilation values of ingested cyanobacteria can vary substantially among cladocerans. For example, Schindler (1968) and Sorokin (1968) generally found cladoceran assimilation of *Anabaena* (and other cyanobacteria) to be consistently lower than assimilation of chlorophyceans (*Chlamydomonas* and *Chlorella*) even when cladocerans exhibited equal feeding rates on each food type. On the other hand, Arnold (1971) showed that *Daphnia pulex* demonstrated generally high cyanobacterial assimilation efficiencies ranging from 16% for *Anacystis nidulans* to 100% for *Anabaena flos-aquae.* Lampert (1977) reported variable cyanobacterial assimilation values for *D. pulex,* including quite low values for *Microcystis.* It can be concluded from these results that even if cladocerans can circumvent mechanical and chemical constraints of feeding, the often low assimilatory values of cyanobacteria strongly suggest that cyanobacteria represent a nutritionally poor phytoplankton group. As a result, cladoceran survivorship and growth based on a cyanobacterial diet has proven inferior in a majority of cases when compared to diets based on eukaryotic (chlorophycean) phytoplankton.

Copepod ingestion and hence assimilation of cyanobacteria also appear inhibited to varying degrees. Both Bogatova (1965) and Infante (1978) reported generally few cyanobacteria in the guts of several calanoid copepods even though surrounding lake waters supported abundant cyanobacterial populations. Furthermore, little cell disruption of ingested cyanobacteria usually takes place, leading to virtually no assimilatory value of ingested food and live passage out of the digestive tract of copepods. Both mechanical and chemical constraints are again most often cited as the chief factors responsible for poor ingestion and assimilation of cyanobacteria. Cyclopoid copepods, being more selective as well as being predatory feeders, appear to more effectively utilize some filamentous and colonial cyanobacteria as food, although this ability is mainly confined to certain specialized taxa. The tropical species *Thermocyclops hyalinus* and *Mesocyclops crassus* can utilize *Lyngbya limnetica* as a food source during blooms (Infante 1978). Diverse temperate herbivorous cyclopoids appear to make little use of cyanobacteria (Fryer 1957). Infante (1981) also found virtually no utilization of *Microcystis* or *Anabaena* by temperate *Mesocyclops crassus*. What *Microcystis* had been ingested appeared to remain intact following gut passage.

In contrast to the generally negative impacts reported for crustacean zooplankton, members of the microzooplankton including rotifers and protozoans often "cobloom" with dominant cyanobacterial species (Braband et al. 1983). Rotifers in particular appear to show little, if any, sensitivity to cyanobacterial bloom species known to induce either avoidance behavior or physical deterioration (starvation and death) among crustaceans. Several investigators have shown common rotifers to be able to subsist, grow, and reproduce on a diet exclusively composed of cyanobacteria. Rotifers appear to make use of cyanobacterial biomass in several ways. Direct feeding on portions of cyanobacterial filaments and colonies can be accomplished through the grasping and breaking off of fragments composed of individual cells or small pieces of filaments followed by ingestion (Starkweather & Kellar 1983). Furthermore, most rotifers are effective consumers of cyanobacteria-associated bacteria. It can be shown that bacteria associated with cyanobacterial bloom species, having derived their biomass from cyanobacterial photosynthate and metabolites, are actively grazed off host filaments and colonies by closely associated rotifers. In a somewhat similar scenario, flagellated and amoeboid protozoans are active grazers of bacterial epiphytes as well as colonial cyanobacteria during bloom events. In fact, *Amoebae* that characteristically colonize *Microcystis* colonies during blooms on the lower Neuse River, North Carolina, eventually establish themselves inside colonies, subsisting entirely on cyanobacterial cells and bacteria that are consumed through phagocytosis.

Such protozoan colonization of cyanobacterial colonies and filaments, as is the case with bacterial associations, does not necessarily signal the onset of senescence or crashes among blooms. More often,

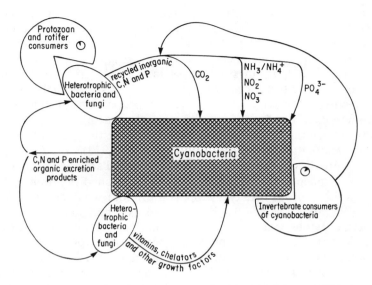

Fig. 7-11. Schematic diagram of heterotrophic bacterial and inverte-brate herbivore interactions with cyanobacteria. Included are routes of cyanobacterial biomass production and utilization, nutrient recycling pathways, and microbial means of supplying growth factors within the cyanobacterial "phycosphere."

protozoan–cyanobacterial associations continue throughout weekly or monthly bloom events. Preliminary evidence indicates that although protozoans and rotifers readily consume cyanobacterial cells during colonization, host cyanobacterial species exhibit enhanced cellular growth rates and greater bloom intensities as a result of such associations. Rotifer and protozoan consumption and digestion of cyanobacterial biomass most likely lead to the release of metabolites and waste products containing inorganic nutrients, such as ammonia, phosphate, and a variety of metals. Intense microzooplankton grazing thus promotes nutrient cycling within the cyanobacterial phycosphere. This is analogous to the nutrient-regneration processes inherent in commonly observed cyanobacterial phycophere associations discussed previously (Fig. 7-11).

MORPHOLOGY AS AN ADAPTIVE FEATURE

It has been proposed that filamentous and colonial habits of cyano-bacteria have in part evolved as structural adaptations effective in minimizing zooplankton grazing pressure (see Holm, Ganf, & Shapiro 1983). However, despite the fact that coloniality clearly represents a barrier to crustacean utilization of cyanobacterial biomass, diverse fish and bird taxa have developed means of utilizing such colonies as food (Fogg et al. 1973; Livingstone & Melack 1984). In tropical waters numerous cichlid fish species *(Tilapia)* are effective colonial cyanobacteria feeders (Moriarty 1973). A large proportion of the protein requirements of flamingos is met through the grazing of surface dwelling cyanobacteria *(Spirulina, Microcystis)*, as are the pigments responsible for feather and bill colors (see Fogg et al. 1973).

There are diverse physiological and ecological advantages to a multicellular existence for cyanobacteria. As mentioned previously, colonial and filamentous taxa can support morphologically and physiologically differentiated cells such as heterocysts and akinetes. Among nonheterocystous taxa, colonies and bundles of filaments lead to the formation of chemical and physical gradients that promote essential physiological processes. Vertical migration is more rapidly accomplished among filaments and colonies as opposed to single cells of gasvacuolate species because multicellular forms are able to rise and sink in the water column more quickly than single cells of comparable density (Reynolds 1975; Reynolds & Walsby 1975).

In summary, several physiological and ecological reasons exist for the development and maintenance of coloniality among cyanobacteria. Clearly, coloniality is advantageous from both growth and reproductive perspectives. Some ecologically successful cyanobacterial genera such as *Synechoccocus* and *Cyanodictyon* never resort to coloniality or filament formation. Instead, these genera are made up of quite small species (1–2-μm cell diameter) that, like numerous eukaryotic phytoplankton, take advantage of high surface to volume ratios, high-affinity nutrient-uptake kinetics and high photosynthetic efficiencies under extremely low light levels in order to occupy low-nutrient and poorly lit waters (Paerl & Mackenzie 1977; Gibson & Smith 1982). Therefore, in terms of morphological specialization, elaboration, and modification, cyanobacterial members of the phytoplankton community appear to have evolved in diverse ways in order to exploit available environmental resources among a range of trophic states.

REPRODUCTION

Cyanobacterial Resting Cells and Propagules

Akinetes are morphologically and physiologically differentiated cells (Geitler 1932) of filamentous colonial and single-celled cyanobacteria that arise from vegetative cells (see Nichols & Adams 1982). Akinetes are normally larger than vegetative cells, having a granular appearance and thickened cell wall. During their differentiation stages akinetes reveal enhanced and altered metabolic activities. In particular, a relatively large proportion of photosynthetic production as compared to vegetative cells is incorporated by akinetes. As a result, akinetes accumulate rich supplies of carbonaceous and nitrogenous storage inclusions as glycogen granules, polyhedral and lipid bodies, and cyanophycin granules (Sutherland, Herdman, & Stewart 1979). There is little doubt that the most ecologically vital function of akinetes is ensuring appropriate "seeding" of both the water column and sediments during environmentally unfavorable growth that includes ice-covered winter months or nutrient-starved periods (Nichols & Adams 1982). Abundant storage products and a relatively impermeable cell wall ensure "overwintering" for at least several years in diverse aquatic habitats (Preston, Stewart, & Reynolds 1980). Sediments of mesotrophic and eutrophic lakes known to support cyanobacteria normally exhibit high numbers of resting akinetes representing numerous species.

Conditions inhibiting continued cyanobacterial growth generally promote differentiation of vegetative cells to akinetes. Aside from previously mentioned adverse changes in weather, other, more specific, factors have been suspected of controlling akinete development. Included are (a) deficiencies in fixed nitrogen supplies (Demeter 1956), principally affecting non-N_2-fixing genera; (b) depletion of available phosphorous (Wolk 1965); (c) deficiencies in iron (Sinclair & Whitton 1977); and (d) light limitation (Sutherland et al. 1979). Organic compounds appear to elicit mixed effects on akinete formation (Nichols & Adams 1982). Enrichment with amino acids, for example, has been shown to increase akinete formation, and certain sugars trigger the opposite effect. Extracellular organic compounds produced by "healthy" cyanobacterial cells have also been shown to stimulate akinete formation (Fisher & Wolk 1976). Clearly, the suite of environmental factors that may control akinete formation are complex and at present are just beginning to be understood (Rother & Fay 1979). Moreover, the mechanisms by which such factors control akinete formation are poorly known (Rother & Fay 1979).

Favorable conditions for cyanobacterial growth lead to germination of akinetes (Rother & Fay 1979). Germination usually leads to the production of at least one and up to 4–5 (unicellular and filamentous genera, respectively) vegetative cells that subsequently form the inocula for actively growing populations (Miller & Lang 1968).

Environmental and evolutionary success has been ensured in part through the production of akinetes. Akinetes are known to be able to withstand extreme fluctuations in temperature, desiccation, pressure, and irradiance, and can accordingly be found under what are normally construed to be "hostile" environmental conditions for algal growth, such as desert and alpine soils, snow and ice, desiccated rocks in Antarctica, deep-sea and lake environments, volcanic lakes, and others.

CONCLUSION: THE CYANOBACTERIAL "STRATEGY"

It is generally agreed that cyanobacteria by virtue of their diverse and unique physiological and ecological traits are an extremely opportunistic freshwater phytoplankton group. Because of their ability to physiologically diversify as oxygenic photoautotrophs, cyanobacteria have impacted and transformed the biosphere in a manner unparalleled by other phytoplankton taxa; the transition from ancient anoxic to current oxic aqueous and atmospheric conditions is largely attributable to the advent and proliferation of cyanobacteria. Furthermore, the impact of cyanobacterial blooms on trophic conditions and resultant recreational and commercial values of affected aquatic ecosystems has been profound.

Despite evolutionary and contemporary biospheric impacts attributable to cyanobacteria, this phytoplankton group is conspicuously sensitive to physical features of the aquatic environment. An evaluation of key ecological criteria underlying cyanobacterial growth and reproductive success suggests that water column stability, adequate irradiation, and consistency in water column temperatures are vitally important. Even if nutrient requirements are satisfied, unfavorable physical conditions can rapidly modify otherwise favorable growth conditions and thereby influence the persistence and bloom potentials for a host of common cyanobacterial genera. Accordingly, a key to the evolutionary and current success of cyanobacteria is the periodic (including seasonal) persistence of physically stable conditions in aquatic ecosystems. Numerous aquatic ecosystems on earth can be considered to be extreme as judged by unusual physical–chemical and resultant biotic conditions. Such extreme conditions can prove advantageous to physiologically and ecologically adaptive cyanobacteria. However, physical and, to a lesser extent, chemical stability must ideally accompany such extremes. Extremeness, within the context of cyanobacterial ecology, therefore signifies marginally inhabitable but environmentally stable habitats. In many respects such conditions probably typified Archean and Precambrian aquatic ecosystems within which cyanobacteria first evolved and flourished.

REFERENCES

Apte, S. K., Rowell, P. and Stewart, W. D. P. (1978). Electron donation to ferredoxin in heterocysts of the N_2 fixing alga *Anabaena cylindrica. Proc. R. Soc. Lond. B Biol. Sci.,* 200, 1–25.

Arnold, D. E. (1971). Ingestion, assimilation, survival, and reproduction by *Daphnia pulex* fed seven species of blue-green algae. *Limnol. Oceanog.,* 16, 906–20.

Baalsrud, K. (1982). The rehabilitation of Norway's largest lake. *Water Sci. Technol.,* 14, 21–30.

Bailey-Watts, A. E., Bindloss, M. E., and Belcher, J. H. (1968). Freshwater primary production by a blue-green alga of bacterial size. *Nature,* 220, 1344–5.

Barghoorn, E. S. and Schopf, J. W. (1965). Microorganisms from the Late Precambrian of central Australia. *Science,* 150, 337–9.

Bautista, M. F. and Paerl, H. W. (1985). Diel N_2 fixation in an intertidal marine cyanobacterial mat community. *Mar. Chem.,* 16, 369–77.

Beeton, A. M. (1965). Eutrophication of the St. Lawrence Great Lakes. *Limnol. Oceanogr.,* 10, 240–54.

Bogatova, I. B. (1965). The food for daphnids and diaptomids in ponds. Trudy vseross. nauchno-issled. *Inst. prud. knozy.,* 13, 165–78. (In Russian, transl. J. F. Haney).

Bothe, H., Distler, E., and Eisbrenner, G. (1978). Hydrogen metabolism in blue-green algae. *Biochemie,* 60, 277–89.

Bothe, H., Yates, M. G., and Cannon, F. C. (1982). Nitrogen fixation. In: *Encyclopedia of Plant Physiology, New Series,* eds. A. Lauchli and R. Bieleski. Berlin: Springer.

Braband, A., Faafeng, B. A., Kallqvist, T., and Nilssen, J. P. (1983). Biological control of undesirable cyanobacteria in culturally eutrophic lakes. *Oecologica,* 60, 1–5.

Brock, T. D. (1967). Life at high temperatures. *Science,* 158, 1012–19.

Brook, A. J. (1959). The waterbloom problem. *Proc. Soc. Water Treat. Exam.,* 8, 133–7.

Brown, R. M., Larson, D. A., and Bold, H. C. (1964). Airborne algae: Their abundance and heterogeneity. *Science,* 143, 583–5.

Bryceson, I. and Fay, P. (1981). Nitrogen fixation in *Oscillatoria (Trichodesmium) erythraea* in relation to bundle formation and trichome differentiation. *Mar. Biol.,* 61, 159–66.

Burns, C. W. and Rigler, F. H. (1967). Comparison of filtering rates of *Daphnia rosea* in lake water in suspensions of yeast. *Limnol. Oceanogr.,* 12, 492–502.

Burris, J. E., Wedge, R., and Lane, A. (1981). Carbon dioxide limitation of photosynthesis of freshwater phytoplankton. *J. Freshwat. Ecol.,* 1, 81–90.

Caldwell, D. E. and Caldwell, S. J. (1978). A *Zoogloea* sp. associated with blooms of *Anabaena flos-aquae. Can. J. Microbiol.,* 24, 922–31.

Carpelan, L. H. (1964). Effects of salinity of algal distribution. *Ecology,* 45, 70–7.

Carpenter, E. J. and Price, C. C., IV (1976). Marine *Oscillatoria (Trichodesmium)*: Explanation for aerobic nitrogen fixation without heterocysts. *Science,* 191, 1278–80.

Carr, N. G. and Whitton, B. A. (1982). *The Biology of Cyanobacteria.* Oxford: Blackwell Scientific.

Clark, N. V. (1978). The food of adult copepods from Lake Kainji, Nigeria. *Freshwat. Biol.,* 8, 321–6.

Cloud, P. (1976). Beginnings of biospheric evolution and their biogeochemical consequences. *Paleobiology,* 2, 351–87.

Collins, M. (1978). Algal toxins. *Microbial. Rev.,* 42, 725–46.

Cronberg, G., Gelin, C., and Larsson, K. (1975). The Lake Trummen restoration project. II. Bacteria, phytoplankton and phytoplankton productivity. *Verh. Int. verin. Limnol.,* 19, 1088–96.

Daft, M. J. and Stewart, W. D. P. (1971). Bacterial pathogens of freshwater blue-green algae. *New Phytol.,* 70, 819–24.

Demeter, O. (1956). Über Modifikationen bei Cyanophycean. *Arch. Mikrobiol.,* 24, 105–33.

Dinsdale, M. T. and Walsby, A. E. (1972). The interrelations of cell turgor pressure, gas vacuolation, and buoyancy in a blue-green alga. *J. Exp. Bot.,* 23, 561–70.

Donze, M., Haveman, J., and Schiereck, P. (1972). Absence of photosystem II in heterocysts of the blue-green alga *Anabaena. Biochim. Biophys. Acta,* 256, 157–61.

Droop, M. R. (1974). Heterotrophy of carbon. In: *Algal Physiology and Biochemistry,* ed. W. D. P. Stewart, pp. 530–59. Oxford: Blackwell, Scientific.

Drouet, F. and Daily, W. A. (1956). Revision of the coccoid Myxophyceae. *Butler Univ. Bot. Stud.,* 12.

Edmondson, W. T. (1969). Eutrophication in North America. In: *Eutrophication: Causes, Consequences, Correctives,* pp. 112–49. *Natl. Acad. Sci. Publ.* No. 1700.

Eloff, J. N., Steinitz, Y., and Shilo, M. (1976). Photooxidation of cyanobacteria in natural conditions. *Appl. Environ. Microbiol.,* 31, 119–26.

Fay, P., Stewart, W. D. P., Walsby, A. E. and Fogg, G. E. (1968). Is the heterocyst the site of nitrogen fixation in blue-green algae? *Nature,* 220, 810–12.

Fisher, R. W. and Wolk, C. P. (1976). Substance stimulating the differentiation of spores of the blue-green alga *Cylindrospermum licheniforme. Nature,* 259, 394–5.

Fogg, G. E. (1969). The physiology of an algal nuisance. *Proc. R. Soc. Lond. B Biol. Sci.,* 173, 175–89.

 (1982). Marine plankton. In: *The Biology of Cyanobacteria,* eds. N. G. Carr and B. A. Whitton. Oxford: Blackwell Scientific.

Fogg, G. E., Steward, W. D. P., Fay, P., and Walsby, A. E. (1973). *The Blue-Green Algae.* London: Academic Press.

Fryer, G. (1957). The food of some freshwater cyclopoid copepods and its ecological significance. *J. Anim. Ecol.,* 26, 263–86.

Gallon, J. R., Kurz, W. G. W., and LaRue, T. A. (1975). The physiology of nitrogen fixation by a *Gloeocapsa* sp. In: *Nitrogen Fixation by Free-living Microorganisms,* ed. W. D. P. Stewart, pp. 159–73. Cambridge: Cambridge University Press.

Gallucci, K. K. and Paerl, H. W. (1983). *Pseudomonas aeruginosa* chemotaxis associated with blooms of N_2-fixing blue-green algae (Cyanobacteria). *Appl. Environ. Microbiol.,* 45, 557–61.

Ganf, G. G. (1975). Photosynthetic production and irradiance-photosynthesis relationships of the phytoplankton from a shallow equatorial lake (Lake George, Uganda). *Oecologia,* 18, 165–83.

Ganf, G. G. and Horne, A. J. (1975). Diurnal stratification, photosynthesis and nitrogen fixation in a shallow equatorial lake (Lake George, Uganda). *Freshwat. Biol.,* 5, 13–39.

Ganf, G. G. and Viner, A. B. (1973). Ecological stability in a shallow equatorial lake (Lake George, Uganda). *Proc. R. Soc. Lond. B. Biol. Sci.,* 184, 321–46.

Geitler, L. (1932). Cyanophyceae. In: *Kryptogamen-Flora von Deutschland, Osterreich and der Schweiz* 14, ed. L. Rabenhorst. Leipzig: Akademische Verlagsgesellschaft.

(1960). Schizophyceen. In: *Encyclopeida of Plant Anatomy,* eds. W. Zimmermann and P. Ozenda. Handbuch der Pflanzenanatomie, 2d ed., pt. 1. Berlin: Bortrager.

Gentile, J. H. (1971). Blue-green and green algal toxins. In: *Microbial Toxins,* Vol. 7. Algal and Fungal, eds. A. Ciegler and S. J. Ajl. New York: Academic Press.

Gerloff, G. C. and Skoog, F. (1957). Nitrogen as a limiting factor for the growth of *Microcystis aeruginosa* in southern Wisconsin lakes. *Ecology,* 38, 556–61.

Gibson, C. E., Wood, R. B., Dickson, E. L., and Jenson, D. H. (1971). The succession of phytoplankton in Lough Neagh, 1968–1970. *Mitt. Int. Verein. Theor. Angew. Limnol.,* 19, 140–60.

Gibson, C. E. and Smith R. V. (1982). Freshwater plankton. In: *The Biology of Cyanobacteria,* eds. N. G. Carr and B. A. Whitton, pp. 463–90. Oxford: Blackwell Scientific.

Goodwin, T. W. (1980). *Biochemistry of the Carotenoids,* 2d ed., vol. 1, London: Chapman and Hall.

Gorham, P. R. and Carmichael, W. W. (1980). Toxic substances from freshwater algae. *Prog. Water Technol.,* 12, 189–98.

Gorham, P. R., McLachlan, J., Hammer, V. T., and Kim, W. W. (1964). Isolation and culture of toxic strains of *Anabaena flos-aquae* (Lyngb.) de Breb. *Verh. Int. Verein. Limnol.,* 15, 796–804.

Grant, N. G. and Walsby, A. E. (1977). The contribution of photosynthate to turgor pressure rise in the planktonic blue-green alga *Anabaena flos-aquae. J. Exp. Bot.,* 28, 409–15.

Griffiths, B. M. (1939). Early references to water-blooms in British lakes. *Proc. Linn. Soc. Lond.,* 151, 12–19.

Grillo, J. F. and Gibson, J. (1979). Regulation of phosphate accumulation in the unicellular cyanobacterium *Synechococcus. J. Bact.,* 140, 508–17.

Harris, G. P. (1980). Temporal and spatial scales in phytoplankton ecology.

Mechanisms, methods, models and management. *Can. J. Fish. Aquat. Sci.,* 37, 877–900.

Harris, G. P., Haffner, G. D., and Piccinin, B. B. (1980). Physical variability and phytoplankton communities. II. Primary productivity by phytoplankton in a physically variable environment. *Arch. Hydrobiol.,* 88, 393–425.

Healey, F. P. (1973). Characteristics of phosphorous deficiency in *Anabaena. J. Phycol.,* 9, 383–94.

(1982). Phosphate. In: *The Biology of Cyanobacteria,* eds. N. G. Carr and B. A. Whitton, pp. 105–24. Oxford: Blackwell Scientific.

Henry, L. E. A., Gogotov, I. N. and Hall, D. O. (1978). Superoxide dismutase and catalase in the protection of the proton-donating systems of nitrogen fixation in the blue-green alga *Anabaena cylindrica. Biochem. J.,* 174, 373–7.

Hoare, D. S., Hoare, S. L., and Moore, R.B. (1967). The photo-assimilation of organic compounds by autotrophic blue-green algae. *J. Gen. Microbiol.,* 49, 351–70.

Holland, H. D. (1978). *The Chemistry of the Atmosphere and Oceans.* New York: Wiley.

Holm, N. P., Ganf, G. G., and Shapiro, J. (1983). Feeding and assimilation rates of *Daphnia pulex* fed *Aphanizomenon flos-aquae. Limnol. Oceanogr.,* 28, 677–87.

Holm-Hansen, O. (1964). Isolation and culture of terrestrial and freshwater algae of Antarctica. *Phycologia,* 4, 43–50.

Hutchinson, G. E. (1961). The paradox of the plankton. *Am. Nat.,* 95, 137–45.

Infante, A. de. (1978). Natural food of herbivorous zooplankton of Lake Valencia (Venezuela). *Arch. Hydrobiol.,* 82, 347–58.

(1981). Natural food of copepod larvae from Lake Valencia, Venezuela. *Verh. Int. Verein. Limnol.,* 21, 709–14.

Jensen, T. E. (1968). Electron microscopy of polyphosphate bodies in a blue-green alga, *Nostoc pruniforme. Arch. Mikrobiol.,* 62, 144–52.

Johnston, R. (1964). Seawater, the natural medium for phytoplankton. II. Trace metals and chelation, and general discussion. *J. Mar. Biol. Ass. U.K.,* 44, 87–109.

Kellar, P. E. and Paerl, H. W. (1980). Physiological adaptations in response to environmental stress during an N_2-fixing *Anabaena* bloom. *Appl. Environ. Microbiol.,* 40, 587–95.

Khoja, T. and Whitton, B. A. (1971). Heterotrophic growth of blue-green algae. *Arch. Mikrobiol.,* 79, 280–2.

King, D. (1970). The role of carbon in eutrophication. *J. Water Pollut. Control Fed.,* 40, 2035–501.

Klemer, A. R., Feuillade, J., and Feuillade, M. (1982). Cyanobacterial blooms. Carbon and nitrogen limitation have opposite effects on buoyancy. *Science,* 215, 1629–31.

Knoll, A. H. (1977). Paleomicrobiology. In: *CRC Handbook of Microbiology,* 2d ed., vol. 1, pp. 8–29. Cleveland: CRC Press.

Konopka, A. (1982). Physiological ecology of a metalimnetic *Oscillatoria rubescens* population. *Limnol. Oceanogr.*, 27, 1154–61.

 (1984). Effect of light-nutrient interactions on buoyancy regulation by planktonic cyanobacteria. In: *Current Perspectives in Microbial Ecology*, eds. M. J. Klug and C. A. Reddy, pp. 41–8. Washington, D.C.: Am. Soc. Microbiol.

Kuentzel, E. L. (1969). Bacteria, carbon dioxide, and algal blooms. *J. Water Pollut. Control Fed.*, 41, 1737–47.

Lampert, W. (1977). Studies on the carbon balance of *Daphnia pulex* de Ceer as related to environmental conditions. II. The dependence of carbon assimilation on animal size, temperature, food concentration and diet species. *Arch. Hydrobiol. Suppl.*, 48, 310–35.

 (1981). Toxicity of the blue-green *Microcystis aeruginosa:* Effective defense mechanism against grazing pressure by *Daphnia. Verh. Int. Verein. Limnol.* 21, 412–27.

Lang, N. J. and Fay, P. (1971). The heterocysts of blue-green algae. II. Details of ultrastructure. *Proc. R. Soc. Lond. B Biol. Sci.*, 178, 193–203.

Lange, W. (1967). Effects of carbohydrates on the symbiotic growth of the planktonic blue-green algae with bacteria. *Nature*, 215, 1277–8.

Lewin, R. A. (1976). Prochlorophyta as a proposed new division of algae. *Nature*, 261, 697–8.

 (1981). The algae of polar bears. *Phycologia*, 20, 303–14.

Likens, G. E. (1972). Nutrients and eutrophication: The limiting nutrient controversy. *Am. Soc. Limnol. Oceanogr. Spec. Symp., vol. 1.*

Livingstone, D. A. and Melack, J. M. (1984). Some lakes of subsaharan Africa. In: *Lakes and Reservoirs: Ecosystems of the World*, vol. 23, ed. F. Tank, pp. 467–97. New York: Elsevier.

Loeblich, A. R. III. (1966). Aspects of the physiology and biochemistry of the Pyrrhophyta. *Phykos. Prof. Iyenger Memorial*, vol. 5, 216–55.

Lorenzen, M. W. and Mitchell, R. (1975). An evaluation of artificial destratification for control of algal blooms. *J. Am. Water Work Ass.*, 67, 373–6.

Lund, J. W. G. (1965). The ecology of the freshwater phytoplankton. *Biol. Rev.*, 40, 231–93.

Lupton, F. S. and Marshall, K. C. (1981). Specific adhesion of bacteria to heterocysts of *Anabaena* spp. and its ecological significance. *Appl. Environ. Microbiol.*, 42, 1085–92.

MacLeod, R. A. (1971). Salinity: Bacteria, fungi, and blue-green algae. In: *Marine Ecology*, vol. 1, pt. 2, ed. O. Kinne, pp. 689–703. London: Wiley.

Margulis, L. (1975). The microbes' contribution to evolution. *Biosystems*, 7, 266–92.

 (1982). *Early Life.* Boston: Science Books International.

Margulis, L., Walker, J. C. G. and Rambler, M. (1976). Reassessment of roles of oxygen and ultraviolet lights in Precambrian evolution. *Nature*, 264, 620–4.

McKnight, D. M. and Morel, F. M. (1979). Release of weak and strong copper complexing agents by algae. *Limnol. Oceanogr.*, 24, 823–37.

Melamed-Harel, H. and Tel-or, E. (1981). Adaptation to salt of the photosynthetic apparatus in cyanobacteria. In: *Photosynthesis,* vol. 6, ed. G. Akoyunogloe, pp. 455–64. Philadelphia: Balaban I.S.

Miller, M. M. and Lang, N. J. (1968). The fine structure of akinete formation and germination in *Cylindrosperum. Arch. Mikrobiol.,* 60, 303–13.

Mohleji, S. C. and Verhoff, F. H. (1980). Sodium and potassium ions effects on phosphorus transport in algal cells. *J. Water Pollut. Control Fed.,* 52, 110–25.

Moriarty, D. J. W. (1973). The physiology of digestion of blue-green algae in the cichlid fish *Tilapia nifotica. J. Zool.,* 171, 25–39.

Munawar, M. and Munawar, I. F. (1975). The abundance and significance of phytoflagellates and nannoplankton in the St. Lawrence Great Lakes. *Verh. Int. Ver. Limnol.,* 19, 705–23.

Murphy, T. P., Lean, D. R. S., and Nalewajko, C. (1976). Blue-green algae: Their excretion of iron-selective chelators enables them to dominate over other algae. *Science,* 192, 900–2.

Nalewajko, C. (1978). Release of organic substances. In: *Handbook of Phycological Methods,* eds. J. A. Hellebust and J. S. Craigie, pp. 389–98. Cambridge: Cambridge University Press.

Nichols, J. M. and Adams, D. G. (1982). Akinetes. In: *The Biology of Cyanobacteria,* eds. N. G. Carr and B. A. Whitton, pp. 387–412. Oxford: Blackwell Scientific.

Niemi, A. (1979). Blue-green algal blooms and N:P ratio in the Baltic Sea. *Acta Bot. Fenn.,* 110, 57–61.

Ohki, K. and Fujita, Y. (1982). Laboratory culture of the pelagic blue-green alga *Trichodesmium thiebautii:* Condition for unialgal culture, *Mar. Ecol. Prog. Ser.,* 1, 185–90.

Paerl, H. W. (1977). Ultraplankton biomass and production in some New Zealand lakes, *N. Z. J. Mar. Freshwat. Res.,* 11, 297–305.

(1978). Role of heterotrophic bacteria in promoting N_2 fixation by *Anabaena* in aquatic habitats. *Microb. Ecol.,* 4, 215–31.

(1982a). Interactions with bacteria. In: *The Biology of Cyanobacteria,* eds. N. G. Carr and B. A. Whitton, pp. 441–61. Oxford: Blackwell Scientific.

(1982b). Environmental factors promoting and regulating N_2 fixing blue-green algal blooms in the Chowan River. North Carolina Resour. Res. Inst. Rep. No. 176.

(1983a). Partitioning of CO_2 fixation in the colonial cyanobacterium *Microcystis aeruginosa:* A mechanism promoting surface scums. *Appl. Environ. Microbiol.,* 46, 252–9.

(1983b). Factors regulating nuisance blue-green algal bloom potentials in the lower Neuse River, North Carolina Water Resour. Res. Inst. Rep. No. 177.

(1984a). Transfer of N_2 and CO_2 fixation products from *Anabaena oscillarioides* to associated bacteria during inorganic carbon sufficiency and deficiency. *J. Phycol.,* 20, 600–8.

(1984b). Alteration of microbial metabolic activities in association with detritus. *Bull. Mar. Sci.,* 35, 393–408.

(1985). Nuisance algal blooms in lakes, rivers, and coastal marine waters. In: *Comparative Ecology of Freshwater and Coastal Marine Ecosystems,* ed. S. W. Nixon. Proc. UNESCO Symp., Am Soc. Limnol. Oceanogr. Publ.

Paerl, H. W. and Bland, P. T. (1982). Localized tetrazolium reduction in relation to N_2 fixation, CO_2 fixation and H_2 uptake in aquatic filamentous cyanobacteria. *Appl. Environ. Microbiol.,* 43, 218–26.

Paerl, H. W., Bland, P. T., Blackwell, H. H., and Bowles, N. D. (1984). The effects of salinity on the potential of blue-green algal *(Microcystis aeruginosa)* bloom. Univ. N.C. Sea Grant Program Working Paper 84–1.

Paerl, H. W., Bland, P. T., Bowles, N. D., and Haibach, M. E. (1985). Adaptation to high-intensity, low wavelength light among surface blooms of the cyanobacterium *Microcystis aeruginosa. Appl. Environ. Microbiol.,* 49, 1046–52.

Paerl, H. W. and Gallucci, K. K. (1985). Chemotaxis: Its role in establishing a N_2 fixing cyanobacterial-bacterial symbiosis. *Science,* 227, 647–9.

Paerl, H. W. and Kellar, P. E. (1978). Significance of bacterial-*Anabaena* (Cyanophyceae) associations with respect to N_2 fixation in freshwater. *J. Phycol.,* 14, 254–60.

(1979). Nitrogen-fixing *Anabaena:* Physiological adaptations instrumental in maintaining surface blooms. *Science,* 204, 620–2.

Paerl, H. W. and Mackenzie, A. L. (1977). A comparative study of the diurnal carbon fixation patterns of nannoplankton and netplankton. *Limnol. Oceanogr.,* 22, 737–8.

Paerl, H. W., Tucker, J., and Bland, P. T. (1983). Carotenoid enhancement and its role in maintaining blue-green algal *(Microcystis aeruginosa)* surface blooms. *Limnol. Oceanogr.,* 28, 847–57.

Paerl, H. W. and Ustach, J. F. (1982). Blue-green algal scums: An explanation for their occurrence during freshwater blooms. *Limnol. Oceanogr.,* 21, 212–17.

Pavoni, M. (1963). Die bedeutung des nannoplanktons in vergleich zum netplankton. *Schwiez. Z. Hydrol.,* 25, 219–341.

Pearsall, W. H. (1932). Phytoplankton in English Lakes. II. The composition of the phytoplankton in relation to dissolved substances. *J. Ecol.,* 20, 241–62.

Porter, K. G. (1977). The plant-animal interface in freshwater ecosystems. *Am. Sci.,* 65, 159–70.

Porter, K. G. and Orcutt, J. D., Jr. (1980). Nutritional adequacy, manageability, and toxicity as factors that determine the food quality of green and blue-green algae for *Daphnia.* In: *The Evolution and Ecology of Zooplankton Communities,* ed. W. C. Kerfoot, pp. 268–81. Spec. Symp. II, Am. Soc. Limnol. Oceanogr.

Postgate, J. R. (1978). *Nitrogen Fixation.* Studies in Biology No. 92. London: Arnold.

Potts, M. and Whitton, B. A. (1979). pH and Eh on Aldabra Atoll. I. Comparison of marine and freshwater environments. *Hydrobiologia,* 67, 11–17.

Prakash, A. (1971). Terrigenous organic matter and coastal phytoplankton fertility. In: *Fertility of the Sea,* ed. J. D. Costlow, pp. 351–8. London: Gordon and Breach.

Preston, T., Stewart, W. D. P., and Reynolds, C. S. (1980). Bloom-forming cyanobacterium *Microcystis aeruginosa* overwinters on sediment surface. *Nature,* 288, 365–7.

Priscu, J. C. and Goldman, C. R. (1983). Seasonal dynamics of the deep-chlorophyll maximum in Castle Lake, California. *Can. J. Fish. Aquat. Sci.,* 40, 208–14.

Reuter, J. E., Loeb, S. C., and Goldman, C. R. (1983). Nitrogen fixation in periphyton of oligotrophic Lake Tahoe. In: *Periphyton of Freshwater Ecosystems,* ed. R. G. Wetzel, pp. 101–9. The Hague: Dr. W. Junk.

Reynolds, C. S. (1972). Growth, gas vacuolation and buoyancy in a natural population of a blue-green alga. *Freshwat. Biol.,* 2, 87–106.

(1975). Interrelations of photosynthetic behavior and buoyancy regulation in a natural population of a blue-green alga. *Freshwat. Biol.,* 5, 323–38.

(1980). Phytoplankton assemblages and their periodicity in stratifying lake systems. *Holarct. Ecol.,* 141–59.

(1984). *The Ecology of Freshwater Phytoplankton.* Cambridge: Cambridge University Press.

Reynolds, C. A. and Walsby, A. E. (1975). Water blooms. *Biol. Rev.,* 50, 437–81.

Rhee, G. Y. (1982). Effects of environmental factors and their interactions on phytoplankton growth. *Adv. Microb. Ecol.,* 6, 33–74.

Richerson, P., Armstrong, R., and Goldman, C. R. (1970). Contemporaneous disequilibrium, a new hypothesis to explain "The Paradox of the Plankton." *Proc. Natl. Acad. Sci. U.S.A.,* 67, 1710–14.

Richerson, P. J., Lopez, M., and Coon, T. (1978). The deep chlorophyll maximum layer of Lake Tahoe. *Verh. Int. Verein. Limnol.,* 20, 426–33.

Robarts, R. D. (1984). Factors controlling primary production in a hypertrophic lake (Hartbeespoort Dam, South Africa). *J. Plank. Res.,* 6, 91–105.

Rother, J. A. and Fay, P. (1979). Blue-green algal growth and sporulation in response to simulated surface bloom conditions. *Br. Phycol. J.,* 14, 59–68.

Rützler, K. (1981). An unusual blue-green alga symbiotic with two new species of *Ulosa* (Porifera: *Hymeniacidonidae*) from Carrie Bow Cay, Belize. *Mar. Ecol.,* 2, 35–50.

Saunders, G. W. (1972). Potential heterotrophy in a natural population of *Oscillatoria agardhii* var. *isothrix* Skuja. *Limnol. Oceanogr.,* 17, 704–11.

Schantz, E. (1971). The dinoflagellate poisons. In: *Microbial Toxins,* vol. 7, eds. S. Kadis, A. Ciegler, and S. Agh, pp. 3–26. New York: Academic Press.

Schiefer, G. E. and Caldwell, D. E. (1982). Synergistic interaction between *Anabaena* and *Zoogloea* spp. in carbon dioxide-limited continuous cultures. *Appl. Environ. Microbiol.,* 44, 84–7.

Schindler, D. W. (1968). Feeding, assimilation and respiration rates of

Daphnia magna under various environmental conditions and their relation to production estimates. *J. Anim. Ecol.,* 37, 369–85.

(1971). Carbon, nitrogen, and phosphorous and the eutrophication of freshwater lakes. *J. Phycol.,* 7, 321–9.

(1975). Whole lake eutrophication experiments with phosphorus, nitrogen and carbon. *Verh. Int. Verein. Limnol.,* 19, 3221–31.

(1977). Evolution of phosphorous limitation in lakes. *Science,* 195, 260–2.

Schopf, J. W. and Walter, M. R. (1982). Origin and early evolution of cyanobacteria: The geological evidence. In: *The Biology of Cyanobacteria,* eds. N. G. Carr, and B. A. Whitton, pp. 543–64. Oxford: Blackwell Scientific.

Shapiro, J. (1973). Blue-green algae: Why they become dominant. *Science,* 179, 382–4.

Schindler, D. B., Paerl, H. W., Kellar, P. E., and Lean, D. R. S. (1981). Environmental constraints on *Anabaena* N_2 and CO_2 fixation. Effects of hyperoxia and phosphate depletion on blooms and chemostat cultures. In: *Developments in Hydrobiology,* vol. 2, eds. J. Barica and L. R. Mur, pp. 221–9. The Hague: Dr. W. Junk.

Simpson, F. B. and Neilands, J. B. (1976). Siderochromes in Cyanophyceae: Isolation and characterization of schizokinen from *Anabaena* sp. *J. Phycol.,* 12, 44–8.

Sinclair, C. and Whitton, B. A. (1977). Influence of nutrient deficiency on hair formation in the Rivulariaceae. *Br. Phycol. J.,* 12, 297–313.

Smith, V. H. (1983). Low nitrogen to phosphorus ratios favor dominance by blue-green algae in lake phytoplankton. *Science,* 221, 669–71.

Sorokin, Y. I. (1968). The use of ^{14}C in the study of nutrition of aquatic animals. *Mitt. Int. Verein. Limnol.,* 16, 1–41.

Stal, L. and Krumbein, W. E. (1985). Nitrogenase activity in the non-heterocystous cyanobacterium *Oscillatoria* sp. grown under alternating light-dark cycles. *Arch. Microbiol.,* 143, 67–71.

Stanier, R. Y. (1977). The position of cyanobacteria in the world of phototrophs. *Carlsberg Res. Commun.,* 42, 77–98.

Stanier, R. Y. and Cohen-Bazire, G. (1977). Phototrophic prokaryotes: The cyanobacteria. *Ann. Rev. Microbiol.,* 31, 225–74.

Starkweather, P. L. and Kellar, P. E. (1983). Utilization of cyanobacteria by *Brachianus calyciflorus: Anabaena flos-aquae* (NRC-44-1) as a sole or complementary food source. *Hydrobiologia,* 104, 373–7.

Stewart, W. D. P. (1974). *Algal Physiology and Biochemistry.* Oxford: Blackwell Scientific.

(1975). *Nitrogen Fixation by Free-living Microorganisms.* London: Cambridge University Press.

Stewart, W. D. P. and Pearson, H. W. (1970). Effects of aerobic and anaerobic conditions on growth and metabolism of blue-green algae. *Proc. R. Soc. Lond. B Biol. Sci.,* 175, 293–311.

Stewart, W. D. P., Pemble, M., and Al-Ugaily, L. (1978). Nitrogen and phosphorous storage and utilization in blue-green algae. *Mitt. Int. Verein. Limnol.,* 21, 224–47.

Stewart, W. D. P., Rowell, P., and Rai, A. N. (1980). Symbiotic nitrogen-fixing cyanobacteria. In: *Proc. Int. Symp. On Nitrogen Fixation,* eds. W. D. P. Stewart and J. R. Gallon, pp. 239–77. Oxford: Oxford University Press.

Stiefel, E. I. (1977). The mechanisms of nitrogen fixation. In: *Recent Developments in Nitrogen Fixation,* eds. W. Newton, J. R. Postgate, and C. Rodriguez-Barrneco, pp. 69–108. London: Academic Press.

Stumm, W. and Morgan, J. J. (1981). *Aquatic Chemistry,* 2d ed. New York: Wiley.

Sunda, W. and Guillard, R. L. (1976). The relationship between cupric ion activity and the toxicity of copper to phytoplankton. *J. Mar. Res.,* 34, 511–29.

Sutherland, J. M., Herdman, M., and Stewart, W. D. P. (1979). Akinetes of the cyanobacterium Nostoc PCC 7524: Macromolecular composition, structure and control of differentiation. *J. Gen. Microbiol.,* 115, 273–87.

Talling, J. F. (1982). Utilization of solar radiation by phytoplankton. In: *Trends in Photobiology,* eds. M. Montenay-Garestier and G. Laustriat, pp. 619–31. New York: Plenum Press.

Throndsen, G. (1973). Motility in some marine nanoplankton flagellates. *Norw. J. Zool.,* 21, 193–200.

Tolbert, N. E. (1974). Photorespiration. In: *Algal Physiology and Biochemistry,* ed. W. D. P. Stewart, pp. 474–504. Oxford: Blackwell Scientific.

Van Liere, L. and Walsby, A. E. (1982). Interactions of cyanobacteria with light. In: *The Biology of Cyanobacteria,* eds. N. G. Carr and B. A. Whitton, pp. 9–45. Oxford: Blackwell Scientific.

Vance, B. D. (1965). Composition and succession of cyanophycean water blooms. *J. Phycol.,* 1, 81–96.

Vollenweider, R. (1976). Advances in defining critical loading levels for phosphorus in lake eutrophication. *Mem. Ist. Ital. Idrobiol.,* 33, 53–85.

Vollenweider, R. A., Munawar, M., and Stadelman, P. (1974). A comparative review of phytoplankton and primary production in the Laurentian Great Lakes. *J. Fish. Res. Board Can.,* 31, 739–62.

Walsby, A. E. (1972). Structure and function of gas vacuoles. *Bacteriol. Rev.,* 36, 1–32.

(1974). The extracellular products of *Anabaena cylindrica* Lemm. II. Fluorescent substances containing serine and their role in extracellular pigment formation. *Br. Phycol. J.,* 9, 383–91.

(1982). Cell-water and cell-solute relations. In: *The Biology of Cyanobacteria,* eds. N. G. Carr and B. A. Whitton, pp. 237–62. Oxford: Blackwell Scientific.

Walsby, A. E. and Booker, M. J. (1980). Changes in buoyancy of a planktonic blue-green alga in response to light intensity. *Br. Phycol. J.,* 15, 311–19.

Waterbury, J. B., Watson, S. W., Guillard, R. R. L., and Brand, L. E. (1979). Widespread occurrence of a unicellular marine planktonic cyanobacterium. *Nature,* 277, 293–4.

Webster, K. E. and Peters, R. H. (1978). Some size-dependent inhibitions of larger cladoceran filters in filamentous suspensions. *Limnol. Oceanogr.,* 23, 1238–45.

Wirth, T. L. and Dunst, R. C. (1967). Limnological changes resulting from artificial destratification and aeration of an impoundment. *Wisconsin Conservation Dept. Res. Dep.* No. 22.

Wolk, C. P. (1965). Heterocyst germination under defined conditions. *Nature,* 205, 201–2.

  (1968). Movement of carbon from vegetative cells to heterocysts in *Anabaena cylindrica. J. Bacteriol.,* 96, 2138–43.

  (1979). Intercellular interactions and pattern formation in filamentous cyanobacteria. In: *Determinants of Spatial Organization,* 37 Symp. Soc. Dev. Biol., eds. S. Sutelny and I. R. Konigsberg, pp. 247–66. New York: Academic Press.

  (1982). Heterocysts. In: *The Biology of Cyanobacteria,* eds. N. G. Carr and B. A. Whitton, pp. 359–86. Oxford: Blackwell Scientific.

Yates, M. G. (1977). Physiological aspects of nitrogen fixation. In: *Recent Developments in Nitrogen Fixation,* eds. W. Newton, J. R. Postgate, and C. Rodriguez-Barrueco, pp. 219–70. London: Academic Press.

Yount, J. L. (1966). Causes and relief of hypereutrophication of lakes. Report, Entomological Research Center, Florida State Board of Health, Vero Beach, Florida.

Chapter 8

PHYSIOLOGICAL MECHANISMS IN PHYTOPLANKTON RESOURCE COMPETITION

DAVID H. TURPIN

INTRODUCTION

In this chapter I will attempt to review the range of phytoplankton adaptations for resource procurement and the role of these adaptations in the physiological basis of resource competition. I use the term *resource* as a synonym for nutrient and will not address other phytoplankton requirements, such as light or space. Although the focus will be on physiological mechanisms, adaptations for resource procurement are found at the biochemical, physiological, and organismal levels of biological organization. Adaptations from these levels are closely integrated and compose the organism's life history strategy. I will argue that the severity of resource limitation is of primary importance in determining which of these levels are employed in response to resource limitation. Where possible, the significance of these adaptations will be evaluated by providing a mechanistic assessment of their effects on the growth of the organism. After providing a link between algal physiology and growth, I will examine the concepts of resource limitation and the models relating resource supply and concentration to algal growth. These models will provide a mechanistic basis for examining resource competition in both homogeneous and fluctuating environments and will be used to explore the mechanisms controlling resource partitioning between phytoplankton populations and the maintenance of community structure.

ADAPTATIONS FOR RESOURCE PROCUREMENT

In a resource-limited environment the successful competitor must exhibit two attributes. First, it must be able to maintain a net population growth rate that is greater than or equal to zero, and second, it must be able to do so at resource levels less than those required by other species. The term *net population growth rate* serves to highlight the importance of loss factors such as grazing, sinking, advection, and cell death in addition to the intrinsic rate of phytoplankton growth (Reynolds 1983, 1984, Chapter 10 this volume; Reynolds et al. 1982). The physiological basis of resource competition lies in the differential ability of species to scavenge essential limiting resources from the environment and to utilize them in the growth process. To this end, algae exhibit a wide variety of adaptations such as changes in structure, form, or function that enhance the procurement, storage, or utilization of limiting resources.

Algae must be viewed as integrated units. Whether unicellular or multicellular, they respond to their environment by internal adjustments that involve biochemical, physiological, and organismal levels of biological organization. In this section adaptations for resource procurement are dealt with sequentially in the following framework: the regulation of transport kinetics, the effect of transport kinetics on the kinetics of assimilation, the relationship between assimilation and growth, selective and nonselective nutrient uptake, the production of extracellular products, morphological changes, development of symbiosis, and life cycle adaptations.

Regulation of Nutrient Transport Kinetics

In their simplest form, transport kinetics can be represented by the Michaellis–Menten relationship [Eq. (1)].

$$\rho_t = \frac{\rho_{t\ max}[S]}{K_t + [S]} \tag{1}$$

In this representation ρ_t is the rate of nutrient transport (Table 8-1 contains a list of all abbreviations), $\rho_{t\ max}$ is the substrate saturated rate of transport, $[S]$ is the substrate concentration, and K_t is the value of $[S]$ at which $\rho_t = \frac{1}{2} \rho_{t\ max}$ (Fig. 8-1). Most researchers use the symbol ρ to denote the cellular rate of uptake $(mol \cdot cell^{-1} \cdot t^{-1})$, whereas V denotes a specific rate (t^{-1}) (Dugdale 1967; Goldman & Glibert 1983). Because ecological success is determined on an organismal basis, the subsequent discussion deals with cellular rates of uptake, ρ. Currently ρ_{max} and K_s are often employed in describing the kinetics of nutrient uptake; however, these values depend on the method and duration of measurement (Goldman & Glibert 1983). For this reason, I have

Table 8-1. *A list of symbols used in the text and their definitions.*

Symbol	Definition	Units
Kinetics of nutrient transport		
ρ_t	Velocity of the nutrient transporter	$mol \cdot cell^{-1} \cdot t^{-1}$
$\rho_{t\ max}$	Substrate-saturated velocity of the nutrient transporter	$mol \cdot cell^{-1} \cdot t^{-1}$
$[S]$	Substrate concentration	$mol \cdot L^{-1}$
K_t	The substrate concentration at which the transporter is half-saturated ($\rho_t = \frac{1}{2}\rho_{t\ max}$)	$mol \cdot L^{-1}$
α_t	The initial slope of the ρ_t-vs-[S] curve defined as $\rho_{t\ max}/K_t$	$L \cdot cell^{-1} \cdot t^{-1}$
Kinetics of nutrient assimilation		
ρ_a	Velocity of whole cell nutrient assimilation	$mol \cdot cell^{-1} \cdot t^{-1}$
ρ_{amax}	Substrate saturated rate of whole cell assimilation	$mol \cdot cell^{-1} \cdot t^{-1}$
K_a	The substrate concentration at which whole cell assimilation is half-saturated ($\rho_a = \frac{1}{2}\rho_{a\ max}$)	$mol \cdot L^{-1}$
α_a	The initial slope of the ρ_a-vs-[S] curve	$L \cdot cell^{-1} \cdot t^{-1}$
Kinetics of steady-state nutrient uptake or assimilation		
ρ	Steady-state rate of uptake or assimilation ($\rho = Q \cdot \mu$)	$mol \cdot cell^{-1} \cdot t^{-1}$
ρ_{max}	Maximal steady-state rate of uptake or assimilation ($\rho_{max} = Q_{max} \cdot \hat{\mu}$)	$mol \cdot cell^{-1} \cdot t^{-1}$
K_s	The substrate concentration at which steady-state assimilation is half-saturated ($\rho = \frac{1}{2}\rho_{max}$)	$mol \cdot L^{-1}$
$[S^*]$	Steady-state external substrate concentrations	$mol \cdot L^{-1}$
α_ρ	The initial slope of the ρ-vs.-[S*] curve	$L \cdot cell^{-1} \cdot t^{-1}$
Kinetics of steady-state nutrient-limited growth		
$\hat{\mu}$	Maximal growth rate	t^{-1}
$\bar{\mu}$	Theoretical growth rate when cell quota (Q) approaches infinity	t^{-1}
$\mu_{i,j}$	Growth rate as supported by resource i or j	t^{-1}
$Q_{i,j}^{A,B}$	Cell quota of the limiting resource i or j for species A or B at a particular growth rate	$mol \cdot cell^{-1}$
Q_{max}	Cell quota at $\hat{\mu}$	$mol \cdot cell^{-1}$
K_Q	Minimum cell quota of the limiting resource (Q when $\mu = 0$)	$mol \cdot cell^{-1}$

Table 8-1 (*cont.*)

Symbol	Definition	Units
Q'	Cell quota of a nonlimiting resource at a particular growth rate	$mol \cdot cell^{-1}$
R	Steady-state coefficient of luxury consumption (Q'/Q)	$mol \cdot cell^{-1}$
K_μ	Steady-state substrate concentration $[S^*]$ at which growth is half-saturated ($\mu = \frac{1}{2}\hat{\mu}$)	$mol \cdot L^{-1}$
α_μ	The initial slope of the μ-vs.-$[S^*]$ curve	$L \cdot mol^{-1} \cdot t^{-1}$
D	Mortality rate or dilution rate	t^{-1}
$R_{i,j}$	Resource i or j	
$[R_j]$	Concentration of resource i or j	$mol \cdot L^{-1}$
R^*	The resource concentration $[S^*]$ at which $\mu = D$	$mol \cdot L^{-1}$
$R^*_{i,j}$	R^* for resource i or j	$mol \cdot L^{-1}$
$R^*_{A,B}$	R^* for species A or B	$mol \cdot L^{-1}$
R_i/R_j	The ratio of resource supply rates (resource ratio)	
$R_c^{A,B}$	The optimum ratio [i.e., the ratio of Q_i to Q_j required for balanced growth; see Eq. (12)] for species A or B	

defined $\rho_{t\ max}$ and K_t to indicate the true kinetics of transport, free of any experimental bias.

Any change in these kinetic parameters that lowers the substrate concentration required for a given rate of transport represents an important adaptation for the procurement of a homogeneously distributed limiting resource. Three such adaptations are possible. The first is an increase in the substrate saturated rate $\rho_{t\ max}$, the second an increase in the affinity of the transporter as reflected by a decrease in K_t, and the third is both a decrease in K_t and an increase in $\rho_{t\ max}$ (Fig. 8-1). From inspection of Fig. 8-1 it is apparent that each adaptation serves to increase the initial slope of the ρ_t-versus-$[S]$ curve (i.e., $\rho_{t\ max}/K_t = \alpha_t$), and thereby each provides a significant adaptation for nutrient procurement in nutrient-limited environments (Healey 1980).

The general response to the onset of nutrient limitation appears to be an increase in $\rho_{t\ max}$ for the limiting resource. This has been shown for phosphate (Perry 1976; Gotham & Rhee 1981a; Riegman & Mur 1984), inorganic carbon (Miller, Turpin, & Canvin 1984a), and ammonium (Eppley & Renger 1974; McCarthy & Goldman 1979; Zevenboom 1980; Goldman & Glibert 1982). Recent work employing short-

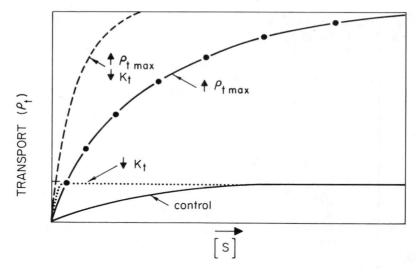

Fig. 8-1. Adaptations to nutrient limitation as reflected in nutrient transport kinetics. Three adaptations are indicated: (1) an increase in the maximum velocity ($\uparrow \rho_{t\ max}$), (2) a decrease in the half-saturation constant of transport ($\downarrow K_t$), and (3) both an increase in $\rho_{t\ max}$ and a decrease in K_t ($\uparrow \rho_{t\ max}$, $\downarrow K_t$). All three adaptations serve to increase α_t over the control condition.

term incubations suggests that such a response may not occur for nitrate (Horrigan & McCarthy 1981; Dortch et al. 1982). The evolutionary significance of this difference between NH_4^+ and NO_3^- procurement may lie in the mechanisms controlling NH_4^+ and NO_3^- supply to phytoplankton populations. In stable, nitrogen-limited environments, readily regenerated NH_4^+ supports the majority of primary productivity (Dugdale & Goering 1967; Eppley et al. 1977; Eppley & Peterson 1979). Consequently, an evolutionarily stable strategy for NH_4^+ procurement would involve an increase in $\rho_{t\ max}$ under NH_4^+ limitation. Nitrate, on the other hand, is not readily available during times of N deficiency and, as a consequence, makes only a minor contribution to the nitrogenous nutrition of phytoplankton during these periods (Dugdale & Goering 1967; Eppley et al. 1977; Eppley & Peterson 1979). Hence there is no selective pressure favoring the development of an increased uptake rate for nitrate in response to NO_3^- limitation. In fact, under conditions of NO_3^- depletion the nature of N recycling has selected for organisms that induce an enhanced NH_4^+ uptake capacity (Horrigan & McCarthy 1981; Dortch et al. 1982).

The effects of nutrient limitation on K_t are ambiguous. This confusion is due in part to methodological problems associated with its measurement (see Goldman & Glibert 1983). The few studies that have examined this problem have shown that measured values of the half-saturation constant vary, depending on incubation times (Turpin 1980; Goldman & Glibert 1983; Parslow, Harrison, & Thompson 1985a,b). Consequently, any changes in K_t in response to nutrient limitation may be masked by analytical problems. Kaplan, Badger, and Berry (1980) have employed the silicon oil technique to examine the kinetics of dissolved inorganic carbon transport in algal cells grown on either high or low concentrations of inorganic carbon. Values of K_t were shown to be independent of the degree of dissolved inorganic carbon (DIC) limitation. Similar results were obtained for phosphorus and nitrate half-saturation constants by Gotham and Rhee (1981a,b). If these results are confirmed and shown to be applicable to other nutrients, it suggests that the major transport adaptation for the procurement of a homogeneously distributed resource is an increase in $\rho_{t\ max}$.

Some confusion has arisen from use of specific uptake rates in the study of phytoplankton adaptation to nutrient limitation. The relationship between the maximum specific uptake rate V_{max} (t^{-1}) and the maximum absolute transport rate $\rho_{t\ max}$ is $V_{max} = \rho_{t\ max}/Q$. Consequently, a decrease in cell quota (Q) as a result of nutrient limitation will cause an increase in V_{max} with no change in cellular transport kinetics (Dugdale 1977; Goldman & Glibert 1983). Thus, many reported increases in V_{max} upon nutrient depletion are in part due to changes in cell quota and do not reflect transport adaptations. Quota flexibility is an important adaptation in its own right and will be dealt with later.

The Effect of Transport Kinetics on the Kinetics of Assimilation

Uptake kinetics describe but one component of an organism's competitive ability in a nutrient-limited system. Another major consideration is the relationship between the kinetics of nutrient transport and the kinetics of whole cell nutrient assimilation. It is important to realize that the kinetics of whole cell assimilation may be very different from the kinetics of the enzyme responsible for assimilation. This is a result of the formation of large internal nutrient pools that facilitate enzyme saturation at low external substrate concentration. Consequently, the initial slope of the assimilation (ρ_a)-versus-$[S]$ curve is set by the initial slope of the transport curve $(\rho_t$ versus $[S])$. Because $\rho_{a\ max}$

is usually much lower than $\rho_{t\ max}$ (Dugdale 1977; Miller et al. 1984a), the ρ_a-versus-[S] curve truncates at the maximum rate of assimilation, $\rho_{a\ max}$. The value of $\rho_{a\ max}$ is usually set by the activity of a key assimilatory process (Raven 1984; Parslow et al. 1985a; Turpin et al. 1985b) (Fig. 8-2).

Changes in transport kinetics can therefore result in changes in whole cell assimilation. An increase in $\rho_{t\ max}$ under nutrient limitation not only increases α_t but results in an increase in the initial slope of the ρ_a-versus-[S] curve (α_a) and a decrease in the half-saturation value for assimilation (K_a) (Fig. 8-2a,b,c). Changes in $\rho_{a\ max}$ as a function of nutrient limitation do not affect the initial slope of the ρ_a-versus-[S]curve, but they do determine its point of truncation. Figure 8-2 illustrates the effect of an increase in $\rho_{t\ max}$ on the kinetics of whole cell assimilation when (a) $\rho_{a\ max}$ remains constant under nutrient limitation [i.e., $\rho_{a\ max}$(ltd) = $\rho_{a\ max}$(non-ltd)], (b) $\rho_{a\ max}$ decreases under nutrient limitation [i.e., $\rho_{a\ max}$(ltd) < $\rho_{a\ max}$(non-ltd)], and (c) $\rho_{a\ max}$ increases under nutrient limitation [i.e., $\rho_{a\ max}$(ltd) > $\rho_{a\ max}$(non-ltd)]. Examples of all three of these responses have been observed (see Wheeler 1983 and references therein). Inspection of Fig. 8-2 shows that, in all cases, an increase in $\rho_{t\ max}$ results in an increase in the initial slope of the ρ_a-versus-[S] curve (α_a) and a lowering of K_a.

For each steady-state nutrient-limited growth rate there exists a unique ρ_t-versus-[S] and ρ_a-versus-[S] curve. At steady state, there is a coincident point of operation for these two curves defined by the point ρ, [S^*], where ρ is the steady-state uptake or assimilation rate and is equal to the product of the growth rate and cell quota ($\mu \cdot Q$). [S^*] is the steady-state nutrient concentration. Consequently at steady-state

Fig. 8-2. The relationship between nutrient transport and assimilation. This figure illustrates the effects of an increase in $\rho_{t\ max}$ on whole cell assimilation under conditions when (A) the maximum rate of assimilation ($\rho_{a\ max}$) is independent of nutrient limitation; (B) the maximum rate of assimilation decreases under nutrient limitation; and (C) the maximum rate of assimilation increases under nutrient limitation. The rate of transport is indicated by the dashed curves (---) under limiting [ρ_t (ltd)] and nonlimiting [ρ_t (non-ltd)] conditions. The rate of steady-state assimilation is indicated by the solid curves (————) under limiting [ρ_a (ltd)] and nonlimiting [ρ_a (non-ltd)] conditions. The resulting values of K_t, K_a, $\rho_{a\ max}$, and $\rho_{t\ max}$ are given. In all cases, increases in $\rho_{t\ max}$ under nutrient limitation results in an increase in the initial slope of the ρ_a-vs.-[S] curve (α_a) and serves as a significant adaptation for nutrient procurement.

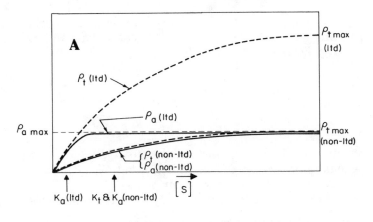

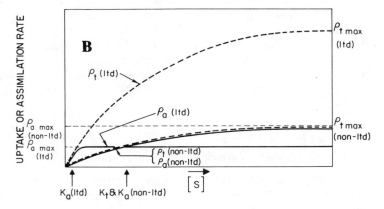

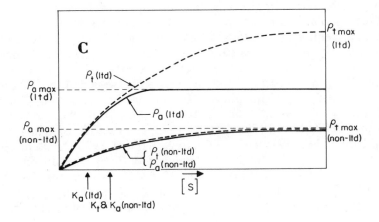

$\rho_t = \rho_a = \rho$. The determination of this point at all growth rates $(0 < \mu \leq \hat{\mu})$ defines a steady-state ρ-versus-$[S^*]$ curve that can be described by a rectangular hyperbola such that

$$\rho = \frac{\rho_{max}[S^*]}{K_s + [S^*]} \tag{2}$$

where

ρ = steady-state uptake or assimilation rate $(\mu \cdot Q)$
ρ_{max} = maximum steady-state assimilation rate $(\hat{\mu} \cdot Q_{max})$
$[S^*]$ = steady-state substrate concentrations
K_s = half-saturation constant for steady-state assimilation or uptake.

As previously demonstrated, an increase in $\rho_{t\ max}$ results in an increase in the initial slope (α_a) and a decrease in the half-saturation value (K_a) of the ρ_a-versus-$[S]$ curve. As a result, increases in $\rho_{t\ max}$ in response to nutrient deficiency are also reflected in a greater initial slope (α_ρ) and a lower value of K_s than would have occurred in the absence of transport adaptations. This demonstrates that adaptations for transport are also reflected in an organism's steady-state assimilation kinetics.

The Relationship between Assimilation and Growth

In order to relate the significance of these adaptations in nutrient uptake and assimilation to the growth process, we must first define the relationship between steady-state nutrient assimilation or uptake (ρ) and growth. In the simplest case, the limiting nutrient complement per cell (cell quota) is constant, and the relative kinetics of nutrient assimilation are equivalent to those of growth (Fig. 8-3a,b,c). Although this situation occurs for inorganic carbon-limited growth (Goldman, Oswald, & Jenkins 1974; Goldman & Graham 1981; Miller et al. 1984a; Turpin et al. 1985b) the more general case is that the cell quota decreases with increasing nutrient limitation (Caperon 1968; Caperon & Meyer 1972; Droop 1968, 1970, 1974; Fuhs 1969; Williams 1971; Thomas & Dodson 1972; Paasche 1973; Rhee 1973; Bienfang 1975; Feuillade & Feuillade 1975; Tilman & Kilham 1976; Nyholm 1976, 1977; Goldman 1977; Sneft 1978; Panikov & Pirt 1978; Ahlgren 1980; Terry 1980; Gotham & Rhee 1981a,b; Terry, Laws, & Burns 1985; Button 1985; Elrifi & Turpin 1985). Much of this work has focused on nitrogen, phosphorus, silicate, and vitamin B-12, and in each case a hyperbolic relationship between growth rate and cell quota was observed. The general form of this relationship was given by Droop (1968) as

$$\mu = \bar{\mu}[1 - K_Q/Q] \tag{3}$$

where

μ = growth rate
$\bar{\mu}$ = growth rate as Q approaches ∞
Q = cell quota for the limiting resource
K_Q = minimum cell quota for the limiting resource.

The relationship between the maximal growth rate ($\hat{\mu}$) and $\bar{\mu}$ is

$$\hat{\mu} = \bar{\mu}\left(1 - \frac{K_Q}{Q_{\max}}\right) \tag{3a}$$

where $Q_{\max}$ = maximum cell quota. Organisms exhibiting a great deal of quota flexibility have $K_Q/Q_{\max}$ values less than 1 or, conversely, a value of $\bar{\mu}$ approaching $\hat{\mu}$ (see Goldman & McCarthy 1978).

Flexibility in the cell quota is a major adaptation to nutrient limitation. In cases where the rate of steady-state resource procurement (ρ) decreases due to substrate limitation, an organism with a constant cell quota would suffer a similar decrease in growth rate. On the other hand, an organism that is capable of decreasing its cellular requirements for the nutrient will be able to offset much of the decrease in resource availability, thus minimizing the effects of nutrient depletion on growth rate. The effect of this adaptation on growth can be illustrated by coupling the hyperbolic kinetics of steady-state nutrient assimilation (ρ) [Eq. (2)] with the kinetics of growth as represented by Droop's equation [Eq. (3)]. During steady-state growth the rate of cell quota increase due to uptake will equal the dilution of cell quota due to growth. Hence Eqs. (2) and (3) can be combined, yielding Eq. (4):

$$\frac{\rho_{\max}[S^*]}{K_s + [S^*]} = \bar{\mu}\left(1 - \frac{K_Q}{Q}\right)Q \tag{4}$$

Equation (4) can be solved for Q and substituted into Eq. (3). Rearrangement results in the expression of growth rate as a function of $\rho_{\max}$, K_s, $\bar{\mu}$, K_Q, and $[S^*]$. The resulting equation [eq. (5)] was derived originally by Droop (cf. 1983) and is of the Monod form (Burmaster

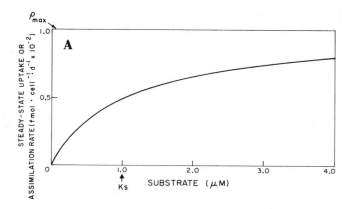

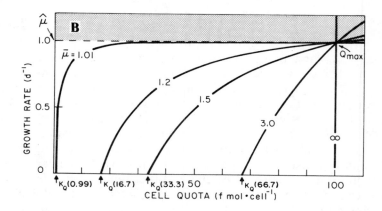

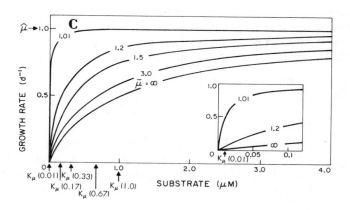

1979; Droop 1973a,b; Kilham 1978; Turpin, Parslow, & Harrison 1981):

$$\mu = \frac{\hat{\mu}[S^*]}{K_\mu + [S^*]} \tag{5}$$

The maximal growth rate $\hat{\mu}$ and the half-saturation constant for growth K_μ are therefore given by the following equations:

$$\hat{\mu} = \frac{\bar{\mu}\rho_{max}}{\bar{\mu}K_Q + \rho_{max}} \tag{6}$$

$$K_\mu = \frac{K_s\bar{\mu}K_Q}{\bar{\mu}K_Q + \rho_{max}} \tag{7}$$

From evaluation of Eq. (7) it is apparent that under conditions of low quota variability (i.e., $\bar{\mu} \to \infty$; $K_Q/Q_{max} \to 1$), the half-saturation constant for growth (K_μ) approaches the half-saturation constant for steady-state uptake or assimilation (i.e., $K_\mu \to K_s$). However, with increases in quota variability (i.e., $\bar{\mu} \to \hat{\mu}$; $K_Q/Q_{max} \ll 1$), the half-saturation constant for growth (K_μ) decreases relative to K_s (i.e., $K_\mu \ll K_s$). This phenomenon is illustrated in Fig. 8-3.

Figure 8-3a represents a single steady-state uptake or assimilation relationship, and Fig. 8-3b represents a family of Droop curves with a common value of Q_{max}. These curves represent different degrees of quota variability in response to nutrient limitation, ranging from a

Fig. 8-3. The effect of cell quota variability on growth kinetics. (A) The relationship between steady-state uptake or assimilation rate and $[S^*]$, the steady-state substrate concentration ($K_s = 1.0$ μM; $\rho_{max} = 100$ fmol·cell^{-1}·d^{-1}). (B) A family of relationships between growth rate and cell quota as described by the Droop relationship [Eq. (3)] where $\bar{\mu} = 1.01$, 1.2, 1.5, 3.0, or ∞ and values of K_Q of 0.99, 16.7, 33.3, 66.7, or 100 fmol·cell^{-1}, respectively. Q_{max} was established at 100 fmol·cell^{-1} and $\hat{\mu} = 1.00$ d^{-1}. (C) The kinetics of growth as calculated using Eq. (5) and the values for ρ_{max}, K_s, $\bar{\mu}$, and K_Q represented in panels A and B. The resulting values of K_μ are indicated with respect to the value of $\bar{\mu}$ used in each calculation. This panel indicates the adaptive significance of quota variability under nutrient limitation. The greater the quota variability (i.e., the smaller the $\bar{\mu}$), the greater the decrease in the half-saturation constant for growth (K_μ) relative to the half-saturation constant for steady-state uptake or assimilation (K_s).

constant Q ($\bar{\mu}/\hat{\mu} = \infty$) to a situation in which Q varies by a factor of 100 ($\bar{\mu}/\hat{\mu} = 1.01$). Equations (5)–(7) were then used to calculate the resulting saturation constants for growth (K_μ) given the common steady-state uptake or assimilation curve (Fig. 8-3a) but different degrees of quota flexibility (Fig. 8-3b). The results illustrate the significance of quota flexibility in the growth of phytoplankton in nutrient-limited environments (Fig. 8-3c). Organisms with no quota flexibility ($\bar{\mu}/\hat{\mu} = \infty$) show $K_\mu = K_s$. The greater the degree of quota flexibility the lower K_μ becomes relative to K_s and the greater the initial slope of the μ-versus-[S^*] curve (α_μ).

Experimental evidence supports this conclusion. Nitrogen is a resource that exhibits quota flexibility. Half-saturation constants for ammonium uptake/assimilation tend to be in the low micromolar range (Eppley, Rogers, & McCarthy 1969b; MacIsaac & Dugdale 1969, 1972; McCarthy 1981; Glibert, Goldman, & Carpenter 1982; Wheeler, Glibert, & McCarthy 1982; Goldman & Glibert 1983), whereas half-saturation constants for growth have been shown to be below our current limits of detection (0.03 μM, Steeman-Nielsen 1978; Goldman & McCarthy 1978). Carbon, on the other hand, exhibits little quota variability, and the values for K_s and K_μ are similar (Turpin et al. 1985b).

If quota flexibility is adaptive, it must be a characteristic on which natural selection has acted. A comparison of the degree of quota variability for a particular resource with the range of concentrations at which the resource is found suggests this is the case. Resources that are routinely found in high concentrations (i.e., C and Na, Wetzel 1975; Talling 1985) show little quota flexibility ($\bar{\mu}/\hat{\mu} \rightarrow \infty$, Goldman et al. 1974; Goldman & Graham 1981; Miller et al. 1984b; Turpin et al. 1985b). On the other hand, resources found at low concentrations (P, Fe, and B$_{12}$; Wetzel 1975) exhibit a high degree of quota flexibility ($\bar{\mu}/\hat{\mu} = 1.01 \rightarrow 1.03$, McCarthy 1981). Nitrogen and silicon, resources found at intermediate concentrations (Wetzel 1975; Lobban, Harrison, & Duncan 1985), exhibit an intermediate degree of quota flexibility ($\bar{\mu}/\hat{\mu} = 1.25$, Goldman & McCarthy 1978; McCarthy 1981). This suggests that decreases in cell quota below Q_{max} as a result of nutrient deficiency may have evolved for resources found in low concentrations. Although quota variability appears to be adaptive, there are physical constraints placed on the degree of quota variability. A resource such as carbon, which constitutes the greatest share of cellular biomass, will only exhibit quota variability with corresponding changes in cell size. Such constraints may impose limits on the degree of quota variability that has evolved in response to resource limitation.

It is apparent that quota flexibility is an adaptation that allows K_μ to decrease below K_s, thereby facilitating an increase in α_μ. In the previous section it was shown that an increase in the maximum rate of nutrient transport $\rho_{t\ max}$ served to increase the initial slope of the ρ_a-versus-$[S^*]$ curve and ultimately decrease K_s. Because of this relationship, any increase in $\rho_{t\ max}$ in response to nutrient limitation will also be reflected in a decrease in K_μ and an increase in the initial slope of the μ-versus-$[S^*]$ curve (α_μ). Consequently, adaptations for nutrient uptake not only enhance limiting nutrient assimilation but are reflected in a more efficient growth process at low substrate concentrations. Button (1985) provides an excellent review of the kinetics of nutrient uptake and assimilation.

Quota variability of the limiting nutrient is often confused with the concept of luxury consumption. An increase in the cell quota for a limiting resource corresponding to an increased nutrient-limited growth rate is anything but luxurious; it is an absolute requirement for an increase in growth rate. Luxury consumption represents the accumulation of a nonlimiting resource above the levels required to maintain the current growth rate. In effect, luxury consumption occurs if the cell quota for a nonlimiting resource (Q') is greater than the cell quota required to maintain that particular growth rate if that resource were limiting (i.e., $Q' > Q$). The coefficient of luxury consumption for resource i is defined as $R = Q'/Q$ (Elrifi & Turpin 1985). A detailed discussion of luxury consumption is given by Elrifi and Turpin (1985) and will not be reiterated here.

Selective and Non-Selective Nutrient Uptake

The environment contains many different forms of potentially limiting elements. Nitrogen, for example, is found as NH_4^+, NO_2^-, NO_3^-, urea, amino acids, and numerous other compounds. With this variety, it is not surprising that phytoplankton exhibit a preference. Under conditions of nitrogen excess the preferred source is usually the most reduced form of inorganic nitrogen, ammonium (Eppley, Coatsworth, & Solorzano 1969a; McCarthy, Taylor, & Taft 1975, 1977; Conway 1977; Serra, Llama, & Cadenas 1978). In situations where phytoplankton are grown on nitrate, the addition of ammonium results in an inhibition of nitrate uptake (Conway 1977; Eppley et al. 1969a). The time required for ammonium inhibition of nitrate uptake increases with increasing N limitation (Conway 1977).

Organic forms of nitrogen (amino acids or urea) may also serve as a nitrogen source for algal growth (Antia et al. 1975), but usually only after inorganic forms have been depleted (Healey 1977; Rees & Syrett 1979). The rate of transport of these compounds increases with increasing N limitation (North & Stephens 1971, 1972; McCarthy

1972; Wheeler, North, & Stephens 1974; Wheeler 1977; Rees & Syrett 1979). The ability of a species to utilize organic forms of nitrogen is affected not only by the degree of nutrient limitation but by the habitat for which the organism is adapted. In-shore, littoral, and benthic marine species are better able to utilize amino acids and other organic nitrogen forms than are offshore species (Wheeler et al. 1974).

Further examples of selective uptake are seen in the cases of sulfur and phosphorus procurement. In *Chlorella vulgaris* evidence exists suggesting that cysteine and methionine can be utilized to offset sulfate deficiency (Passera & Ferrari 1975). Soldatini, Ziegler, and Ziegler (1978) have shown that *C. vulgaris* will preferentially take up sulfite (SO_3^{2-}) over sulfate (SO_4^{2-}), once again exhibiting a preference for the reduced inorganic form. Phosphorus is also found in a wide variety of organic and inorganic forms. Although there has been no work on the preferred forms of inorganic phosphorus, orthophosphate is preferred to the organic forms (Cembella, Antia, & Harrison 1984). In general, the reduced forms of inorganic nutrients are preferred under nutrient sufficiency, and organic forms are utilized only when the inorganic sources are depleted. This response may have evolved due to selective pressures favoring the minimization of energetic expenditures associated with nutrient uptake and assimilation.

Production of Extracellular Products

In many cases nutrient limitation results in the production of extracellular products that faciliate uptake of the limiting resource. These products mediate chemical transformations of the element in question and fall into two distinct classes. In the first case, the compound secreted is capable of chelating and mobilizing insoluble forms of the limiting nutrient. Iron is a highly insoluble nutrient required for algal growth [$K_{sp}(Fe(OH)_3) = 1.1 \times 10^{-36}$]. In many cases iron deficiency triggers the production of extracellular iron (Fe^{3+}) chelators (siderophores) by cyanobacteria (Simpson & Neilands 1976; Murphy, Lean, & Nalewajko 1976) and some eukaryotic algae (Trick et al. 1983). The siderophore produced by *Anabaena* (schizokinen) has been characterized as a dihydroxamic acid exhibiting very high affinity for iron ($K = 10^{30}$, Simpson & Neilands 1976). It has been demonstrated that the production of these siderophores by cyanobacteria enables them to outcompete other algae for available iron (Murphy et al. 1976).

The second class of extracellular products are enzymes that are produced in response to a nutrient stress and are capable of structurally modifying compounds containing the limiting element. The production of extracellular phosphatases in response to phosphorus deficiency has been well documented (Brandes & Elston 1956; Eppley 1962; Healey 1973) and takes place almost immediately upon phosphate depletion (Healey 1973). Although the presence of phosphatases has been used as an indicator of phosphate limitation (Perry 1976; Rhee 1973), interpretation of such results should be done with great care (Cembella et al. 1984). In addition to phosphatases, Galloway and Krauss (1963) demonstrated that pyrrophosphate induced the production of an extracellular pyrrophosphatase in *Chlorella pyrenoidosa*. The production of extracellular enzymes is not restricted to phosphorus procurement. It appears that under nitrogen deficiency some microalgae produce extracellular deaminases that allow utilization of the amino group of some amino acids (Saubert 1957; Belmont & Miller 1965).

Although the ecological significance of extracellular enzymes has yet to be conclusively demonstrated, their presence in natural systems suggests they may be important in determining nutrient availability (Cembella et al. 1984). Species differences in the production of these enzymes may provide a basis for competitive exclusion. In order to constitute an evolutionarily stable strategy, the species producing the extracellular product must be the major beneficiary of its action. This may be ensured either through retaining the extracellular product at the cell surface, as in the case of some extracellular phosphatases (Brandes & Elston 1956) or through the production of species-specific chelators (Murphy et al. 1976).

Morphological Changes

Both size and shape play a major role in determining the cell's surface-to-volume (S/V) ratio. Lewis (1976) has shown that microalgal S/V ratios fall within a narrow range out of the spectrum of geometric possibilities, thus suggesting that natural selection must be operating on this relationship. This in turn suggests that S/V must be an important factor in the adaptation of these organisms to their environment. The work of Munk and Riley (1952) coupled with that of Pasciak and Gavis (1974, 1975) and Gavis (1976) has indicated that S/V ratios and general surface morphology have a significant influence on the ability of phytoplankton to absorb limiting nutrients. There is some indication that nutrient limitation may result in changes to a cell's S/V characteristics; for species with such flexibility this would appear to be adaptive (Adamich, Gibor, & Sweeney 1975; Yarish 1976; Leppard, Massalski, & Lean 1977; Sinclair & Whitton 1977). For a detailed discussion of the morphological basis of competition and succession, readers are referred to Margalef (1978), Sournia (1981), and Reynolds (1984; Chapter 10 this volume).

Development of Symbiosis

Several potentially important symbiotic relationships involving marine diatoms have been identified. Some members of the diatom genus *Rhizosolenia* found in the nutrient-depleted central North Pacific gyre have been shown to harbor an endosymbiotic nitrogen fixing blue-green alga, *Richelia* (Mague, Weare, & Holm-Hansen 1974). The endosymbiont has been shown to fix atmospheric nitrogen, which presumably can be subsequently assimilated by the host. Martinez and co-workers (1983) have identified another endosymbiotic association involving *Rhizosolenia* and N_2-fixing bacteria. The implications of such symbiotic relationships on resource competition are presently unknown. In the preceding cases they may serve to allow *Rhizosolenia* to remain in a nitrogen-limited environment in the presence of superior competitors. Analogous nutritionally beneficial algal symbioses have yet to be clearly identified from freshwater systems. It may be that lakes are too young geologically for such associations to have evolved.

Life Cycle Adaptations

It is well documented that nutrient limitation serves to induce life cycle changes in many phytoplankton species. In many cases these changes appear to be adaptive because they result in production of resting stages capable of surviving the period of stress and providing seed stock when favorable conditions return. Nutrient limitation can induce life cycle changes when the capacity of biochemical and physiological adaptations are exceeded. Numerous examples of life cycle adaptations including the roles of nutrient limitation are covered in other chapters of this book.

The Hierarchy and the Time Scale of Adaptation

Figure 8-4 represents a hierarchy of adaptations to nutrient limitation that are manifested in phytoplankton life history strategies. This hierarchy extends from the biochemical to the organismal levels of biological organization. Adaptations at any level of organization are not independent of the other levels in the hierarchy. In fact, it is the inter-

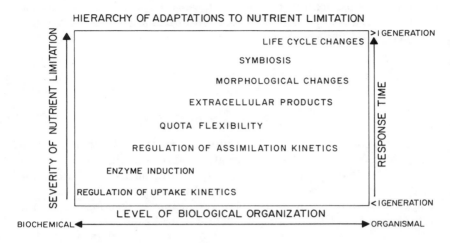

Fig. 8-4. The hierarchy of adaptations to nutrient limitation composing a phytoplankter's life history strategy. These adaptations span the biochemical through to the organismal level of organization. Mild nutrient stress is usually dealt with at the biochemical level of organization, whereas a severe stress often elicits life cycle changes. Corresponding to this continuum is an increase in the response time for adaptation.

action between these levels that enables optimization for a particular environment.

In most cases, nutrient limitation of low severity elicits adaptations at the biochemical and physiological levels. As the severity of nutrient limitation increases, adaptations from higher levels of the hierarchy are employed. The most severe stresses require life cycle adaptations. Corresponding to this continuum is an increase in the response time required for the adaptations to take place. Biochemical adaptations usually occur in less than a generation time, whereas the induction of life cycle events changes may take much longer.

Every species exhibits a unique set of life history adaptations. Some of this variation among species is a result of the emphasis each level of biological organization receives during adaptation to nutrient limitation. Organisms that are *r*-selected tend to emphasize adaptations at the whole organism level of organization (Fig. 8-4). They are fast growers and, upon the onset of nutrient limitation, often undergo life cycle changes and produce resting stages (Kilham & Kilham 1980). The *K*-selected competitors adapt to low-nutrient environments through many of the biochemical and physiological adaptations previously outlined in this chapter (Fig. 8-4). The general morphological adaptations that seem to covary with these strategies are discussed by Reynolds (1984; Chapter 10 this volume).

THE ROLE OF RESOURCES IN LIMITING ALGAL GROWTH

The Limitation of Rates and Yields

When dealing with resource limitation of growth, it is imperative that one define two concepts of limitation. It is possible to have one resource limit the algal biomass in a system while a second limits the rate at which this biomass is attained. In many instances it has been shown that phosphorus is the element primarily responsible for the limitation of algal biomass or production in freshwater systems (Schindler 1977; Schindler & Fee 1974). In some cases the rate at which this biomass is attained may be controlled by an entirely different resource. An extreme example is in the case of inorganic carbon, which has been identified as a rate-limiting resource in some poorly buffered low-pH systems (Burris, Wedge, & Lane 1981). In these cases the rate-limiting factor is the transfer of atmospheric CO_2 to the lake (Schindler & Fee 1973), but the large supply of atmospheric CO_2 ensures that carbon will never become a biomass-limiting nutrient in aquatic systems (Schindler 1977). In other situations the same nutrient may limit both algal biomass and growth rate. The dynamics of phytoplankton community structure is a result of differential net growth rates attained by different species in a phytoplankton community. Consequently, it is through the process of competition for the rate-limiting resources that nutrients play a role in shaping phytoplankton community structure.

Models of Resource-Limited Growth

Steady-state phytoplankton growth rates have been described as a function of external resource concentrations using the Monod (1950) equation [Eq. (5)]. Two other models currently employed in describing phytoplankton rate processes are the Michaelis–Menten uptake model [Eq. (2)] and the Droop relationship [Eq. (3)]. It has already been shown that Eqs. (2) and (3) can be combined mathematically to produce an equation of Monod form. These three models are mathematically equivalent at steady state and can be considered submodels of a "supramodel" of algal growth (Burmaster 1979). Although more complex models have also been constructed (Davis, Breitner, & Harrison 1978), it appears that in steady-state systems the Monod model enjoys a strong physiological underpinning, provided there are no major changes in K_s with growth rate (Turpin et al. 1985b). In cases where a single Monod equation is physiologically invalid due to changes in K_s, a forced Monod fit still provides an empirical description of nutrient-limited algal growth (Turpin et al. 1985b). The situation for non-steady-state growth will be discussed subsequently.

RESOURCE COMPETITION IN HOMOGENEOUS ENVIRONMENTS

A Single Limiting Resource

In order for a population to be maintained in a system, it must exhibit a growth rate equal to or greater than its mortality rate D (i.e., net growth rate ≥ 0). The resource concentration at which the net growth equals zero has been termed R^* (Tilman 1977, 1982; Tilman, Kilham, & Kilham 1982). The value of R^* for a given species is a variable depending on the population's mortality rate and can be predicted from the Monod equation (5) if growth rate (μ) is equated to the mortality rate (D) such that:

$$D = \mu = \frac{\hat{\mu} R^*}{K_\mu + R^*} \tag{8}$$

where

D = mortality rate
R^* = resource concentration at which $\mu = D$.

Two general cases of competition for a single resource are defined by this model. In the first case, one of two species exhibits the lowest R^* value regardless of the mortality rate. Such a case is illustrated as species A in Fig. 8-5. In such a configuration, species A will always lower the resource concentration below that required for species B to maintain a positive net growth rate (i.e., $R_A^* < R_B^*$). Consequently, A is the successful competitor, and B is excluded from the system. Such a relationship between two species means that there is no potential for resource partitioning along a concentration gradient of the limiting nutrient.

In the second case the Monod curves for species A and B cross (Fig. 8-6). The result is that when the mortality rate is less than D', species A exhibits the lowest R^*, and when $D > D'$, species B has the competitive advantage. Crossing μ-versus-$[S^*]$ curves (Fig. 8-7) are appealing from a theoretical perspective (Dugdale 1967) because they permit

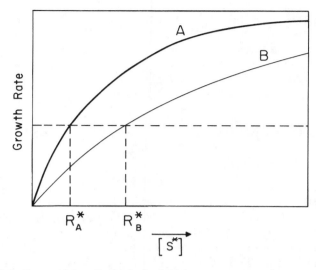

Fig. 8-5. Competition for a single limiting resource in which species A exhibits a lower R^* than does species B at all growth/mortality rates. The dotted line represents a mortality rate experienced by both species and the resulting R^* values for species A and B.

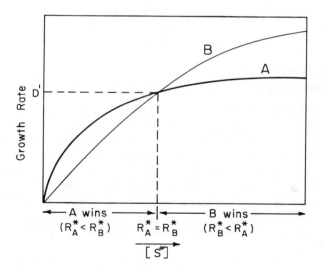

Fig. 8-6. Competition for a single limiting resource in which species A has the lowest R^* value ($R^*_A < R^*_B$) when mortality/growth rates are less than D', and species B has the lowest R^* value when mortality/growth rates are greater than D'.

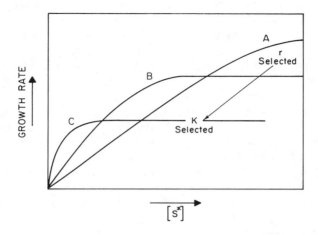

Fig. 8-7. Resource partitioning along a single resource supply/concentration continuum illustrating the necessary tradeoffs between species and the relative positions of *r*- and *K*-selected organisms.

partitioning of a single resource along a resource supply continuum. This continuum can refer to a concentration, dilution rate, growth rate, or specific flux continuum. Although each term has a slightly different connotation, at steady-state they are all correlated. Although resource partitioning along such gradients is a common phenomenon for bacteria (Jannasch 1967; Harder & Veldkemp 1971; Veldkemp & Jannasch 1972; Meers 1973; Veldkemp & Kuenen 1973; Matin & Veldkemp 1974; Kuenen et al. 1977), no unequivocal examples have been reported for phytoplankton. The results of continuous culture experiments employing natural phytoplankton assemblages have been viewed by some authors as support for this hypothesis (Dunstan & Tenore 1974; Harrison & Davis 1979; Mickelson, Maske, & Dugdale 1979; S. S. Kilham pers. comm.). Other workers (Smith & Kalff 1983) provide evidence suggesting that such partitioning does not occur in freshwater systems, although their interpretation has been the subject of some discussion (Sommer & Kilham 1985; Smith & Kalff 1985).

If species partition along a single resource supply continuum (Fig. 8-7), what are the tradeoffs in competitive ability that must occur? Organisms dominating at high substrate concentration must exhibit high maximal growth rates ($\hat{\mu}$), high values of K_μ, and low initial slopes (α_μ) relative to the other species. At the other end of the continuum are the better resource competitors with low values of K_μ, low maximal growth rates ($\hat{\mu}$), and correspondingly high initial slopes. These tradeoffs result in the organisms assorting along this resource concentration or supply continuum in a $K \rightarrow r$ type fashion (Fig. 8-7). The fast-growing r-selected organisms are capable of rapid colonization of new environments. K-selected organisms are slow growers that are more efficient at resource utilization and, hence, better competitors. The latter tend to dominate in stable, relatively closed systems such as the epilimnia of lakes during the later part of summer.

Cell size may also be an important consideration in a successional sequence. Typical patterns of phytoplankton seasonal periodicity begin with r-selected species growing in a nutrient-rich environment (Margalef 1958; Kilham & Kilham 1980; Sommer 1981), and as the available nutrient concentrations decrease, K-selected species begin to dominate due to their superior competitive ability. The strong inverse correlation between growth rate and cell size (Banse 1976; Chan 1978; Smith & Kalff 1983) would suggest that r-selected species would be smaller than K-selected ones.

Support for this generalization can be found in Sommer's (1981) phytoplankton periodicity studies from Lake Constance. Sommer abstracted three functional groups of phytoplankton based on maximum in situ net growth rates. The first group to appear in the temporal sequence exhibited net growth rates greater than 0.4 d^{-1} and consisted of small Chlorococcales, Volvocales, centric diatoms, and all of the Cryptophyceae with the exception of *C. rostratiformis*. The second group to appear was composed of those organisms that exhibited net growth rates between 0.2 and 0.4 d^{-1}, and it was represented by pennate diatoms, medium-sized centric diatoms, Cyanophyta, *Dinobryon, Cryptomonas rostratiformis, Mougeotia,* and *Ulothrix.* The final group in the temporal sequence consisted of those species that exhibited in situ net growth rates less than 0.2 d^{-1}, and it was represented by the dinoflagellates, desmids, and the large centric diatom *Stephanodiscus astrea.* With the exception of the flagellates there was a strong inverse correlation between cell size and growth rate (Sommer 1981). The higher growth rates of small cells can in part be explained by surface-to-volume considerations (Pasciak & Gavis 1974, 1975; Gavis 1976). Flagellates, however, may be exempt from such size-dependent constraints due to their motility (Gavis 1976).

Resource partitioning along a single resource supply continuum does not enable coexistence of potentially competing species (Taylor & Williams 1975) and hence does little to solve the "paradox of the plankton" (Hutchison 1961). This is a major difference when compared to partitioning along a resource ratio continuum in which stable steady-state coexistence between potential competitors may occur (Petersen 1975; Taylor & Williams 1975; Tilman 1977; Tilman 1982).

Development of zones of stable coexistence along a single resource supply continuum, however, may be facilitated through interactions with a resource ratio continuum (see "Interactions between Resource Continua").

Although evidence in support of phytoplankton resource partitioning along a single resource supply continuum is equivocal, this concept is based on a strong conceptual foundation. The assumption that it does occur allows for the construction of models consistent with a wide variety of seasonal periodicity studies. Although such models are attractive, detailed examination is required to determine their relative importance in comparison to other factors, such as resource ratios and variability, grazing, sinking, and irradiance.

Several Potentially Limiting Resources

In natural systems many resources are capable of limiting phytoplankton growth rates. Consequently, the simple theory of the preceding section must be extended to enable us to deal with several potentially limiting resources. The elegant work of Droop (1974), Rhee (1974), and Panikov (1979) has shown that nutrient limitation can be modeled most appropriately by a threshold model. In such a case an organism's growth rate may be limited only by a single resource and interactions with other potentially limiting resources are minimal. The need then arises to identify the resource that is limiting different populations in a community. At the theoretical level this can be accomplished by examining external or internal resource concentrations by employing either the Monod (4) or Droop (3) equations, respectively.

Identification of the Limiting Resource from External Concentrations. In a steady-state system, the transition between limitation by resources R_i and R_j occurs when the external substrate concentrations of the two resources are such that each in isolation would produce an identical nutrient-limited growth rate. The ratio of $[R_i]/[R_j]$ at which this occurs can be determined by substituting $[R_i]$ and $[R_j]$ into Eq. (8) and solving for the case $\mu_i = \mu_j$, giving

$$\frac{[R_i]}{K_{\mu i} + [R_i]} = \frac{[R_j]}{K_{\mu j} + [R_j]} \tag{9}$$

Rearranging we obtain

$$\frac{[R_i]}{[R_j]} = \frac{K_{\mu i}}{K_{\mu j}} \tag{10}$$

The result provides that when $[R_i]/[R_j] > K_{\mu i}/K_{\mu j}$ the population is limited by R_j, and when $[R_i]/[R_j] < K_{\mu i}/K_{\mu j}$ limitation is by R_i (Titman 1976; Tilman 1977).

Identification of the Limiting Resource from Internal Concentrations and the Concept of Optimum Ratios. The use of external resource concentrations gives an indication of the concentrations required for balanced growth but provides no information on either the cellular requirements for the resources or the rates of their consumption. By examining resource supply rates and the cellular nutrient ratio at which transition between nutrient limitations occurs, we can determine both the limiting resource and the consumption rates of each resource.

The optimum ratio is the cellular ratio of resources required such that no resource is in short supply relative to any other (Rhee & Gotham, 1980). It therefore represents the cellular nutrient ratio at which the transition from limitation by one resource to limitation by another will occur. If the ratio of resource supply rates (R_i/R_j) is greater than the optimum ratio $(R_c = Q_i/Q_j)$, then R_j will be the limiting resource. If the supply ratio is less than the optimum ratio, R_i will be limiting. The optimum ratio and the ratio of resource supply rates therefore determine the resource that will be of importance in determining the outcome of any competitive interaction.

The optimum ratio was originally defined as the ratio of subsistence quotas for the respective nutrient pair $(K_{Qi}/K_{Qj}$, Rhee & Gotham 1980). Such a growth-rate-independent formulation is valid only if the cell quotas of the resources in question do not change with growth rate or at least that they do so in a parallel fashion. This is unlikely because the relationship between nutrient cell quota and growth rate is extremely variable and depends on both the nutrient and the organism in question [see Eq. (3); Droop 1968, 1973a, 1974, 1983; Paasche 1973; Rhee 1973; Goldman et al. 1974; Tilman & Kilham 1976; Goldman 1977; Goldman & McCarthy 1978; Terry 1980; Goldman & Graham 1981; Elrifi & Turpin 1985; Layzell, Turpin, & Elrifi 1985; Turpin et al. 1985b]. As a result, the optimum ratio is best viewed as a growth-rate-dependent variable (Terry 1980; Terry et al. 1985; Elrifi & Turpin 1985; Turpin 1986). This can be demonstrated by rearranging Eq. (3) such that Q is represented as a function of μ [Eq. (11)]:

$$Q = \frac{K_Q}{1 - \mu/\overline{\mu}} \tag{11}$$

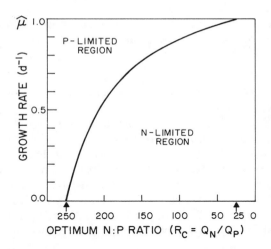

Fig. 8-8. An example of the growth-rate dependence of the optimum N:P ratio for a hypothetical species exhibiting a $\bar{\mu}_N$ of 1.25 and a $\bar{\mu}_P$ of 1.01 as in the general case suggested by McCarthy (1981). The formulation assumes a K_{QN} of 50 fmol·cell^{-1}, K_{QP} of 0.2 fmol·cell^{-1} and $\hat{\mu}$ of 1.0 d^{-1}. This figure illustrates how the optimum ratio of an ecologically important nutrient pair (N:P) can vary dramatically with growth rate. In this case, the relative change in R_c was a factor of 10. If resource supply ratios (R_N/R_P) fall to either side of the optimum ratio curve, the resulting limitation is as indicated.

The growth-rate-dependent optimum ratio ($R_c = Q_i/Q_j$) can then be expressed by Eq. (12):

$$R_c = \frac{Q_i}{Q_j} = \frac{K_{Qi}/(1 - \mu/\bar{\mu}_i)}{K_{Qj}/(1 - \mu/\bar{\mu}_j)} \tag{12}$$

If Q_i is invariant, it may be substituted directly into the numerator of this equation. It can be shown that if $\bar{\mu}_i < \bar{\mu}_j$ (correspondingly, $K_{Qi}/Q_{max\ i} < K_{Qj}/Q_{max\ j}$), then the cell's requirements for nutrient i increase with growth rate at a rate greater than the requirements for nutrient j. The optimum ratio will therefore increase with increasing growth rate. Likewise if $\bar{\mu}_i > \bar{\mu}_j$ (correspondingly, $K_{Qi}/Q_{max\ i} > K_{Qj}/Q_{max\ j}$), the optimum ratio will decrease with increasing growth rate. An example of this phenomenon is given in Fig. 8-8. The further the ratio of $\bar{\mu}_i/\bar{\mu}_j$ deviates from 1, the greater the growth-rate-dependent change in the optimum ratio. Although theoretical evidence has been provided suggesting a growth-rate dependence of the optimum ratio (Terry 1980; Elrifi & Turpin 1985), it was only recently that experimental support for this hypothesis was obtained (Terry et al. 1985; Turpin 1986).

The Range of Growth-Rate-Dependent Optimum Ratios. The magnitude of growth-rate-dependent changes in R_c depends on the nutrient pair in question. In previous studies the optimum N:P ratios were shown to vary only slightly (Terry 1980; Terry et al. 1985; Elrifi & Turpin 1985), whereas Turpin (1986) has shown that optimum C:P ratios can vary by at least a factor of 20. Figure 8-9 reports the theoretical range of R_c changes as a function of all possible values of $\bar{\mu}_i$ and $\bar{\mu}_j$. This figure clearly illustrates that the greatest growth-rate-dependent changes in R_c occur for nutrient pairs in which one nutrient shows high cell quota variability with growth rate ($\bar{\mu}_i \to \hat{\mu}$) while the other remains constant ($\bar{\mu}_j/\bar{\mu}_i \to \infty$). As $\bar{\mu}_i$ increases and/or the ratio of $\bar{\mu}_j/\bar{\mu}_i$ decreases, the growth-rate-dependent variability in R_c decreases. Figure 8-9 also illustrates the approximate domains of various nutrient pairs based on the range of $\bar{\mu}$ values reported for various nutrients (McCarthy 1981; Elrifi & Turpin 1985 and refs. therein). It is apparent that if C is one nutrient of the nutrient pair, R_c will vary greatly, whereas for nutrient pairs such as N:P the expected change is less but may still be ecologically significant. These predictions are consistent with measured changes in R_c (Terry et al. 1985; Turpin 1986).

Optimum Ratios and Their Implications for Competition, Coexistence, and Stability. The field of resource-based competition theory has recently been advanced by the superb work of Tilman (1977, 1981, 1982), in which resource partitioning has been shown to occur along resource ratio continua. Determination of the outcome of competition, whether it results in dominance by a single species or the formation of zones of stable or unstable coexistence, can be determined using simple graphical techniques (Tilman 1982). These well-described techniques make use of zero-net-growth isoclines and resource consumption vectors. The "resource ratio" hypothesis has been used to explain a number of spatial and temporal successional sequences in natural systems and is currently gaining wide acceptance. This theory and the experimental support thereof has been reviewed in recent years (Tilman 1982; Tilman et al. 1982). It will therefore not be reiterated here but will instead be used as a basis from which to examine the potential ecological implications for phytoplankton of growth-rate-dependent optimum ratios.

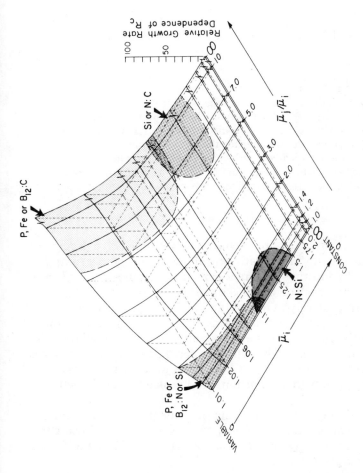

Fig. 8-9. Growth-rate-dependent changes in the optimum Q_i/Q_j ratio (R_c), expressed as the ratio of R_c at $\hat{\mu}$ to R_c when $\mu = 0$ [i.e., $(Q_{maxi}/Q_{maxj}) \cdot (K_{Qj}/K_{Qi})^{-1}$], as a function of $\bar{\mu}_i$, and the ratio of $\bar{\mu}_j/\bar{\mu}_i$. In this example $\bar{\mu}_i$ and $\bar{\mu}_j$ are normalized such that $\hat{\mu} = 1$. The largest growth-rate-dependent change in R_c occurred for nutrient pairs in which Q_i exhibits great variability with growth rate ($\bar{\mu}_i$ approaches 1) and Q_j is constant ($\bar{\mu}_j$ approaches infinity and hence a large value of $\bar{\mu}_j/\bar{\mu}_i$). As μ_j increases (i.e., Q_i becomes less variable) and/or $\mu_j/\bar{\mu}_i$ decreases to 1, the growth-rate dependence of R_c decreases. The approximate domains for a variety of nutrient pairs have been illustrated.

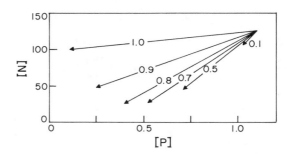

Fig. 8-10. The growth-rate-dependent slopes of N:P consumption vectors for the hypothetical species in Fig. 8-8 illustrated on a surface delineated by N and P concentration. The consumption vectors, as drawn, represent the direction and instantaneous magnitude of relative N and P concentration changes that would occur as a result of cells growing at the indicated steady-state growth rate. The exact position of these vectors on the N:P surface would be determined by the organism's zero net growth isoclines and the resource supply point (Tilman 1982).

At steady state, the rate of nutrient consumption ρ is the product of the steady-state growth rate μ and the cell quota Q. The consumption of two potentially limiting resources can be represented by consumption vectors **C** on a surface where the ordinate and abscissa represent $[R_i]$ and $[R_j]$, respectively (Tilman 1982). The slope of the consumption vector in the absence of luxury consumption represents the uptake rate of R_i relative to R_j and is equivalent to the optimum ratio R_c. Because optimum ratios are growth-rate dependent, so are the slopes of consumption vectors (Fig. 8-10). The instantaneous rates of resource consumption are proportional to the steady-state growth rate (Fig. 8-10).

The potential effects of growth-rate-dependent optimum ratios on competition, coexistence, and stability can best be discussed using two hypothetical species. If Q_i and Q_j for both species follow the Droop relationship and given that $\bar{\mu}_i > \bar{\mu}_j$ (i.e., $K_{Qi}/Q_{\max\,i} > K_{Qi}/Q_{\max\,j}$), then the optimum ratios ($R_c = Q_i/Q_j$) will decrease with increasing growth rate (Terry et al. 1985; Elrifi & Turpin 1985). Given the variability in optimum ratios between species (Rhee & Gotham 1980; Terry 1980; Tilman 1982; Elrifi & Turpin 1985), a situation could arise in which the growth-rate-dependent optimum ratio curves of the two species cross (Fig. 8-11). This crossover point, termed the *optimum ratio equivalance point*, occurs at the growth rate where the R_c of each species is

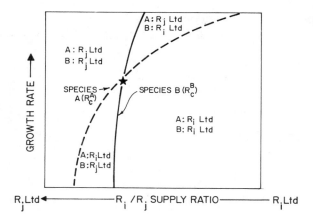

Fig. 8-11. Interactions between optimum ratios (R_c) and zones of nutrient limitation. The optimum ratios of hypothetical species (A and B) are represented on a surface delineated by growth rate and the R_i/R_j supply ratio. The growth-rate-dependent optimum ratio curves cross at the optimum ratio equivalence point indicated by *. The limitation experienced by each species in each of the four zones is as indicated.

identical (marked by a star in Fig. 8-11). At supply ratios greater than the optimum ratios of both species ($R_i/R_j > R_c^A$ and R_c^B), both would be limited by R_j. At supply ratios less than both optimum ratios ($R_i/R_j < R_c^A$ and R_c^B), both species would be limited by R_i. In the region between the growth-rate-dependent optimum ratio curves, each species is limited by a different resource and there is the potential for coexistence (Fig. 8-11).

If species A is the superior competitor for R_j and species B is the superior competitor for R_i (see Monod kinetics of Table 8-2), then the outcome of competition can be determined using the techniques outlined previously. When both species are R_j limited (Fig. 8-11), species A would win (Fig. 8-12). In the zone where both species are R_i limited (Fig. 8-11), species B would win (Fig. 8-12). The nature of the coexistence established in the zones where both species are limited by different resources can be evaluated using the graphical approach of Tilman (1982). Zero-net-growth-isoclines and the associated consumption vectors represent the regions in Fig. 8-12a corresponding to growth rates below, equivalent to, and above the optimum ratio equivalence point (Fig. 8-12b,c,d, respectively). At the two species equilibrium point on the zero-net-growth isoclines (open circles in Fig. 8-12b,c,d), species A

Table 8-2. *The relative kinetics of nutrient-limited growth for two*
hypothetical species grown under either R_i or R_j limitation. Kinetic
values are those defined by the Monod equation [Eqs. (5) and (8)].

	Half-saturation constant for growth (K_μ)		Relative maximal growth rate ($\hat{\mu}$)
	R_i	R_j	
Species A	High	Low	1.0
Species B	Low	High	1.0

is R_i limited and species B is R_j limited. In the zone where $R_c^A > R_i/R_j > R_c^B$ (Fig. 8-12a), each species consumes more of the resource limiting its growth rate than do the other species. This is illustrated by the consumption vectors in Fig. 8-12b. In this configuration, resource levels will eventually converge on the two species equilibrium point, and coexistence will be stable (Tilman 1982). This zone of stable coexistence is indicated by the stippled surface in Fig. 8-12a.

As growth rates increase and the optimum ratios change, the consumption vectors pivot around the two species equilibrium point. The consumption vector for species A (C_A) moves down and to the right relative to C_B, and C_A and C_B converge when growth rates reach the optimum ratio equivalence point (Fig. 8-12c). At growth rates above the equivalence point, in the region where $R_c^B > R_i/R_j > R_c^A$, the consumption vectors are such that each species consumes more of the resource limiting the other species (Fig. 8-12d). This results in a zone of unstable coexistence (Tilman 1982) and is represented in Fig. 8-12a as the hatched volume. If the Monod kinetics of each species were reversed (Table 8-2), the result would be a zone of unstable coexistence up to the optimum ratio equivalence point, and a zone of stable coexistence at growth rates above this value.

Some new potential implications of growth-rate-dependent optimum ratios for phytoplankton competition, coexistence, and community stability are now apparent. The first is that the breadth of the zone of stable coexistence between two species may depend on the specific flux of the limiting resources and, hence, the organisms' growth rate (Fig. 8-12a). Furthermore, changes in dilution rate or specific flux

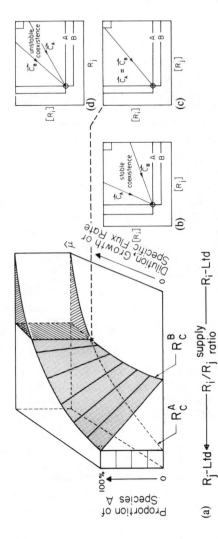

Fig. 8-12. The same surface as in Fig. 8-11 but expanded in the third dimension to represent the outcome of competition given the Monod kinetics outlined in Table 8-2. This figure shows the potential effects of growth-rate-dependent optimum ratios on competition, coexistence, and stability in a hypothetical two-species system. (a) The growth-rate-dependent optimum ratios ($R_c = Q_i/Q_j$) of hypothetical species A and B are represented on a surface delineated by the R_i/R_j supply ratio and dilution rate or specific flux. The point at which the optimum ratios curves cross is termed the optimum ratio equivalence point and is indicated by *. The outcome of competition over this surface is indicated in the vertical dimension. At R_i/R_j supply ratios resulting in R_j limitation to both species (i.e., $R_i/R_j > R_c^A$ and R_c^B), species A dominates due to its superior competitive ability under R_j limitation (Table 8-2). At supply ratios resulting in R_i limitation to both species (i.e., $R_i/R_j < R_c^A$ and R_c^B), species B dominates due to its superior competitive ability under R_i limitation (Table 8-2). In the regions between R_c^A and R_c^B; there is the potential for coexistence. The nature of this coexistence can be determined graphically using zero-net-growth isoclines (ZNGI) and consumption vectors as outlined by Tilman (1982). (b) The ZNGI for species A and B and the consumption vectors for species A (C_A) and B (C_B) at dilution rates below the optimum ratio equivalence point. The two-species equilibrium point is indicated by ○. This configuration results in stable coexistence of both species when $R_c^A > R_i/R_j > R_c^B$ (see text), and is indicated by the stippled surface in (a). (c) The ZNGI and accompanying consumption vectors for both species at the optimum ratio equivalence point. (d) The ZNGI and accompanying consumption vectors for both species at dilution rates greater than the optimum ratio equivalence point. This configuration results in unstable coexistence of both species when $R_c^B > R_i/R_j > R_c^A$ and is indicated by the hatched area in (a) (see text).

may completely eliminate a stable zone of coexistence. This points to a potentially important mechanism by which dilution rate or specific flux may affect the outcome of competition and the nature of two species equilibria in planktonic systems.

Resource Partitioning and Competitive Tradeoffs along a Resource Ratio Continuum. To this point I have dealt primarily with the mechanistic basis of resource partitioning along resource ratio continua and the implications of dilution rate or specific flux on the integrity of this partitioning. Although the importance of growth-rate-dependent optimum ratios in phytoplankton ecology has yet to be demonstrated, the importance of optimum ratios in general in allowing resource partitioning along a resource ratio continuum is well established. In the following section I wish to examine some tradeoffs that occur between species assorted along such continua.

Tilman et al. (1982) compiled data and reported tradeoffs in the competitive abilities for phosphate and silicate in several species of freshwater diatoms. In general, there was an inverse correlation between the competitive abilities for Si and P. Those species that were good Si competitors were poor at competing for P, and vice versa (see Sommer, Chapter 6). These tradeoffs suggested that a decline in Si:P ratios over the season should result in a progression from *Synedra* to *Asterionella* to *Fragilaria* to *Diatoma* to *Stephanodiscus*. In the view of Tilman et al. (1982) this is consistent with observations of natural Lake Michigan phytoplankton dynamics made by Kilham (1971).

Another classic example of phytoplankton partitioning along a resource ratio continuum is that of blue-green algal dominance as a function of N:P. The ability of some blue-green algae to fix nitrogen enables these organisms to dominate many natural systems when nitrogen is limiting to other species (see Paerl, Chapter 7). In a detailed assessment of species composition of numerous lakes, Smith (1983) was able to demonstrate a dramatic increase in the abundance of cyanobacteria when epilimnetic N:P ratios fell below 29:1 by weight (13:1 by atoms). Other workers have also provided convincing evidence showing blue-green algal dominance is strongly correlated with low N:P conditions (Schindler 1977; Tilman and Kiesling 1984). This observation not only demonstrates the advantages that N_2-fixation confers on these organisms but suggests that the blue-green algae are poor competitors for phosphorus. This tradeoff could be viewed as the evolutionary cost associated with N_2-fixing capability although the rationale for such a tradeoff is not yet apparent.

There are more tangible examples of the costs associated with N_2-

fixation. Zevenboom et al. (1981) isolated a nonheterocystous mutant of *Aphanizomenon flos-aquae* that exhibited a higher growth rate at all light intensities than did the wild type. The cost of N_2-fixation for *Aphanizomenon flos-aquae* therefore appears to be reflected as a decrease in maximal growth rate. It has been well established that heterocystous blue-green algae have lower maximal growth rates when grown on N_2 than when grown on NO_3^- or NH_3 (Ward & Wetzel 1980; Rhee & Lederman 1983; Layzell et al. 1985; Turpin et al. 1985a). This decrease in growth rate can be primarily attributed to the costs associated with the establishment and maintenance of the heterocysts (Turpin et al. 1985a; see also Paerl's comments on symbiotic bacteria, Chapter 7). Their ability to maintain high surface densities through use of gas vacuoles and other physiological adaptations have enabled them to ameliorate the cost of N_2-fixation by maintaining themselves in a high-light, rapid-growth environment (Paerl & Keller 1979; Van Liere & Walsby 1982). In some hypereutrophic systems, however, the costs of N_2-fixation can restrict their success (Zevenboom 1980; Zevenboom & Mur 1980).

Interactions between Resource Continua

In the preceding sections I have examined resource partitioning along supply rate and ratio continua as well as interactions between these continua mediated by growth-rate-dependent optimum rates. Other interactions may also occur. Changes in the supply rate of a single resource alters resource supply ratios. Similarly, a given resource ratio can be attained at an infinite number of resource supply rates (Fig. 8-13). What then are the potential effects of these interactions beyond those associated with growth-rate-dependent optimum ratios?

Potential species interactions are maximal when partitioning occurs along both resource supply rate and resource ratio continua. The nature of these interactions can be illustrated by choosing two hypothetical species that exhibit Monod kinetics such that partitioning occurs along these continua (Fig. 8-7). The kinetic parameters of two such species and the resulting growth curves are given in Table 8-3 and Fig. 8-14, respectively. The outcome of competition between these species can be calculated as a function of both supply rate (dilution or specific flux rate) and the resource supply ratio (Fig. 8-15). For simplicity, optimum ratios are considered constant. The conclusion is that the nature of the partitioning along one of the continua depends on the position on the other (Fig. 8-15). The implications are that alterations

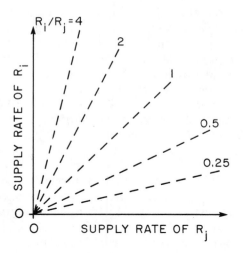

Fig. 8-13. The interactions between the supply rates of R_i and R_j and the resource supply ratio R_i/R_j. The dashed lines represent lines of constant resource ratio.

Table 8-3. *Monod model kinetic parameters ($\hat{\mu}$ and K_μ) and the optimum ratio (R_c) of two hypothetical species (A and B) with respect to R_i and R_j. The growth kinetics of these species are outlined in Fig. 8-14 and the theoretical outcome of competition as a function of supply rate and R_i/R_j is given in Fig. 8-15.*

	$\hat{\mu}$	$K_{\mu i}$	$K_{\mu j}$	R_c
Species A	1.0	4.0	0.1	40
Species B	0.85	3.0	0.05	60

in either the magnitude of resource supply or the ratio of resource supply rates may change the outcome of competition or the nature of species coexistence. Interactions between supply rates and ratios may therefore allow zones of stable coexistence to develop along the resource supply continuum where prior to this added complexity such a phenomena was impossible (see previous section).

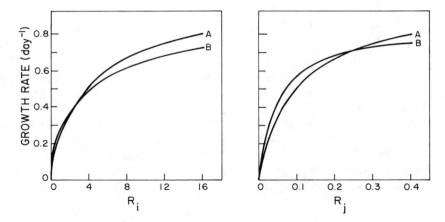

Fig. 8-14. The growth kinetics of two hypothetical species growing under R_i or R_j limitation that exhibit the ability to partition along either an R_i or an R_j supply continuum. The kinetics of growth are as given in Table 8-3.

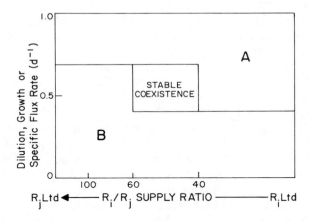

Fig. 8-15. A dominance plane representing the outcome of competition between species A and B (see Table 8-3 and Fig. 8-14) as a function of steady-state dilution rate (specific flux) and the resource supply ratio R_i/R_j. The interaction of these two continua will theoretically allow for the establishment of a zone of stable coexistence along the resource supply rate (growth rate, dilution rate) continuum. In the present example this could occur only when R_i/R_j is between 40 and 60.

COMPETITION IN FLUCTUATING ENVIRONMENTS: IMPLICATIONS OF RESOURCE VARIABILITY

A homogeneous environment is but a single point on the continuum of resource fluctuation frequencies. Fluctuations of the supply environment range from low-frequency seasonal fluctuations to fluctuations of such ultrahigh frequency that the environment approaches true homogeneity. Adequate consideration of a broad range of frequencies by use of the mechanistic simplicity employed for homogeneous environments is all but impossible (Grenney, Bella, & Curl 1973; Turpin et al. 1981; Powell & Richardson 1985). The coverage in this section will therefore emphasize strategies employed in the optimization of environments exhibiting varying degrees of resource variability.

Partitioning of Variable Resources

The role of resource variability in determining phytoplankton community structure was assessed by Turpin and Harrison (1979). To evaluate the effects of resource fluctuations, these authors established chemostat cultures containing a natural assemblage of marine phytoplankton as an inoculum. The temporal patchiness of the limiting nutrient (ammonium) was controlled so that addition was either continual or pulsed at eight times a day or once a day. Culture dilution rates and the flux of all nonlimiting nutrients were held constant, and the successional sequence was monitored for three weeks. At the end of the experiment it was shown that community structure differed between the treatments. The homogeneous environment favored the diatom *Chaetoceros*, whereas the patchy environment (once a day addition) favored a second diatom, *Skeletonema*. The intermediate regime resulted in an assemblage codominated by these two species. The major features of these experiments were subsequently replicated (Turpin & Harrison 1980), and, coupled with the recent work of Robinson & Sandgren (1983), Scavia et al. (1984), and Sommer (1984, 1985), they confirm that resource partitioning can occur along a continuum of resource variability.

Prior to examining the mechanisms controlling the partitioning and adaptations to resource variability, it is important that we assess the relationship between the frequency, duration, and magnitude of resource fluctuations (also see Reynolds, Chapter 10). High-frequency, low-duration resource fluctuations can be viewed as environments that approach homogeneity. An example would be a chemostat culture where the new medium is being added dropwise every few seconds, or possibly bacteria regeneration in natural systems. The duration and magnitude of these high-frequency resource fluctuations is low, and the significance of an individual fluctuation is small (i.e., lowering a chemostat flow rate by one drop per day is relatively insignificant). On the other hand, a single low-frequency resource fluctuation event is of major significance to community resource supply and tends to be of much greater duration (i.e., a semicontinuous culture that is being diluted once a day or week, or periodic storm events in natural systems). Therefore there exists an inverse correlation between the frequency of resource fluctuation, its duration, and the magnitude of its significance to community resource supply. Because of these frequency-dependent characteristics, it is reasonable to assume that different adaptations would be employed to optimize different frequency resource fluctuations.

In recent years the concepts of resource fluctuations and limiting-nutrient patchiness have been linked to zooplankton excretion and macroaggregates (McCarthy & Goldman 1979; Goldman, McCarthy, & Peavey 1979; Lehman & Scavia 1982a,b). Many of the ecological implications of resource variability, however, do not require that they occur at a microscale. Other episodic events such as storms, seiches, and wind mixing or seasonal turnovers all produce resource fluctuations of varying frequency and duration. Nevertheless a great deal of energy has been expended on the assessment of nutrient microzones in the sea. Since the occurrence of microzones was first proposed (McCarthy & Goldman 1979), their potential significance has been the subject of much discussion (Turpin & Harrison 1979, 1980; Jackson 1980; Turpin et al. 1981; Williams & Mur 1981; Lehman & Scavia 1982a,b; Scavia et al. 1984; Currie 1984). The exciting work of Lehman and Scavia (1982a,b) has shown that phytoplankton can utilize nutrients in the microzones produced by zooplankton. The question of how important microscale nutrient distribution is to phytoplankton nutrition still remains unanswered. Although such an answer will not be forthcoming in this contribution, I would suggest that the significance of this phenomenon will depend on the species in question and the adaptations it employs in dealing with the wide range of resource fluctuation frequencies.

Adaptations to Resource Variability

What are the physiological adaptations for resource procurement in fluctuating environments, and to which frequencies are they best suited? Adaptations to resource variability must serve to increase the growth of the organism over that which would occur in the absence of the adaptation. As we have already discussed, optimization of a homogeneous environment may occur through maintaining a low K_μ for growth by increases in $\rho_{t\ max}$ and decreases in the cell quota under nutrient limitation. As the resource concentration begins to fluctuate, the relevance of this "steady-state" solution decreases. Nutrient-limited algae encountering a pulse of nutrient will initially consume the resource at a rate equivalent to a point on the nutrient transport curve (ρ_t, S); see Fig. 8-2. As uptake proceeds, internal pools will fill and assimilation will become limiting (Conway, Harrison, & Davis 1976; Wheeler, Glibert, & McCarthy 1982a; Wheeler & McCarthy 1982). Uptake rate will then decrease to a point on the assimilation curve (ρ_a, S); see Fig. 8-2. In pulses of significant duration, the cells' physiological state and its rate of assimilation will change. Therefore the time course of nutrient disappearance and resource procurement is nonlinear (Conway et al. 1976; Conway & Harrison 1977; Collos 1980; Goldman & Glibert 1983; Parslow et al. 1985b); consequently, both the kinetics of transport ($\rho_{t\ max}$ and K_t) and the kinetics of assimilation ($\rho_{a\ max}$ and K_t) are important components in optimization of fluctuating environments. For example, Fig. 8-16 illustrates resource consumption by two species (A and B) exhibiting different ρ_t and ρ_a components of uptake. Species A has a higher value of ρ_t but a lower value of ρ_a. The result is that each species is able to optimize resource encounters of different frequencies and duration. In high-frequency environments (nutrient encounters on the order of seconds to minutes) a high ρ_t would enable maximum nutrient procurement. Hence, species A would obtain a greater proportion of the resource in high-frequency environments where patch encounters were of a duration less than t' (Fig. 8-16). As the frequency of resource fluctuation decreases and the duration of encounters increases, the significance of the short-lived ρ_t component of nutrient procurement decreases and the longer-term assimilation component (ρ_a) becomes more significant. Hence, when resource encounters are of duration greater than t', procurement of the resource is maximized by species B (Fig. 8-16). The fact that $\rho_{t\ max}$ is in part proportional to the cells' S/V ratio (Gavis 1976; Raven 1984) indicates that small cells may have an advantage in environments dominated by high-frequency resource fluctuations. Large cells, which exhibit low specific rates of respiration (Laws 1975), may have an advantage during lower-frequency resource fluctuations. This is consistent with the observations of Turpin and Harrison (1980), in which small cells dominated under conditions of high-frequency resource fluctuations, but large cells dominated during low-frequency resource encounters.

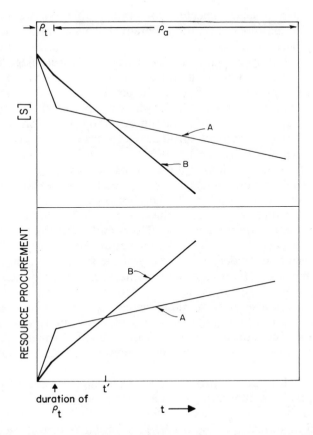

Fig. 8-16. The tradeoffs between the ρ_t and ρ_a components of uptake and assimilation in the optimization of transient nutrient availability. This figure relates the amount of resource procured to the duration of nutrient encounter. The short-lived nature of the ρ_t component ensures that organisms with high values of $\rho_{t\ max}$ will optimize resource procurement during encounters of short duration ($< t'$). Species A represents such an example. During encounters of longer duration, the ρ_a component plays an important role in the optimization of nutrient procurement, as illustrated by species B. When pulse duration exceeds t', species B will monopolize the limiting resource.

There is some experimental evidence showing that different adaptations are required to optimize resource fluctuations of different frequencies. Turpin and Harrison (1979) and Quarmby, Turpin, and Harrison (1982) showed that the nutrient uptake capacity of phytoplankton assemblages selected for under different frequencies of resource fluctuation were quite different. Each assemblage was physiologically best suited for the procurement of the limiting nutrient in the temporal supply regime under which that assemblage had been selected. Optimization of high-frequency fluctuations appeared to occur through an increase in $\rho_{t\ max}$ (Turpin & Harrison 1979; Turpin et al. 1981; Quarmby et al. 1982), whereas species dominating low-frequency environments (one encounter per day) were shown to optimize nutrient procurement through high rates of long-term assimilation, ρ_a (Turpin & Harrison 1979).

Lower-frequency resource fluctuations occurring on a monthly or yearly basis (i.e., much longer than a generation time) elicit responses at the organismal level of phytoplankton organization. These adaptations include such responses as the development of high growth rates during long-duration nutrient encounters and the formation of resting cysts during the times of nutrient deprivation. Sommer (1985) has provided convincing evidence that many of the successful competitors in low-frequency fluctuating environments exhibit a marked growth response over the period of resource oscillation. Examples of life cycle changes in response to nutrient deficiency are common and have been well reviewed in other chapters in this volume. Intermediate degrees of resource fluctuation can be accommodated through adaptations such as luxury consumption and behavioral and morphological changes.

From the limited existing evidence, there appears to be a hierarchy of phytoplankton adaptations to resource procurement from fluctuating environments that spans the biochemical and organismal levels of biological organization (Fig. 8-17). The similarity of this hierarchy to that which occurs in response to increasing severity of limitation (Fig. 8-4) is not unexpected. When the limiting resource is homogeneously distributed, the intensity of resource competition will be constant and those organisms able to dominate the environment will be K-selected and exhibit adaptations serving to increase α_μ. Some of these adaptations, such as an increase in $\rho_{t\ max}$ and cell quota flexibility, coupled with high values of ρ_a, are also very important in coping with high-frequency resource fluctuations. Environments predominated by low-frequency resource fluctuations would be dominated by r-selected organisms exhibiting the ability to attain high growth rates during the long-duration nutrient encounters, yet capable of withstanding the periods of nutrient deprivation encountered in such environments by producing resting stages (Fig. 8-17). Different life history strategies would be reflected in the assortment of species along a continuum of resource variability. Organisms dominating under homogeneous con-

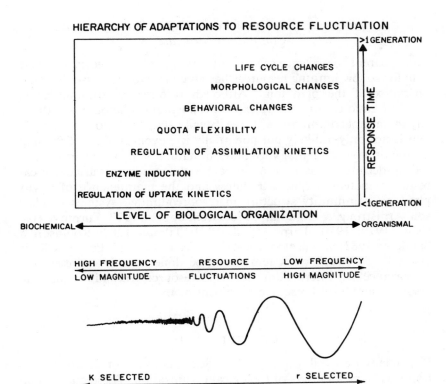

Fig. 8-17. The hierarchy of adaptations to resource fluctuations. These adaptations span the biochemical through to the organismal levels of organization. Homogeneously distributed resources and high-frequency resource fluctuations are adapted to by modifications at the biochemical and physiological levels of organization. Low-frequency fluctuations are often adapted to by life cycle changes. Species emphasizing adaptations at different points along this continuum represent intermediates along a $K \longrightarrow r$ type contimuum.

ditions would obtain most of their nutrient requirements from the homogeneous background levels or very-high-frequency resource encounters. Species that dominate low-frequency environments would obtain the majority of their annual nutrient ration from low-frequency episodic events such as those associated with turnovers and storm events. Between these two extremes would be organisms capable of maximizing nutrient procurement at intermediate frequencies of nutrient availability. In essence, each species would be expected to exhibit a unique "power spectrum" of resource acquisition.

Implications of Resource Variability to Phytoplankton Community Structure

The apparent coexistence of so many phytoplankton species that compete for so few limiting resources has given rise to Hutchison's (1961) "paradox of the plankton." Although resource partitioning along resource ratio continua may explain a portion of the observed diversity in natural communities, a true "steady-state" solution to the problem is not only unlikely but potentially irrelevent in light of the high degree of environmental variability in natural systems. Over the preceding decade it has become apparent that resource variability can result in increased species richness and the maintenance of phytoplankton community structure in resource-limited systems (Richerson, Armstrong, & Goldman 1970; Grenney et al. 1973; Turpin & Harrison, 1979, 1980; Turpin et al. 1981; Tilman 1982; Robinson & Sandgren 1983; Scavia et al. 1984; Sommer 1984, 1985; Powell & Richerson 1985). Much work remains to be done in order to assess the importance of interactions between resource variability and the resource supply and resource ratio continua.

CONCLUSIONS

Phytoplankton exhibit a hierarchy of adaptations to resource limitation involving all levels of biological organization. These adaptations function in an integrative fashion and serve either to enhance growth rates above those that would occur in the absence of any adaptation or, through life cycle responses, to enable the organisms to survive during times of nutrient stress. These mechanisms provide the basis of resource competition and facilitate resource partitioning along resource supply and resource ratio continua. The interactions between these continua may occur directly as outlined in Fig. 8-15 or indirectly through the phenomena of growth-rate-dependent optimum ratios (Fig. 8-12). These interactions have the potential to contribute to the maintenance of community structure. Superimposed upon these continua and their patterns of interaction is the spectrum of resource fluctuation. This provides yet another continuum for resource partitioning serving to maintain species diversity. Adaptations that serve to optimize resource procurement from fluctuating environments depend on the frequency of resource fluctuation. High-frequency fluctuations elicit adaptations based in the biochemical and physiological levels of organization, whereas adaptations to low-frequency events are found at the organismal level. The resulting hierarchy of adaptations is similar to that occurring in response to increasing severity of nutrient limitation (Figs. 8-4 & 8-17).

REFERENCES

Adamich, M., Gibor, A., and Sweeney, B. M. (1975). Effects of low nitrogen levels and various nitrogen sources on growth and whorl development in *Acetabularia* (Chlorophyta). *J. Phycol.*, 11, 364–7.

Ahlgren, G. (1980). Effects on algal growth rates by multiple nutrient limitation. *Arch. Hydrobiol.*, 89, 43–53.

Antia, N. J., Berland, B. R., Bonin, D. J., and Maestrinni, S. Y. (1975). Comparative evaluation of certain organic and inorganic sources of nitrogen for phototrophic growth of marine microalgae. *J. Mar. Biol. Ass. U. K.*, 55, 519–39.

Banse, K. (1976). Rates of growth, respiration and photosynthesis of unicellular algae as related to cell size – a review. *J. Phycol.*, 12, 135–40.

Belmont, L. and Miller, D. A. (1965). The utilization of glutamate by algae. *J. Exp. Bot.*, 16, 318–24.

Bienfang, P. K. (1975). Steady-state analysis of nitrate-ammonium assimilation by phytoplankton. *Limnol. Oceanogr.*, 20, 402–11.

Brandes, D. and Elston, R. N. (1956). An electron microscopical study of the histochemical localization of alkaline phosphatase in the cell wall of *Chlorella vulgaris*. *Nature*, 177, 274–5.

Burmaster, D. E. (1979). The continuous culture of phytoplankton: mathematical equivalence among three steady-state models. *Am. Nat.*, 113, 123–34.

Burris, J. E., Wedge, R., and Lane, A. (1981). Carbon dioxide limitation of photosynthesis of freshwater phytoplankton. *J. Freshwater Ecol.*, 1, 81–96.

Button, D. K. (1985). Kinetics of nutrient-limited transport and microbial growth. *Micro. Rev.*, 49, 270–97.

Caperon, J. (1968). Population growth response of *Isochrysis galbana* to a variable nitrate environment. *Ecology*, 50, 188–92.

Caperon, J. and Meyer, J. (1972). Nitrogen-limited growth of marine phytoplankton. I. Changes in population characteristics with steady-state growth rate. *Deep Sea Res.*, 19, 601–18.

Cembella, A. D., Antia, N. J., and Harrison, P. J. (1984). The utilization of inorganic and organic phosphorous compounds as nutrients by eukaryotic microalgae: a multidisciplinary perspective: Part I. *Critical Rev. Microbiol.*, 10, 317–91.

Chan, A. T. (1978). Comparative physiological study of marine diatoms and dinoflagellates in relation to irradiance and cell size. 1. Growth under continuous light, *J. Phycol.*, 14, 396–402.

Collos, Y. (1980). Transient situations in nitrate assimilation by marine diatoms. I. Changes in uptake parameters during nitrogen starvation. *Limnol. Oceanogr.*, 25, 1075–81.

Conway, H. L. (1977). Interactions of inorganic nitrogen in the uptake and assimilation by marine phytoplankton. *Mar. Biol.,* 39, 221–32.

Conway, H. L. and Harrison, P. J. (1977). Marine diatoms grown in chemostats under silicate or ammonium limitation. IV. Transient response of *Chaetoceros debilis, Skeletonema costatum* and *Thalassiosira gravida* to a single addition of the limiting nutrient. *Mar. Biol.,* 43, 33–43.

Conway, H. L., Harrison, P. J., and Davis, C. O. (1976). Marine diatoms grown in chemostats under silicate or ammonium limitation. II. Transient response of *Skeletonema costatum* to a single addition of the limiting nutrient. *Mar. Biol.,* 35, 187–99.

Currie, D. J. (1984). Microscale nutrient patches: Do they matter to the plankton? *Limnol. Oceanogr.,* 29, 211–14.

Davis, C. O. Breitner, N. F., and Harrison, P. J. (1978). Continuous culture of marine diatoms under silicon limitation. 3. A model for Si-limited diatom growth. *Limnol. Oceanogr.,* 23, 41–52.

Dortch, Q., Clayton, J. R., Thoreson, S. S., Bressler, S. L., and Ahmed, S. I. (1982). Response of marine phytoplankton to nitrogen deficiency: decreased nitrate uptake vs. enhanced ammonium uptake. *Mar. Biol.,* 70, 13–19.

Droop, M. R. (1968). Vitamin B_{12} and marine ecology. IV. The kinetics of uptake, growth and inhibition in *Monochrysis lutheri. J. Mar. Biol. Ass. U. K.,* 48, 689–733.

   (1970). Vitamin B_{12} and marine ecology. V. Continuous culture as an approach to nutritional kinetics. *Helgolander Wiss. Meeresunters,* 20, 629–36.

   (1973a). Some thoughs on nutrient limitation in algae. *J. Phycol.,* 9, 264–72.

   (1973b). Nutrient limitation in osmotophic protista. *Amer. Zool.,* 13, 209–14.

   (1974). The nutrient status of algal cells in continuous culture. *J. Mar. Biol. Ass. U. K.,* 54, 825–55.

   (1983). 25 years of algal growth kinetics: a personal view. *Bot. Mar.,* 26, 99–112.

Dugdale, R. C. (1967). Nutrient limitation in the sea: Dynamics, identification and significance. *Limnol. Oceanogr.,* 12, 685–95.

   (1977). Modelling. In: *The Sea,* vol. 6, eds. E. D. Goldberg, I. N. McCave, J. J. O'Brien, and J. H. Steele, pp. 789–806. New York: Wiley.

Dugdale, R. C. and Goering, J. J. (1967). Uptake of new and regenerated forms of nitrogen in primary productivity. *Limnol. Oceanogr.,* 12, 196–206.

Dunstan, W. W. and Tenore, W. T. (1974). Control of species composition in enriched mass cultures of natural phytoplankton populations. *J. Appl. Ecol.,* 11, 529–36.

Elrifi, I. R. and Turpin, D. H. (1985). Steady-state luxury consumption and the concept of optimum nutrient ratios: a study with phosphate and nitrate limited *Selenastrum minutum* (Chlorophyta). *J. Phycol.,* 21, 592–602.

Eppley, R. W. (1962). Hydrolosis of polyphosphates by *Porphyra* and other seaweeds. *Physiol. Plant.,* 15, 246–51.

Eppley, R. W., Coatsworth, J. L., and Solorzano, L. (1969a). Studies of nitrate reductase in phytoplankton. *Limnol. Oceanogr.*, 14, 194–205.

Eppley, R. W. and Peterson, B. J. (1979). Particulate organic matter flux and planktonic new production in the deep ocean. *Nature*, 282, 677–80.

Eppley, R. W. and Renger, E. H. (1974). Nitrogen assimilation of an oceanic diatom in nitrogen-limited continuous culture. *J. Phycol.*, 10, 15–23.

Eppley, R. W., Rogers, J. N., and McCarthy, J. J. (1969b). Half-saturation constants for uptake of nitrate and ammonia by marine phytoplankton. *Limnol. Oceanogr.*, 14, 912–20.

Eppley, R. W., Sharp, J. H., Renger, E. H., Perry, M. Y., and Harrison, W. G. (1977). Nitrogen assimilation by phytoplankton and other microorganisms in the surface waters of the central North Pacific Ocean. *Mar. Biol.*, 39, 111–20.

Feuillade, J. B. and Feuillade, M. G. (1975). Etude des besoins en azote et en phosphore d'*Oscillatoria rubescens* D.C. a l'adie de cultures en chemostats. *Verh. Internat. Verein. Limnol.*, 19, 2698–708.

Fuhs, G. W. (1969). Phosphorus content and rate of growth in the diatoms *Cyclotella nana* and *Thalassiosira fluviatilis*. *J. Phycol.*, 5, 312–21.

Galloway, R. A. and Krauss, R. W. (1963). Utilization of phosphorous sources by *Chlorella*. In: *Microalgae and Photosynthetic Bacteria. Plant Cell Physiol. Spec. Issue*, pp. 569–74.

Gavis, J. (1976). Munk and Riley revisited: nutrient diffusion transport and rates of phytoplankton growth. *J. Mar. Res.*, 34, 161–79.

Glibert, P. M., Goldman, J. C., and Carpenter, E. J. (1982). Seasonal variations in the utilization of ammonium and nitrate by phytoplankton in Vineyard Sound, Massachusetts. *Mar. Biol.*, 70, 237–49.

Goldman, J. C. (1977). Steady-state growth of phytoplankton in continuous culture: comparison of internal and external nutrient equations. *J. Phycol.*, 13, 251–8.

Goldman, J. C. and Glibert, P. M. (1982). Comparative rapid ammonium uptake by four marine phytoplankton species. *Limnol. Oceanogr.*, 27, 814–27.

   (1983). Kinetics of inorganic nitrogen uptake by phytoplankton. In: *Nitrogen in the Marine Environment*, eds. E. J. Carpenter and D. G. Capone, pp. 233–74. New York: Academic Press.

Goldman, J. C. and Graham, S. J. (1981). Inorganic carbon limitation and chemical composition of two freshwater green microalgae. *Appl. Environ. Microbiol.*, 41, 60–70.

Goldman, J. C. and McCarthy, J. J. (1978). Steady-state growth and ammonium uptake of a fast-growing marine diatom. *Limnol. Oceanogr.*, 23, 695–703.

Goldman, J. C., McCarthy, J. J., and Peavey, D. G. (1979). Growth rate influence on the chemical composition of phytoplankton in oceanic waters. *Nature.*, 279, 210–15.

Goldman, J. C., Oswald, W. J., and Jenkins, D. (1974). The kinetics of inorganic carbon limited algal growth. *J. Water Poll. Control Fed.*, 46, 554–74.

Gotham, I. J. and Rhee, G. Y. (1981a). Comparative kinetics studies of phosphate-limited growth and phosphate uptake in phytoplankton in continuous culture. *J. Phycol.,* 17, 257–65.

    (1981b). Comparative kinetic studies of nitrate-limited growth and nitrate uptake in phytoplankton in continuous culture. *J. Phycol.,* 17, 309–14.

Grenney, W. J., Bella, D. A., and Curl, H. C. (1973). A theoretical approach to interspecific competition in phytoplankton communities. *Amer. Nat.,* 107, 405–25.

Harder, W. and Veldkamp, H. (1971). Competition of marine psychrophilic bacteria at low temperatures. *Antonie Van Leeuwenhock,* 37, 51–63.

Harrison, P. J. and Davis, C. O. (1979). The use of outdoor phytoplankton continuous cultures to analyze factors influencing species succession. *J. Exp. Mar. Biol. Ecol.,* 41, 9–23.

Healey, F. P. (1973). Characteristics of phosphorus deficiency in *Anabaena. J. Phycol.,* 9, 383–94.

    (1977). Ammonium and urea uptake by some freshwater algae. *Can. J. Bot.,* 55, 61–9.

    (1980). Slope of the Monod equation as an indicator of advantage in nutrient competition. *Microb. Ecol.,* 5, 281–6.

Horrigan, S. G. and McCarthy, J. J. (1981). Urea uptake by phytoplankton at various stages of depletion. *J. Plank. Res.,* 3, 403–14.

Hutchison, G. E. (1961). The paradox of the plankton. *Am. Nat.,* 95, 137–45.

Jackson, G. A. (1980). Phytoplankton growth and zooplankton grazing in oligotrophic oceans. *Nature,* 284, 439–41.

Jannasch, H. W. (1967). Enrichment of aquatic bacteria in continuous culture. *Archiv. Mikrobiol.,* 59, 165–73.

Kaplan, A., Badger, M. R., and Berry, J. A. (1980). Photosynthesis and the intracellular inorganic carbon pool in the blue-green alga *Anabaena variabilis:* Response to external CO_2 concentration. *Planta,* 149, 219–26.

Kilham, P. (1971). A hypothesis concerning silica and the freshwater planktonic diatoms. *Limnol. Oceanogr.,* 16, 10–18.

Kilham, S. S. (1978). Nutrient kinetics of freshwater planktonic algae using batch and semicontinuous methods. *Mitt. Internat. Verein. Limnol.,* 21, 147–57.

Kilham, P. and Kilham, S. S. (1980). The evolutionary ecology of phytoplankton. In: *The Physiological Ecology of Phytoplankton, Studies of Ecology,* Vol. 7, ed. I. Morris, pp. 571–97. London: Blackwell.

Kuenen, J. G., Boonstra, J., Schroder, H. G. I., and Veldkamp, H. (1977). Competition for inorganic substrates among chemoorgantrophic and chemolithotrophic bacteria. *Micro. Ecol.,* 3, 119–30.

Laws, E. A. (1975). The importance of respiration losses in controlling the size distributions of marine phytoplankton. *Ecology,* 56, 419–26.

Layzell, D. B., Turpin, D. H., and Elrifi, I. R. (1985). Effect of N source on the steady-state growth and N assimilation of P-limited *Anabaena flosaquae. Plant Physiol.,* 78, 739–45.

Lehman, J. T. and Scavia, D. (1982a). Microscale patchiness of nutrients in plankton communities. *Science,* 216, 729–30.

(1982b). Microscale nutrient patches produced by zooplankton. *Proc. Natl. Acad. Sci.,* 79, 5001–5.

Leppard, G. G., Massalski, A., and Lean, D. R. S. (1977). Electron-opaque microscopic fibrils in lakes: Their demonstration, their biological derivation and their potential significance in the redistribution of cations. *Protoplasma,* 92, 289–309.

Lewis, W. M. Jr. (1976). Surface/volume ratio: implications for phytoplankton morphology. *Science,* 192, 419–26.

Lobban, C. S., Harrison, P. J., and Duncan, M. J. (1985). *The Physiological Ecology of Seaweeds.* Cambridge: Cambridge University Press.

MacIsaac, J. J. and Dugdale, R. C. (1969). The kinetics of nitrate and ammonia uptake by natural populations of marine phytoplankton. *Deep-Sea Res.,* 16, 45–57.

(1972). Interactions of light and inorganic nitrogen in controlling nitrogen uptake in the sea. *Deep-Sea Res.,* 19, 209–32.

Mague, T. H., Weare, N. M., and Holm-Hansen, O. (1974). Nitrogen fixation in the North Pacific Ocean. *Mar. Biol.,* 24, 109–19.

Margalef, R. (1978). Life-forms of phytoplankton as survival alternatives in an unstable environment. *Oceanologica Acta,* 1, 493–509.

(1958). Temporal succession and spacial heterogeneity in phytoplankton. In: *Perspectives of Marine Biology,* ed. A. A. Buzzati-Traverso, pp. 323–49. Los Angeles: University of California Press.

Martinez, L., Silver, M. W., King, J. M., and Alldredge, A. L. (1983). Nitrogen fixation by floating diatom mats: A source of new nitrogen to oligotrophic ocean waters. *Science,* 221, 152–4.

Matin, A. and Veldkamp, H. (1974). Physiological basis of high substrate affinity. *ASM Abstracts.,* p. 50.

McCarthy, J. J. (1972). The uptake of urea by natural populations of marine phytoplankton. *Limnol. Oceanogr.,* 17, 738–48.

(1981). The kinetics of nutrient utilization. In: *Physiological Basis of Phytoplankton Ecology,* ed. T. Platt, pp. 211–33. *Can. Bull. Fish. Aqua. Sci.,* 210.

(1981). Uptake of major nutrients by estuarine plants. In: *Estuaries and Nutrients,* eds. B. J. Neilson and L. E. Cronin, pp. 139–63. Clifton NJ: Humana Press.

McCarthy, J. J. and Goldman, J. C. (1979). Nitrogenous nutrition of marine phytoplankton in nutrient-depleted waters. *Science,* 203, 670–2.

McCarthy, J. J., Taylor, W. R., and Taft, J. L. (1975). The dynamics of nitrogen and phosphorus cycling in the open waters of the Chesapeake Bay. In: *Marine Chemistry in the Coastal Environment,* ed. T. M. Church, pp. 664–81. *ACS Symp. Ser.,* 18.

(1977). Nitrogenous nutrition of the phytoplankton of Chesapeake Bay. I. Nutrient availability and phytoplankton preference. *Limnol. Oceanogr.,* 22, 996–1011.

Meers, J. L. (1973). Growth of bacteria in mixed cultures. *Critical Rev. Microbiol.,* 2, 39–184.

Mickelson, M. J., Maske, H., and Dugdale, R. C. (1979). Nutrient-determined dominance in multispecies chemostat cultures of diatoms. *Limnol. Oceanogr.,* 24, 298–315.

Miller, A. G., Turpin, D. H., and Canvin, D. T. (1984a). Growth and photosynthesis of the cyanobacterium *Synechococcus leopoliensis* in HCO_3^--limited chemostats. *Plant Physiol.,* 75, 1064–70.

(1984b). Na^+ requirement for growth, photosynthesis, and pH regulation in the alkalotolerant cyanobacterium *Synechococcus leopoliensis. J. Bacteriol.,* 159, 100–6.

Monod, J. (1950). La technique de la culture continue: Theorie et applications. *Ann. Inst. Pasteur Lille,* 79, 390–410.

Munk, W. H. and Riley, G. A. (1952). Absorption of nutrients by aquatic plants. *J. Mar. Res.,* 11, 215–40.

Murphy, T. P., Lean, D. R. S., and Nalewajko, C. C. (1976). Blue-green algae: Their excretion of iron selective chelators enables them to dominate other algae. *Science,* 192, 900–2.

North, B. B. and Stephens, G. C. (1971). Uptake and assimilation of amino acids by *Platymonas.* II. Increased uptake in nitrogen deficient cells. *Biol. Bull.,* 140, 242–54.

(1972). Amino acid transport in *Nitzschia ovalis* Arnott. *J. Phycol.,* 8, 64–8.

Nyholm, N. (1976). A mathematical model for microbial growth under limitation by conservative substrates. *Biotech. Bioeng.,* 18, 1043–56.

Nyholm, N. (1977). Kinetics of phosphate limited growth. *Biotech. Bioeng.,* 19, 467–92.

Paasche, E. (1973). Silicon and the ecology of marine plankton diatoms. 1. *Thalassiosira pseudonana (Cyclotella nana)* grown in a chemostat with silicate as limiting nutrient. *Mar. Biol.,* 19, 117–26.

Paerl, H. W. and Keller, P. E. (1979). Nitrogen fixing *Anabaena:* physiological adaptations instrumental in maintaining surface blooms. *Science,* 204, 620–2.

Panikov, N. (1979). Steady-state growth kinetics of *Chlorella vulgaris* under double substrate (urea and phosphate) limitation. *J. Chem. Tech. Biotech.,* 29, 442–50.

Panikov, N. and Pirt, S. J. (1978). The effects of co-operativity and growth yield variation on the kinetics of nitrogen or phosphate limited growth of *Chlorella* in a chemostat culture. *J. Gen. Microbiol.,* 108, 295–303.

Parslow, J. S., Harrison, P. J., and Thompson, P. A. (1985a). Interpreting rapid changes in uptake kinetics in the marine diatom *Thalassiosira pseudonana* (Hustedt). *J. Exp. Mar. Biol. Ecol.,* 91, 53–64.

(1985b). Ammonium uptake by phytoplankton cells on a filter: a new high-resolution technique. *Mar. Ecol. Prog. Ser.,* 25, 121–9.

Pasciak, W. J. and Gavis, J. (1974). Transport limitation of nutrient uptake in phytoplankton. *Limnol. Oceanogr.,* 19, 881–8.

(1975). Transport limited nutrient uptake rates in *Ditylum brightwellii. Limnol. Oceanogr.,* 20, 604–17.

Passera, C. and Ferrari, G. (1975). Sulphate uptake in two mutants of *Chlorella vulgaris* with high and low sulphur amino acid content. *Physiol. Plant.*, 35, 318–21.

Perry, M. J. (1976). Phosphate utilization by an oceanic diatom in phosphorus-limited chemostat culture and in the oligotrophic waters of the central North Pacific. *Limnol. Oceanogr.*, 21, 88–107.

Petersen, R. (1975). The paradox of the plankton: An equilibrium hypothesis. *Am. Nat.*, 109, 35–49.

Powell, T. and Richerson, P. J. (1985). Temporal variation, spatial heterogeneity, and competition for resources in plankton systems: A theoretical model. *Am. Nat.*, 125, 431–64.

Quarmby, L. Q., Turpin, D. H., and Harrison, P. J. (1982). Physiological responses of two marine diatoms to pulsed additions of ammonium. *J. Exp. Mar. Biol. Ecol.*, 63, 173–81.

Raven, J. A. (1984). *Energetics and Transport in Aquatic Plants.* New York: Alan R. Liss.

Rees, T. A. V. and Syrett, P. J. (1979). The uptake of urea by the diatom, *Phaeodactylum. New Phytol.*, 82, 169–78.

Reynolds, C. S. (1983). *The Ecology of Freshwater Phytoplankton.* Cambridge: Cambridge University Press.

 (1984). Phytoplankton periodicity: the interactions of form, function and environmental variability. *Freshwat. Biol.* 14, 111–42.

Reynolds, C. S., Thompson, J. M., Ferguson, A. J. D., and Wiseman, S. W. (1982). Loss processes in the population dynamics of phytoplankton maintained in closed systems. *J. Plank. Res.*, 4, 561–600.

Rhee, G-Y. (1973). A continuous culture study of phosphate uptake, growth rate and polyphosphate in *Scenedesmus* sp. *J. Phycol.*, 9, 495–506.

 (1974). Phosphate uptake under nitrate limitation by *Scenedesmus* and its ecological implications. *J. Phycol.*, 10, 470–5.

Rhee, G-Y. and Gotham, I. J. (1980). Optimum N:P ratios and coexistence of planktonic algae. *J. Phycol.*, 16, 486–9.

Rhee, G-Y. and Lederman, T. C. (1983). Effects of nitrogen sources on P-limited growth of *Anabaena flos-aquae. J. Phycol.*, 19, 179–85.

Richerson, P., Armstrong, P., and Goldman, C. R. (1970). Contemporaneous disequilibrium, a new hypothesis to explain the "paradox of the plankton." *Proc. Natl. Acad. Sci. U.S.A.*, 67, 1710–14.

Riegman, R. and Mur, L. R. (1984). Regulation of phosphate uptake kinetics in *Oscillatoria aghardhii. Arch. Microbiol.*, 139, 28–32.

Robinson, J. V. and Sandgren, C. D. (1983). The effect of temporal environmental heterogeneity on community structure: a replicated experimental study. *Oecologia*, 57, 98–102.

Saubert, S. (1957). Amino acid utilization by *Nitzschia thermalis* and *Scenedesmus bijugatus. South Afr. J. Sci.*, 53, 335–9.

Scavia, D., Fahnenstiel, G. L., Davis, J. A., and Kreis, R. G., Jr. (1984). Small-scale nutrient patchiness: some consequences and a new encounter mechanism. *Limnol. Oceanogr.*, 29, 785–93.

Schindler, D. W. (1977). Evolution of phosphorus limitation in lakes. *Science*, 195, 260–2.

Schindler, D. W. and Fee, E. J. (1973). Diurnal variation of dissolved inorganic carbon and its use in estimating primary production and CO_2 invasion in Lake 227. *J. Fish. Res. Bd. Can.,* 30, 1501–10.

(1974). Experimental lakes area: whole lake experiments in eutrophication. *J. Fish. Res. Bd. Can.,* 31, 937–53.

Serra, J. L., Llama, M. J., and Cadenas, E. (1978). Nitrate utilization by the diatom *Skeletonema costatum.* II. Regulation of nitrate uptake. *Plant Physiol.,* 62, 991–4.

Simpson, F. B. and Neilands, J. B. (1976). Siderochromes in Cyanophyceae: isolation and characterization of schizokinen from *Anabaena* sp. *J. Phycol.,* 12, 44–8.

Sinclair, C. and Whitton, B. A. (1977). Influence of nutrient deficiency on hair formation in the Rivulariaceae. *Br. Phycol. J.,* 12, 297–313.

Smith, R. E. H. and Kalff, J. (1983). Competition for phosphorus among co-occurring freshwater phytoplankton. *Limnol. Oceanogr.,* 28, 448–64.

(1985). Phosphorus competition among phytoplankton: a reply. *Limnol. Oceanogr.,* 30, 440–4.

Smith, V. H. (1983). Low nitrogen to phosphorus ratios favor dominance by blue-green algae in lake phytoplankton. *Science,* 221, 669–71.

Sneft, W. H. (1978). Dependence of light-saturated rates of photosynthesis on intracellular concentrations of phosphorus. *Limnol. Oceanogr.* 23: 709–18.

Soldatini, G. F., Zieglier, I., and Ziegler, H. (1978). Sulfite: Preferential sulphur source and modifier of CO_2 fixation in *Chlorella vulgaris. Planta,* 143, 225–31.

Sommer, U. (1981). The role of r- and K-selection in the succession of phytoplankton in Lake Constance. *Acta Ecologica,* 2, 327–42.

(1984). The paradox of the plankton: Fluctuations of phosphorus availability maintain diversity of phytoplankton in flow-through cultures. *Limnol. Oceanogr.,* 29, 633–6.

(1985). Comparison between steady-state and non-steady state competition: Experiments with natural phytoplankton. *Limnol. Oceanogr.,* 30, 335–46.

Sommer, U. and Kilham, S. S. (1985). Phytoplankton natural competition experiments: A reinterpretation. *Limnol. Oceanogr.,* 30, 436–40.

Sournia, A. (1981). Morphological bases of competition and succession. In: *Physiological Basis of Phytoplankton Ecology,* ed. T. Platt, pp. 339–46. *Can. Bull. Fish. Aquat. Sci.,* 210.

Steeman-Nielsen, E. (1978). Growth of plankton algae as a function of N-concentration measured by means of a batch technique. *Mar. Biol.,* 46, 185–9.

Talling, J. (1985). Inorganic carbon reserves of natural waters and ecophysiological consequences of their photosynthetic depletion: Microalgae. In: *Inorganic Carbon Uptake by Photosynthetic Organisms,* eds. W. J. Lucas and J. A. Berry, pp. 403–42. Baltimore: Waverly Press.

Taylor, P. A. and Williams, P. J. (1975). Theoretical studies on the coexistence of competing species under continuous flow conditions. *Can. J. Microbiol.,* 21, 90–8.

Terry, K. L. (1980). Nitrogen and phosphorus requirements of *Pavlova lutheri* in continuous culture. *Bot. Mar.,* 23, 757–64.

Terry, K. L., Laws, E. A., and Burns, D. J. (1985). Growth rate variation in the N:P requirement ratio of phytoplankton. *J. Phycol.,* 21, 323–9.

Thomas, W. H. and Dodson, A. N. (1972). On nitrogen deficiency in tropical Pacific Oceanic phytoplankton. II. Photosynthetic and cellular characteristics of a chemostat-grown diatom. *Limnol. Oceanogr.,* 17, 515–23.

Tilman, D. (1977). Resource competition between planktonic algae: An experimental and theoretical approach. *Ecology,* 58, 338–48.

   (1981). Test of resource competition theory using four species of Lake Michigan algae. *Ecology,* 62, 802–15.

   (1982). *Resource Competition and Community Structure.* Princeton, NJ: Princeton University Press.

Tilman, D. and Kiesling, R. L. (1984). Freshwater algal ecology: Taxonomic tradeoffs in the temperature dependence of nutrient competitive abilities. In: *Current Perspectives in Microbial Ecology,* eds. M. J. Klug and C. A. Reddy, pp. 314–19. Washington, D.C.: Amer. Soc. Microbiol.

Tilman, D. and Kilham, S. S. (1976). Phosphate and silicate growth and uptake kinetics of the diatoms *Asterionella formosa* and *Cyclotella meneghiniana* in batch and semi-continuous cultures. *J. Phycol.,* 12, 375–83.

Tilman, D., Kilham, S. S., and Kilham, P. (1982). Phytoplankton community ecology: the role of limiting nutrients. *Ann. Rev. Ecol. Syst.,* 13, 349–72.

Titman, D. (1976). Ecological competition between algae: Experimental confirmation of resource based competition theory. *Science,* 192, 463–5.

Trick, C. G., Andersen, R. J., Gillam, A., and Harrison, P. J. (1983). Prorocentrin: An extracellular siderophore produced by the marine dinoflagellate *Prorocentrum minimum. Science,* 219, 306–8.

Turpin, D. H. (1980). Processes in nutrient based phytoplankton ecology. Ph.D. Thesis. University of British Columbia, Vancouver.

   (1986). Growth rate dependent optimum ratios in *Selenastrum minutum:* Implications for competition, coexistence and stability in phytoplankton communities. *J. Phycol.,* 22, 94–101.

Turpin, D. H. and Harrison, P. J. (1979). Limiting nutrient patchiness and its role in phytoplankton ecology. *J. Exp. Mar. Biol. Ecol.,* 39, 151–66.

   (1980). Cell size manipulation in natural marine planktonic diatom communities. *Can. J. Fish. Aquat. Sci.,* 7, 1193–5.

Turpin, D. H., Layzell, D. B., and Elrifi, I. R. (1985a). Modeling the C economy of *Anabaena flos-aquae:* Estimates of establishment, maintenance and active costs associated with growth on NH_3, NO_3^- and N_2. *Plant Physiol.,* 78, 746–52.

Turpin, D. H., Miller, A. G., Parslow, J. S., Elrifi, I. R., and Canvin, D. T. (1985b). Predicting the kinetics of DIC-limited growth from the short-term kinetics of photosynthesis in *Synechococcus leopoliensis* (Cyanophyta). *J. Phycol.,* 21, 409–18.

Turpin, D. H., Parslow, J. S., and Harrison, P. J. (1981). On limiting nutrient patchiness and phytoplankton growth: a conceptual approach. *J. Plank. Res.,* 3, 421–31.

Van Liere, L. and Walsby, A. E. (1982). Interactions of cyanobacteria with light. In: *The Biology of Cyanobacteria, Botanical Monograph,* vol. 19, eds. N. G. Carr and B. A. Whitton, pp. 9–45. Berkley: University of California Press.

Veldkamp, H. and Jannasch, H. W. (1972). Mixed culture studies with the chemostat. *J. Appl. Chem. Biotech.,* 22, 105–23.

Veldkamp, H. and Kuenen, J. G. (1973). The chemostat as a model system for ecological studies. *Bull. Ecol. Res. Comm. (Stockholm),* 17, 347–55.

Ward, A. K. and Wetzel, R. G. (1980). Interactions of light and nitrogen sources among planktonic blue-green algae. *Arch. Hydrobiol.,* 90, 1–25.

Wetzel, R. G. (1975). *Limnology.* Philadelphia: Saunders.

Wheeler, P. A. (1977). Effects of nitrogen source on *Platymonas* (Chlorophyta) cell composition and amino acid uptake rates. *J. Phycol.,* 13, 301–3.

(1983). Phytoplankton nitrogen metabolism. In: *Nitrogen in the Marine Environment,* eds. E. J. Carpenter and D. G. Capone, pp. 309–46. New York: Academic Press.

Wheeler, P. A., Glibert, P. M., and McCarthy, J. J. (1982). Ammonium uptake and incorporation by Chesapeake Bay phytoplankton: Short-term uptake kinetics. *Limnol. Oceanogr.,* 27, 1113–28.

Wheeler, P. A. and McCarthy, J. J. (1982). Methylammonium uptake by Chesapeake Bay phytoplankton: evaluation of the use of the ammonium analogue for field uptake measurements. *Limnol. Oceanogr.,* 27, 1129–40.

Wheeler, P. A., North, B. B., and Stephens, G. C. (1974). Amino acid uptake by marine phytoplankton. *Limnol. Oceanogr.,* 19, 249–59.

Williams, F. M. (1971). Dynamics of microbial populations. In: *Systems Analysis,* ed. B. C. Patten, pp. 197–267. New York: Academic Press.

Williams, P. J. and Mur, L. R. (1981). Diffusion as a constraint on the biological importance of microzones in the sea. In: *Echohydrodynamics,* ed. J. C. Nihoul, pp. 209–18. *Elsevier Ocean Sci. Ser.,* 5(32).

Yarish, C. (1976). Polymorphism of selected marine Chaetophoraceae (Chlorophyceae). *Br. Phycol. J.,* 11, 29–38.

Zevenboom, W. (1980). Growth and nutrient uptake kinetics of *Oscillatoria aghardhii.* Ph.D. Thesis. University of Amsterdam.

Zevenboom, W. and Mur., L. R. (1980). N₂-fixing cyanobacteria: why they do not become dominant in Dutch hypertrophic lakes. In: *Hypertrophic Ecosystems,* eds. J. Barica and L. R. Mur, pp. 123–30. The Hague: Dr. W. Junk.

Zevenboom, W., Van der Does, J., Bruning, K., and Mur, L. R. (1981). A non-heterocystous mutant of *Aphanizomenon flos-aquae,* selected by competition in light-limited continuous culture. *FEMS Microbiol. Lett.,* 10, 11–16.

Chapter 9

SELECTIVE HERBIVORY AND ITS ROLE IN THE EVOLUTION OF PHYTOPLANKTON GROWTH STRATEGIES

JOHN T. LEHMAN

INTRODUCTION

Analyses of herbivory in plankton communities have historically emphasized the role of zooplankton as a vital link in a food chain that stretches from algae to fish. Regarded in this way, the zooplankton are represented as sets of populations that collectively transfer mass and energy from one trophic level to the next. Empirical studies that explore this vein have emphasized ingestion rates, assimilation efficiencies, growth efficiencies, and ecological transfer efficiencies. The frame of reference is fixed on the animals, which are known to exhibit behavioral idiosyncrasies of food acquisition, complexities of metabolic regulation, and diverse styles of somatic growth and reproduction. The quantitative link between zooplankton and their phytoplankton food is generally expressed as a mortality rate, generated as the product of animal abundance and individual or mass-specific ingestion rates.

In this chapter I explore the role of herbivory in plankton communities by placing the frame of reference among the phytoplankton rather than among the planktonic grazers. Many of the discoveries made in the course of zooplankton studies during the last decade have helped to define the selection regimes experienced by planktonic algae in natural waters. These recent advances provide an opportunity to

This work was supported by ONR N00014-84-K-0671 and by grants from the College of Literature, Science and Arts, The University of Michigan. Keith Kennedy performed the video analyses, and J. R. Strickler provided helpful advice about optical design.

speculate about the constraints placed by grazing on algal morphology, physiology, and growth strategies.

With the adoption of this approach it becomes important to establish whether herbivory occurs at sufficient magnitude and regularity to warrant regarding it as an agent of natural selection, and also whether the effects are distributed differentially with respect to algal species and with respect to genetically based variants within species. Other authors in this volume have mentioned herbivory in their discussions of individual algal groups, and there is indeed an array of morphological and physiological attributes that may be regarded as adaptations against grazing pressure. These include aggregation of cells into colonies or filaments of macroscopic dimensions, adoption of attenuated cell shapes, and enclosure of cells within layers of siliceous scales with elongate bristles. But embracing the existence of coloniality, exuberant morphology, or large cell size as de facto evidence of strong grazing pressure merely evades the questions that most demand attention:

1. Does grazing pressure exert significant loss rates on phytoplankton species in nature?
2. Is there enough variability among species and genotypes in their susceptibility to grazing pressure to cause significant differential mortality?
3. What compensatory mechanisms can phytoplankton use to escape or overcome the losses they experience?

In these pages I describe some of the ways these questions have been studied empirically, and I present a model framework for exploring tradeoffs between algal growth strategies and mortality regimes. Before doing so, however, a brief introduction to the planktonic herbivore community is necessary.

PLANKTONIC HERBIVORE ARRAY

The phylogenetic composition of planktonic herbivores in freshwater ecosystems is considerably less diverse than that of marine environments. The groups that dominate lakes include protozoans, a few families of copepods: Calanoida of the families Diaptomidae, Pseudocalanidae, Centropagidae, and Temoridae (Wilson 1959), Cyclopoida of the Cyclopidae (Yeatman 1959), and two groups that have arisen and radiated primarily in fresh water, the Cladocera and the Rotifera. Cladocerans and copepods are both crustacean arthropods, whereas rotifers comprise a distinct and separate phylum. Rotifers exhibit the remarkable characteristic called eutely, or cell constancy, which they share with only a few other invertebrate phyla. The number of somatic cells and nuclei in these animals is fixed at hatching, and only the germ cell line is capable of continued division in the adults. Postembryonic development of rotifers is thus very brief, and the animals are capable of generation times as short as one day.

Patterns of reproduction differ among the three main groups of rotifers, but among the Monogononta, which is the group most common in the freshwater plankton, parthenogenesis is the rule. Diploid amictic females produce diploid eggs mitotically, and these develop into genetic replicates of their mothers. Under the influence of environmental stimuli that are still poorly understood, some eggs may develop into mictic females. These females are morphologically and genetically similar to amictic females, but they produce haploid eggs through meiosis. Unfertilized, the eggs develop into haploid, functional males. Fertilized, the eggs become specialized resting stages that can withstand dessication and cold. The resting eggs hatch into amictic females after considerable delay.

Cladocera also reproduce primarily by parthenogenesis, and males occur during episodes of environmental stress. Sex is determined environmentally in these animals, and the stimulus appears to be a rapid decline in food available to females during ovigenesis (D'Abramo 1980). Fertilized gametogenetic eggs are enclosed in specialized resistant structures, and, as with the rotifers, these eggs are the resting stage.

Copepods exhibit obligate gametogenetic reproduction. The females of both the Diaptomidae and Cyclopidae carry their fertilized eggs in external clusters or clutches, whereas the other families broadcast their eggs into the water. Postembryonic development involves up to 12 instars punctuated by molts. At hatching the earliest instars, called nauplii, bear only three pairs of appendages, but at each molt more appendages are added and the original ones grow more specialized. Whereas generation times among the Rotifera and Cladocera vary from one to several days, copepods have generation times from weeks to months. In all cases, development times are strongly dependent on water temperatures.

Planktonic protozoa are extremely diverse and include amoebae, ciliates, and flagellates. It is not unusual to find both autotrophic and heterotrophic nutritional features in the same individual. Ciliates with endosymbiotic algae, for instance, are sometimes very abundant (e.g., Hecky & Kling 1981), and some algal flagellates can ingest bacteria (e.g., Pascher 1943; Bird & Kalff 1986). Because their nutrition is independent of light intensity and photoperiod, these microzooplankton can have generation times as short as hours and can potentially outgrow many algae.

The seasonality of zooplankton populations in temperate lakes causes a seasonal variation in grazing rates. Many Cladocera and rotifers overwinter as resting eggs. Many copepods, as well, are able to form resting eggs or to enter diapause, although many univoltine or bivoltine species overwinter as adults. Microzooplankton may become abundant during the winter (Bamforth 1958), and that is likely when algal species with mixotrophic forms of nutrition have their greatest advantage. As metazoan populations of zooplankton increase in the spring and summer, they apply grazing pressure not merely to the algae but to the microheterotrophs as well. It seems likely that the few algae present during winter have evolved in the constant presence of microheterotrophy and may even have adopted it themselves. Summer species of phytoplankton have coevolved with larger metazoan grazers as well.

GRAZING PRESSURE FROM PLANKTONIC HERBIVORES

Several methods have been used to investigate the quantitative role of herbivory in plankton communities, but in general the accumulated evidence can be sorted into two main categories: (a) experimental studies of enclosed populations, and (b) inferences from the dynamics of natural phytoplankton populations.

Enclosure Studies

This category includes measurements of ingestion rates of tracer particles (Rigler 1971; Haney 1973), sequential radiotracer labeling of natural phytoplankton and then their grazers (Roman & Rublee 1981), and direct counts of phytoplankton abundance in the presence and absence of potential herbivores (Gliwicz 1975; Weers & Zaret 1975; Porter 1976). Only the latter method has been used successfully to estimate the rates at which different species of algae are cropped by herbivores. Tracer-based grazing estimates are extremely useful for discovering how different herbivore species graze a single particle type, but apportioning the tracer-derived rates among different algal species is a formidable task. Investigators usually assign algal species to "edible" or "inedible" ranks based on gut residues found among field populations of the herbivores of interest, and then apply the tracer-derived rates to those edible species. Alternatively, algal particles are assigned to one category or another based on their mean or maximum linear dimensions. Commonly 30 μm or 50 μm is used to segregate smaller edible taxa from larger inedible ones. Dual-tracer techniques have evolved recently in an effort to overcome some of this ambiguity. A tracer algal species, for instance, may be labeled with radiocarbon, and a tracer bacterium with tritium. This can be very useful for ascertaining the availability of particular particles to different grazers, but it is still not particularly helpful for deciphering how mortalities are distributed across many different algal species in situ.

One method to measure grazing rates on individual algal species directly was introduced by Lehman (1980) and was applied to studies of a natural community by Lehman and Sandgren (1985). Phytoplankton are exposed inside enclosures to a range of herbivore densities, and net growth rates (r, d^{-1}) of the individual algal species are then plotted against herbivore abundance. The slope of the resulting relation [liters·(mg zooplankton DW)$^{-1}$·d^{-1}] measures the strength of grazing pressure on different co-occurring species simultaneously. Where the method has been applied, rates have differed substantially among species, and they have varied through time for single algal species (Figs. 2 and 3, Lehman & Sandgren 1985). These findings are consistent with the discovery, discussed later, that zooplankton possess a complex repertoire of grazing behavior and that rates of particle capture may vary with size, shape, taste, or maybe even surface charge. For instance, although significant grazing was occasionally detected by Lehman and Sandgren (1985) for algal particles as large as 200 μm, most grazed species had unit sizes no larger than 25 μm.

Experimental studies reveal that grazing rates on tracer particles can be large at times, equivalent to loss rates of 0.35 to 0.7 d^{-1} or more (Haney 1973; Knoechel & Holtby 1986). The species-specific grazing studies, however, reveal that the high rates are confined to relatively few algal taxa, and that for many species grazing by planktonic crustaceans has a relatively minor effect on net population growth rate. The effect of grazing varies with season, and it is maximal in summer when zooplankton are most active and abundant.

Population Dynamics Studies

The second approach to assessing the effects of grazers on phytoplankton involves interpretations of the dynamics of plankton populations in situ. If algal populations decline at times when grazers are at peak or increasing abundance, such evidence is consistent with an hypothesis of grazer control. Such is the case for the "clear-water phase" of Lake Constance and elsewhere (Lampert and Schober 1978).

The log-transformed abundances of the chrysophyte flagellate *Dinobryon divergens* in Egg Lake (from Lehman & Sandgren 1985) are plotted in Fig. 9-1. Time periods when the population was in balanced exponential growth or decline are characterized by straight lines fit to the logarithms of cell counts. During these periods the slopes of the lines are proportional to net intrinsic population growth rates (r):

$$r = \mu - k \tag{1}$$

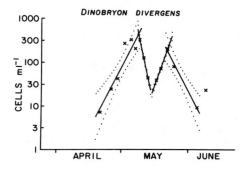

Fig. 9-1. Log-transformed abundances of *Dinobryon divergens* in Egg Lake. The abundances are fit piecewise by linear regression, and 95% confidence intervals of the regressions are shown. The slopes of the fitted lines are proportional to net population growth rates *r*.

where μ is cell division rate (d^{-1}) and k is the composite loss rate from all sources (d^{-1}). An unknown proportion of k can be ascribed to grazing. Changes in r mean that either μ, k, or possibly both, changed in magnitude. The time series data on algal and zooplankton abundances can consequently help to interpret the possible influences of herbivory in natural systems.

Algal population maxima, for instance, might result from increasing loss rates because peak abundances represent a point of transition from net positive growth to net decline. Peak algal abundances should thus occur most frequently when zooplankton biomass is increasing if grazing rates are an important component of overall losses. To investigate this notion quantitatively, I plot (Fig. 9-2) the frequency distribution of peak algal abundances for more than 20 taxa in two lakes. The distributions differ significantly from uniform ones, meaning that changes in growth rate are not random, but they also do not correlate with the changing biomass of herbivorous zooplankton in the lakes ($R = 0.05$ for Egg Lake and $R = -0.38$ for Sportsman Lake). This finding supports the results of enclosure experiments conducted simultaneously and downplays the importance of grazing to these phytoplankton communities (Lehman & Sandgren 1985).

Such evidence as this points to the fact that many algal species are spared from high rates of herbivory. The influence of grazing rather appears to be concentrated on taxa such as cryptomonads and other small flagellates or unicells. These taxa have only recently come under intense physiological and ecological investigation (see chapters by Klaveness and Sandgren in this volume), so the mechanisms by which

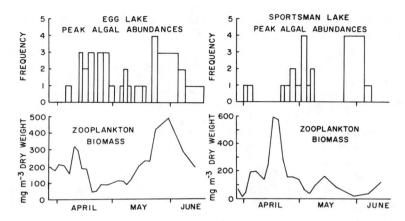

Fig. 9-2. Frequency distributions of the occurrence of algal population maxima during sampling intervals, and concurrent mean zooplankton biomass. Peak abundances were identified as shown in Fig. 9-1. Of the 32 peak abundances identified for Egg Lake, 27 were for species grazed in enclosure experiments; in Sportsman Lake, 18 of the 19 cases were for grazed species.

these species survive occasionally intense grazing pressure are only imperfectly understood. The fact that the risk of mortality is not shared equally by all phytoplankton is consistent with the notion that existing algal diversity has been shaped in large measure by selection for grazer avoidance. Mortality by grazing is, however, only one of the pressures faced by the algae, and some of the alternative selection forces conflict.

CONFLICTING SELECTION PRESSURES

Phytoplankton that are susceptible to grazing face the risk of mortality from not one grazer alone but from the entire grazer array described earlier (Table 9-1). Because many microzooplankton have short generation times, rapid numerical population responses, and because they are free of dependence on photoperiods for nutrition, these unicellular heterotrophs can expand their populations dramatically in the presence of adequate food. It is not impossible that the proliferation of colonial and filamentous growth habits in many algal groups arose originally as an adaptation against microheterotrophs. Large cell size can help algae escape the influence of the smaller grazers, as can aggregation of cells in colonies or filaments.

Table 9-1. *Size spectrum of micro-herbivores and consequences for the microspatial distribution of nutrient release.*

Grazer array		Generation times	Remineralization
Microheterotrophs:	Flagellates	Hours	Homogeneous
	Ciliates	Hours	
Micrometazoa:	Rotifers	Days	
Juvenile stages of larger metazoa:	Nauplii	Days	
Crustaceans:	Cladocera	Days	
	Copepods	Weeks+	Heterogeneous

Large metazoan grazers have longer generation times than do the algae and thus exhibit slower numerical population responses. When these large grazers dominate the dynamics of a community, which they might typically do in early summer in temperate regions, it is more likely that algal growth was first slowed, perhaps by nutrient limitation, than that the herbivores won an unfettered growth rate contest.

Although the potential exists to reduce susceptibilities to grazing mortality through changes in growth habit, there are evidently sources of selection pressure that oppose the morphological solution of gigantism. Herbivory is not constant, and there are economies of scale related to small size and high surface-to-volume ratio that favor a diminutive morphology (see Turpin, Chapter 8). Chief among these advantages may be the light-absorbing advantages of small unicells (Kirk 1983) and the advantages of surface area for nutrient acquisition. These conflicting allometries of selection pressure, where large sizes are favored to avoid grazers but small cells are favored for energy and nutrient acquisition, are the types of conflicts that probably generated the morphological and physiological diversities of natural phytoplankton. Individual species represent individual novel solutions to a conflicting set of natural challenges. The solutions are sometimes morphological, sometimes physiological, and often both.

The times when grazing exerts its greatest influence are also times when ambient reservoirs of nutrient concentrations have become exhausted, and algal production proceeds with recycled nutrients. Just as the algae face an array of grazers, they experience an array of nutrient availabilities (Table 9-1). Nutrients remineralized by numerous microheterotrophs represent a relatively uniform and homogeneous supply of nutrients to the water. The small heterotrophs are so numerous that the nutrients they release can be regarded as a diffuse background supply rate (Jackson 1980). Nutrients released from the larger herbivores are much more heterogeneous. Not only are the larger animals less numerous, but they represent more concentrated "point sources" of nutrients on the microscale. These are the animals that have been best studied in regard to grazing behavior and for which the existence of microscale nutrient patches has been documented (Lehman & Scavia 1982). In the case of these herbivores, encounters with algae may lead to mortality or to nutrient enrichment.

RECENT DEVELOPMENTS FROM STUDIES OF ZOOPLANKTON FEEDING

A small revolution in recent attention to mechanisms of food acquisition by zooplankton can be traced to visualization of capture and ingestion phenomena by microcinematography (Alcarez, Paffenhofer, & Strickler 1980; Koehl & Strickler 1981; Strickler 1984). The approaches to date have involved the application of behavioral observations at the scale of individual algal cells. The visual evidence showing discriminatory feats and detailing the capture events have been complemented with evidence that copepods, at least, employ chemosensation as a way to identify potential food (Poulet & Marso 1978). This means that zooplankton exhibit complex behavior, including event processing abilities and use of simple decision rules. Thus the intense interest in mechanisms of particle selection by zooplankton has shown that features other than cell size alone may regulate whether a cell is susceptible to being grazed. It also means, however, that specific adaptations by phytoplankton crafted to escape one grazer may be ineffective against another that uses alternative capture methods or decision rules.

Equally important, the microscopic observations of grazing behavior have provided vivid evidence that algal cells form an integral association with their surrounding water, owing to small-scale viscosity effects at low Reynolds numbers (Purcell 1977; Zaret 1980). Phytoplankton and the water in their immediate vicinities move as single units, and this forces capture mechanisms to rely on cell surface properties (Rubenstein & Koehl 1977; Gerritsen & Porter 1982; Porter, Feig, & Vetter 1983) or on peculiar shear-based processes (Koehl 1984; Strickler 1984). This close association between cells and their surroundings means, however, that algal cells might experience microenvironments that differ from average chemical conditions in the water.

DEMOGRAPHIC MODEL OF GRAZING AND NUTRIENT ACQUISITION

My aim in this chapter is to explore some of the behavioral phenomena that have been documented among zooplankton with respect to their possible effect on the evolution of algal growth strategies. The preceding discussion has focused attention on conditions in midsummer when algal production relies almost entirely on recycled nutrients and when zooplankton can graze the daily net production of some species. Others have argued that taxa that have adopted morphologies or growth strategies that make them immune to grazing will progressively dominate the phytoplankton community under these conditions (Porter 1977). What deserve attention here are the characteristics of cells that persist in the face of grazing pressure, and the nature of demographic tradeoffs that favor their persistence.

The general model developed here proceeds from two principal assumptions. The first is that frequencies at which grazing herbivores encounter algae are greater than the mortalities. That is, capture efficiencies for different algae may vary with the physical or chemical attributes of the cells. The second assumption is that rates of nutrient acquisition govern the cell division rates of the phytoplankton. This means that changes in nutrient uptake characteristics of the cells could lead to changes in cellular growth rates. As the details of the model are developed, it will become evident that this second assumption is not a harsh one. Basically, the assumption requires that a cell must at least double its nutrient content from one cell division to the next, but the mechanisms and kinetics of acquisition are left undefined.

In order to investigate the implicit tradeoffs between nutrient uptake abilities and the ability to escape mortality from grazers, the model was designed with the following specific characteristics:

1. Cellular nutrient contents, or cell quotas, are used to generate demographic population projections.
2. Nutrient uptake proceeds by two mechanisms. The first is a constant rate from uniform background concentrations, and the second is pulse uptake from episodic enrichments.
3. Encounters between algae and herbivores lead either to mortality for the algal cell or to nutrient enrichment of the cell.

The model is conveniently represented in the form of a projection matrix and a vector of cell quotas (Table 9-2). It is similar in structure to a size-based demographic model introduced by Gage, Williams, and Horton (1984) for continuous algal cultures. Cells are assumed to divide by equal binary fission, so after cell cleavage each daughter cell

Table 9-2. *Formulation of the demographic model described in the text.*[a]

$$
\begin{bmatrix}
0 & 0 & 0 & 0 & 0 & 0 & 2(E-D) & E-D & 2(1-E) \\
1-E & 0 & 0 & 0 & 0 & 0 & 0 & E-D & 2(E-D) \\
0 & 1-E & 0 & 0 & 0 & 0 & 0 & 0 & 0 \\
E-D & 0 & 1-E & 0 & 0 & 0 & 0 & 0 & 0 \\
0 & E-D & 0 & 1-E & 0 & 0 & 0 & 0 & 0 \\
0 & 0 & E-D & 0 & 1-E & 0 & 0 & 0 & 0 \\
0 & 0 & 0 & E-D & 0 & 1-E & 0 & 0 & 0 \\
0 & 0 & 0 & 0 & E-D & 0 & 1-E & 0 & 0 \\
0 & 0 & 0 & 0 & 0 & E-D & 0 & 1-E & 0 \\
0 & 0 & 0 & 0 & 0 & 0 & E-D & 0 & 1-E
\end{bmatrix}
\times
\begin{bmatrix}
Q_1 \\ Q_2 \\ Q_3 \\ Q_4 \\ Q_5 \\ Q_6 \\ Q_7 \\ Q_8 \\ Q_9 \\ Q_{10}
\end{bmatrix}
$$

Projection matrix   Cell quotas

[a] E is the probability of an encounter with a zooplankter in the time step Δt; D is the probability of mortality ($D < E$); Q_i is the vector of cell nutrient quotas. This example shows a case where cells gain a 20% nutrient boost from episodic encounters ($\Delta Q = 20\%$). Only 10 nutrient categories are shown, for simplicity.

contains exactly one-half of the parent's nutrient quota. Alternative growth patterns, like coenobial growth whereby a cell divides to form four daughter cells, can be modeled by doubling the values in the upper right triangle of the matrix. The example of binary fission is used here for simplicity.

Ambient nutrient conditions are presumed to support a basal cell generation time based on cell physiology and growth relations. Cell generation time under uniform nutrient conditions was divided into 20 equal intervals. This means that model time step is an implicit function of cell growth rate, such that

$$\Delta t = \frac{\ln(2)}{\mu/20} \tag{2}$$

where μ is the growth rate (d^{-1}) that would be sustained independently of any episodic enrichments. Cell quotas of nutrient $(mass \cdot cell^{-1})$ were divided into 20 categories so that model cells could proceed from Q to $2Q$ in 5% increments. The model departs from traditional Leslie projection matrix models because opportunities also exist for cells to advance by larger, episodic enrichments.

As shown in Table 9-2, the major subdiagonal elements of the matrix report the probability that cells will not experience an encounter with a grazer during the interval Δt. If the encounter were to occur, there are two outcomes: death or nutrient enhancement. The conditional probability of death for the cell depends on the efficiency of the grazer, and the magnitude of nutrient gain depends on physiological properties of the alga. Survivors of an encounter might gain some nutrient in the process, and that increment to cellular nutrient quota may be larger than the cell might otherwise gain from bulk solution. These transitions or "jumps" in cellular nutrient quota are represented by minor subdiagonal elements in the projection matrix (Table 9-2), positioned farther and farther from the main diagonal as the magnitude of the transition increases. Finally, terms in the upper right triangle of the matrix represent recruitment of cells through cell division. Some of the cells may enter with cell quotas of more than Q because of large increments gained immediately before cell division.

By designing the model this way, two features of algal growth strategies related to grazing could be explored. First, it is possible to examine the effects of changes in grazer efficiency. Algae may exhibit morphological or chemical traits that can decrease the efficiency with which an herbivorous zooplankter can extract the cells from the water. Lower capture efficiencies would permit higher net growth rates, enabling a clone to survive in the presence of grazers. Secondly, the model helps make explicit a tradeoff between risk of mortality and

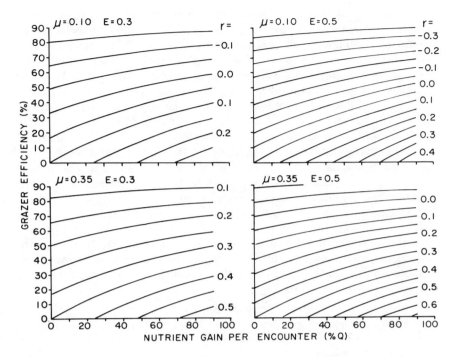

Fig. 9-3. Contour diagram for model-generated net intrinsic growth rate (r, d^{-1}). μ is growth rate under uniform nutrient conditions (d^{-1}); E is encounter rate ($d^-;1$). Contour interval is 0.05 d^{-1}.

potential nutrient gain by dealing in one common currency: population growth rate. For the model calculations reported here, cell generation times under uniform nutrient conditions were arbitrarily set to either two or seven days ($\mu = 0.35$ or 0.1 d^{-1}), and the presumed encounter frequencies with zooplankton were set to either 0.3 or 0.5 d^{-1}. Matrices were generated with permutations of grazer efficiencies from 0 to 100% and with nutrient gain per encounter (ΔQ) from 0 to 90% of Q in 10% increments.

The projection matrices were subjected to numerical eigenanalysis. All real and complex parts of the eigenvalues and eigenvectors were identified, and the dominant real values and vectors were extracted. The dominant eigenvalue corresponds to the steady-state growth rate under conditions defined by the projection matrix. The dominant eigenvector represents the steady-state frequency distribution of cell quotas.

Fig. 9-3 shows that net population growth rate, r (d^{-1}), is influenced both by changes in grazer efficiency and nutrient increment. Results are qualitatively similar for all combinations of basal growth rates and encounter frequencies. Steepness of the response surface increases markedly with increasing encounter rates. Analysis of the response surface reveals that changes are, on average, four times larger with respect to grazer efficiency than with respect to nutrient increment. In other words, a 1% change in grazer efficiency is equivalent to roughly a 4% change in ΔQ. The results show, not surprisingly, a simple and direct tradeoff between grazer efficiency and net growth rate.

In light of the immediate growth reward associated with diminished availability to grazers, it is evident that the demographic model predicts continual selection pressure for grazer resistance. In nature this force may be countered by the constraints thus placed on nutrient uptake and maximum growth rates. If the growth strategies used to resist grazers result in lengthened cell generation times, the solution may produce no net release from herbivory. Mortality varies in proportion to the product of grazer efficiency and Δt, so unless an adaptation decreases grazer efficiency by more than it costs in increased generation time the probability that a cell gets eaten before it can divide remains unchanged.

DIRECT OBSERVATION OF GRAZER EFFICIENCY AND NUTRIENT GAINS

It should be obvious from Figure 9-3 that the efficiency at which grazers harvest the cells they encounter has a major influence on model results. Whether the same can be said for the effect of real grazers on real algae requires some knowledge of actual grazing efficiencies. The new techniques of microscopic behavioral analysis have been adapted for quantitative studies in only a few cases, but the results are encouraging. Scavia et al. (1984), for instance, used a microvideographic technique to chart the paths of algal cells around a feeding zooplankter. They reported that *Daphnia magna* ingest only 20–70% of the *Chlamydomonas* cells they encounter by means of their feeding currents. The efficiency declined with increasing cell abundance. Cells entrained in the feeding current but not ingested were drawn into the carapace gape of the animal and were subsequently released unharmed.

I recently confirmed these observations in my laboratory using *Daphnia pulex* and *Cryptomonas erosa* v. *reflexa*. Even with a cryptomonad flagellate that is acknowledged as excellent food for a variety of zooplankton (Stemberger 1981; Infante & Abella 1985), particle capture is far from perfect. Figure 9-4 shows the general regions of incurrent and excurrent flow defined by tracing the trajectories of single algal cells revealed in video recordings. The released cells swim freely

Fig. 9-4. Incurrent and excurrent flow regions established from microvideo recordings of *Daphnia pulex* feeding on *Cryptomonas erosa* v. *reflexa*. The regions were defined by tracing the trajectories of algal cells entrained and released with the feeding currents.

Table 9-3. *Grazing efficiency of* Daphnia pulex *feeding on* Cryptomonas erosa v. reflexa *(3000 cells mL^{-1}, 15°C; microvideo technique).*

	Cells observed in 5 min		
Replicate	Entrained	Released	Efficiency %
1	345	122	65
2	440	136	69
3	281	60	79
4	225	38	83
5	275	78	72
Mean	313	87	74
SE:	37	18	3

and without any evident harm. By counting the numbers of cells entrained in the incurrent flow and released with the excurrent flow, a general estimate of retention efficiency can be obtained. At *Cryptomonas* abundances of 3000 cells·mL^{-1}, *D. pulex* retained 74% of the cells it encountered (Table 9-3). Considering these observations along with those of Scávia et al. (1984), it appears that capture efficiencies may indeed be variable among algal taxa. The issue merits closer attention to determine the importance to capture efficiency of various algal cell attributes, such as shape, texture, size, and surface chemistry.

The other major feature of the demographic model described here that needs empirical corroboration is the magnitude of nutrient gain potentially achieved by algal cells. Lehman & Scavia (1982) showed that algae could gain episodic enrichments in the presence of large grazers, but there are as yet no data reporting the magnitudes of nutrient gains by different taxa. At the moment, therefore, it is clear that the modeled mechanism exists, but it is uncertain whether algae have evolved to exploit it. The same uncertainty can be attached to the arguments about cell morphology and the supposed tradeoffs between grazer resistance and maximum division rates. In order for algal cells to gain maximal benefit from encounters, they would have to be selected for rapid nutrient acquisition over time scales of several seconds. This means they must rely on rapid surface adsorption of nutrients and their subsequent transport into the cell. It is known that lake phytoplankton can at times double their phosphorus contents in 2 min or less (Lehman and Sandgren 1982), but it is not known if the trait exists to exploit nutrient patches.

CONCLUSIONS

The few empirical studies available to date suggest that the efficiencies with which grazers capture their food and the nutrient uptake characteristics of freshwater phytoplankton exhibit values that are theoretically interesting from the viewpoint of natural selection of growth strategies. Variants that can exploit changes in either grazer resistance or nutrient uptake without compromising the other may gain a selective advantage. Grazing pressure is not, however, a constant force on the many diverse species that compose phytoplankton communities, and it is unlikely that antiherbivore compromises are directed against single zooplankton species. The realistic challenge to phytoplankton in nature is to resist mortality from a complex array of grazers and to exploit nutrients on many spatial and temporal scales. The demographic model developed here has much heuristic value, but it is not a quantitative analytical tool. It does permit different growth strategies to be interrelated and compared on common grounds. The results invite empirical comparisons of growth rates and nutrient uptake rates among taxa that differ in resistance to grazers. The advent of such a systematic empirical relation, backed by theory, may help to explain one of the balances of selection forces that have shaped phytoplankton communities by evolution.

REFERENCES

Alcarez, M., Paffenhofer, G-A., and Strickler, J. R. (1980). Catching the algae: a first account of visual observations on filter-feeding calanoids. In: *The Evolution and Ecology of Zooplankton Communities*, ed. W. C. Kerfoot. Amer. Soc. Limnol. Oceanogr. Spec. Symp. 3. Hanover, NH: University Press of New England.

Bamforth, S. S. (1958). Ecological studies on the planktonic protozoa of a small artificial pond. *Limnol. Oceanogr.*, 3, 398–412.

Bird, D. F. and Kalff, J. (1986). Bacterial grazing by planktonic lake algae. *Science*, 231, 493–5.

D'Abramo, L. R. (1980). Ingestion rate decrease as the stimulus for sexuality in populations of *Moina macrocopa*. *Limnol. Oceanogr.*, 25, 422–9.

Gage, T. B., Williams, F. M., and Horton, J. B. (1984). Division synchrony and the dynamics of microbial populations: a size specific model. *Theor. Pop. Biol.*, 26, 296–314.

Gerritsen, J. and Porter, K. G. (1982). The role of surface chemistry in filter feeding by zooplankton. *Science*, 216, 1225–7.

Gliwicz, Z. M. (1975). Effect of zooplankton grazing on photosynthetic activity and composition of phytoplankton. *Proc. Int. Assoc. Theor. Appl. Limnol.*, 19, 1490–7.

Haney, J. F. (1973). An in situ examination of the grazing activities of natural zooplankton communities. *Arch. Hydrobiol.*, 72, 87–132.

Hecky, R. E. and Kling, H. J. (1981). The phytoplankton and protozooplankton of Lake Tanganyika: species composition, biomass, chlorophyll content, and spatio-temporal distribution. *Limnol. Oceanogr.*, 26, 548–64.

Infante, A. and Abella, S. E. B. (1985). Inhibition of *Daphnia* by *Oscillatoria* in Lake Washington. *Limnol. Oceanogr.*, 30, 1046–52.

Jackson, G. A. (1980). Phytoplankton growth and zooplankton grazing in oligotrophic oceans. *Nature*, 284, 439–41.

Kirk, J. T. O. (1983). *Light and Photosynthesis in Aquatic Ecosystems*. Cambridge: Cambridge University Press.

Knoechel, R. and Holtby, L. B. (1986). Construction and validation of a body-length-based model for the prediction of cladoceran community filtering rates. *Limnol. Oceanogr.*, 31, 1–16.

Koehl, M. A. R. (1984). Mechanisms of particle capture by copepods at low Reynolds numbers: Possible modes of selective feeding. In: *Trophic Interactions within Aquatic Ecosystems*, eds. D. G. Meyers and J. R. Strickler, pp. 135–66. AAAS Selected Symp 85. Boulder, CO: Westview Press.

Koehl, M. A. R. and Strickler, J. R. (1981). Copepod feeding currents: food capture at low Reynolds number. *Limnol. Oceanogr.*, 26, 1062–73.

Lampert, W. and Schober, U. (1978). Das regelmässige Auftreten von Frühjahrsalgenmaximum und "Klarwasserstadium" im Bodensee als Folge von klimatischen Bedingungen und Wechselwirkungen zwischen Phyto- und Zooplankton. *Arch. Hydrobiol.,* 82, 364–86.

Lehman, J. T. (1980). Release and cycling of nutrients between planktonic algae and herbivores. *Limnol. Oceanogr.,* 25, 620–32.

Lehman, J. T. and Sandgren, C. D. (1982). Phosphorus dynamics of the prokaryotic nannoplankton in a Michigan lake. *Limnol. Oceanogr.,* 27, 828–38.

(1985). Species-specific rates of growth and grazing loss among freshwater algae. *Limnol. Oceanogr.,* 30, 34–46.

Lehman, J. T. and Scavia, D. (1982). Microscale nutrient patches produced by zooplankton. *Proc. Nat. Acad. Sci. USA,* 79, 5001–5.

Pascher, A. (1943). Zur Kenntnis verschiedener Ausbildungen der planktonischen *Dinobryon. Int. Rev. Hydrobiol.,* 43, 110–23.

Porter, K. G. (1976). Enhancement of algal growth and productivity by grazing zooplankton. *Science,* 192, 1332–4.

(1977). The plant-animal interface in freshwater ecosystems. *Amer. Sci.,* 65, 159–70.

Porter, K. G., Feig, Y. S., and Vetter, E. F. (1983). Morphology, flow regimes, and filtering rates of *Daphnia, Ceriodaphnia,* and *Bosmina* fed natural bacteria. *Oecologia,* 58, 156–63.

Poulet, S. A. and Marsot, P. (1978). Chemosensory grazing by marine calanoid copepods. *Science,* 200, 1403–5.

Purcell, E. M. (1977). Life at low Reynolds number. *Am. J. Phys.,* 45, 3–11.

Rigler, F. H. (1971). Feeding rates: zooplankton. In: *A Manual on Methods for the Assessment of Secondary Productivity in Fresh Waters,* eds. W. T. Edmondson and G. G. Winberg, pp. 228–55. Int. Biol. Program Handbook 17. Oxford: Blackwell Scientific.

Roman, M. R. and Rublee, P. A. (1981). A method to determine in situ zooplankton grazing rates on natural particle assemblages. *Mar. Biol.,* 65, 303–9.

Rubenstein, D. I. and Koehl, M. A. (1977). The mechanisms of filter feeding: some theoretical considerations. *Amer. Nat.,* 111, 981–94.

Scavia, D., Fahnenstiel, G. L., Davis, J. A., and Kreis, R. G. Jr. (1984). Small scale nutrient patchiness: some consequences and a new encounter mechanism. *Limnol. Oceanogr.,* 29, 785–93.

Stemberger, R. S. (1981). A general approach to the culture of planktonic rotifers. *Can. J. Fish. Aquat. Sci.,* 38, 721–4.

Strickler, J. R. (1984). Sticky water: a selective force in copepod evolution. In: *Trophic Interactions within Aquatic Ecosystems,* eds. D. G. Meyers and J. R. Strickler, pp. 187–239. AAAS Selected Symp. 85. Boulder, CO: Westview Press.

Weers, E. T. and Zaret, T. M. (1975). Grazing effects on nannoplankton in Gatun Lake, Panama. *Proc. Int. Assoc. Theor. Appl. Limnol.,* 19, 1480–3.

Wilson, M. S. (1959). Calanoida. In: *Freshwater Biology,* 2d ed., ed. W. T. Edmondson, pp. 738–794. New York: Wiley.

Yeatman, H. C. (1959). Cyclopoida. In *Freshwater Biology,* 2d ed., ed. W. T. Edmondson, pp. 795–815. New York: Wiley.

Zaret, R. E. (1980). The animal and its viscous environment. In: *The Evolution and Ecology of Zooplankton Communities,* ed. W. C. Kerfoot. Amer. Soc. Limnol. Oceanogr. Spec. Symp. 3. Hanover, NH: University Press of New England.

Chapter 10

FUNCTIONAL MORPHOLOGY AND THE ADAPTIVE STRATEGIES OF FRESHWATER PHYTOPLANKTON

COLIN S. REYNOLDS

INTRODUCTION

Lacking formally defined laws or even many potentially falsifiable theories, ecology continues to rely heavily upon concepts for its development. One such concept relates the distributions of organisms to the evolutionary strategies they have adopted to promote their prospects for growth and survival. Strategies may be regarded as groupings of similar morphological, physiological, reproductive, or behavioral traits that have evolved among species or populations and that are better suited to particular sets of given environmental conditions than to others. Accordingly, different organisms that have adopted similar strategies are likely to have similar ecologies (Grime 1979).

The utility of the concept depends upon its demonstrable applicability to a wide range of plants and animals in a variety of ecological systems. This chapter seeks to reconcile established differences among the biologies of individual species of the freshwater phytoplankton to the basic adaptive strategies that have been distinguished among certain other groups of organisms. It then relates these differences to the selective processes influencing their spatial and temporal distributions.

ENVIRONMENTAL CONSTRAINTS UPON PHYTOPLANKTON GROWTH

Planktonic communities are tenuously organized around the availability of the light energy that is required to sustain photosynthesis. Daily integrals of photosynthetic carbon fixation are sensitive to the intensity, duration, and spectral composition of the incident irradiance penetrating the water and, especially, to its hyperbolic attenuation with depth (Talling 1971). Adaptations to gain residence within or, at least, frequent access to the illuminated near-surface layer have an obvious functional relevance and, presumably, have been favored by natural selection (Walsby & Reynolds 1980).

The rate of carbon fixation places an upper capacity on the rate of production of new biomass, but this potential is rarely fulfilled (Talling 1984; Reynolds, Harris, & Gouldney 1985). Two major environmental factors commonly limit the realization of the potential for growth. One factor is that there may be insufficient essential nutrients (e.g., nitrogen, phosphorus) available to match the supply of photosynthate such that the rate of growth is limited by a depleted "cell quota" (Droop 1974) of one or other of these nutrients, in accord with the Monod model (see Tilman, Kilham, & Kilham 1982; Droop 1983). The second factor is the extent of wind-driven turbulent mixing that, above a defined critical range (Reynolds 1984a), entrains and transports planktonic algae vertically through the light gradient and, on occasion, beyond the euphotic layer (wherein net photosynthesis is possible). Under such conditions, the entrained organisms are subjected to rapid and frequent fluctuations in insolation, and, relative to the length of the solar day, they may well experience a shortcoming in the total photoperiod in which photosynthetic gain is possible.

These restrictions imposed on phytoplankton growth may be respectively analogized to the two main categories of external factors said to limit the amount of living material in biological systems (Grime 1979). Nutrient limitation of the rates of cellular growth and replication corresponds to *stress;* frequent involuntary translocations of individuals out of the euphotic layer, especially if these result in the destruction of existing biomass, arguably represent a form of *disturbance.* Among the pelagic habitats provided by the world's freshwaters (lakes, ponds, and rivers of various sizes and geographical locations), the intensity of stress and disturbance factors varies conspicuously. Just as there exists a wide spectrum of nutrient availability, ranging from ultraoligotrophic to hypereutrophic waters, and a broad range of potentially limiting nutrients, so there is a diversity of physical mixing characteristics (Lewis 1983). This diversity ranges from amixis (permanently ice-covered polar lakes) to continuous holomixis (exposed expanses of shallow or riverine waters) and includes systems that are discontinuously polymictic (usually warm-water lakes, mixing, and restratifying at diel frequencies). Moreover, the intensity of stress and disturbance can vary markedly within individual lakes under the influence of seasonal fluctuations in daylength, solar warming and cooling, wind action, frontal activity, rainfall intensity, nutrient loading, and hydraulic flushing. Given the microscopic sizes of planktonic algae and the comparative brevity of successive generations (days or less), it is not surprising that periodic development of nutrient stress and variability in vertical mixing should represent the major selective influence to which natural phytoplankton respond (Lewis 1978; Margalef 1978). Whereas "high-frequency" environmental oscillations (periodicity < 1 day) such as the photosynthetically active radiation received by individual cells must be accommodated within their physiological capabilities, lower-frequency oscillations in daylength, water temperature, and vertical mixing may be experienced by separate generations of phytoplankton (see Turpin, Chapter 8). Depending on the extent to which interspecific differences in algal tolerances and thresholds disadvantage the less well-adapted species, any responses to the variability are likely to be manifest in the dynamics of growth and loss of each species present and, eventually, in the altered species composition of the community (Harris 1980; Ivanovici and Wiebe 1981). Certainly, the literature abounds with descriptions of the conspicuous seasonal cycles of the abundance and specific composition of phytoplankton in particular bodies of freshwater. Many of these have been shown to conform to one or another basic pattern related to seasonal variations in physical mixing and the supply of nutrients (Findenegg 1947, 1966; Kalff & Knoechel 1978; Reynolds 1980, 1984b; Harris 1983). Moreover, experimental manipulations of these components in model systems have evoked analogous dynamic responses among phytoplankton (Reynolds 1986). What has been less clear is *why* individual species respond to imposed conditions of stress and disturbance with such apparent consistency. The hypothesis to be developed here is that the species of phytoplankton have evolved particular strategic mechanisms, each involving adaptations of cell morphology and physiological function, that equip them preferentially to survive under one or another of the permutations of low or high nutrient stress with low or high physical disturbance (Table 10-1).

PRIMARY STRATEGIES OF PHYTOPLANKTON

Of the four possible permutations of stress and disturbance (Table 10-1), the combination of continuous severe stress and high disturbance results in habitats hostile to the reestablishment of communities (Grime 1979; Smith 1985). The three remaining contingencies, however, have respectively favored the evolution of the three primary strategies that have been discerned variously among terrestrial flow-

Table 10-1. *Adaptive strategies (C, S, R) in the evolution of phytoplankton in freshwater pelagic environments (based on Grime, 1979).*

Intensity of disturbance	Intensity of stress	
	Low	High
Low	Competitors: $C(r, \text{I})$	Stress-tolerant spp.: $S(K, \text{III})$
High	Ruderals: $R(w, \text{II})$	No viable strategy

ering plants, fungi, insects, and mammals (for full discussion, see Southwood 1977; Grime 1979). Our hypothesis is that the same basic strategies apply to the freshwater phytoplankton and that they conform to the aquatic analogues of stress and disturbance factors as proposed here. Accordingly, it is proposed that species of phytoplankton must be adapted (a) to exploit environments saturated by light and nutrients, through the investment in rapid growth and reproduction, and to do so before other species (i.e., they are good *competitors, sensu* Grime) or (b) to operate under conditions of severe depletion of the external supply of essential nutrients (i.e., they are *stress tolerant*) or (c) to tolerate frequent or continuous turbulent transport through the light gradient (i.e., they approximate to disturbance-tolerant *ruderals, sensu* Grime).

It should be noted at once that Grime's (1979) terminology differs from established usage. In particular, the term *competitor* is usually applied to species whose superior ability to operate close to the environmental carrying capacity tends to eliminate species with inferior abilities, according to the principle of competitive exclusion (Hardin 1960); they therefore correspond closely with Grime's concept of stress-tolerant species. Grime's competitors are equivalent to species hitherto recognized as being opportunistic, colonist, or fugitive. Disturbance tolerance is generally understood. Thus, the concept of differing strategies is not at issue. To avoid confusion, the three categories are hereinafter designated *C*-, *S*-, and *R*-strategies, respectively.

These three primary strategies are not mutually exclusive. Rather, it is anticipated that individual species might show a number of adaptive features mainly tending toward those consistent with a *C*-, *S*-, or *R*-strategy and that some intergrading should be discernible.

The following attempt to support the hypothesis that the evolutionary ecology of freshwater phytoplankton conforms to this concept of primary strategies embodies notable previous attempts to categorize the commoner organisms on the basis of their distributions in nature or of their physiological responses in the laboratory. Several such attempts have broadly distinguished those planktonic algae capable of developing relatively rapid rates of growth in situ (r) and that can respond quickly to the availability of environmental resources from those species with obligately slower rates of growth (low r) that are saturated at a lower level of available resources (K) and that can better tolerate or accommodate to periods of resource stress. This distinction between r- and K-selected species, initially developed by MacArthur and Wilson (1967), has been applied to photoplankton (Margalef 1978; Kilham & Kilham 1980; Sommer 1981) to account for the dynamic changes in the growth and attrition of individual species characterizing particular stages in observed seasonal cycles. This view implies that there is a gradient, or continuum, between extreme r and K species, which Margalef (1978) has represented against axes describing increasing nutrient resources and increasing turbulence as a diagonal line (see Fig. 10-1a). For freshwaters, among which the range of nutrient resources and the fluctuations in the physical mixing conditions each vary over several orders of magnitude, rather more of Margalef's spatial representation is available for exploitation. Reynolds (1980) suggested that autogenic successions involved transitions from r- to K-strategists along gradients that, in Margalef's matrix, would follow a downward direction, but that stability changes (mixing, restratification) permitted allogenic displacements in the horizontal plane, selecting independently for or against those further species shown to be more tolerant of, or obligately dependent on, physical mixing of the water column. These species required a separate category, which Reynolds et al. (1983b) labeled w (see Fig. 10-1b). To avoid confusion arising from whether w species might be either r- or K-strategists, Reynolds (1984b) introduced the categories I (for fast-growing, erstwhile r-species), II (for mixing-tolerant w-species), and III (for slow-growing, stress-tolerant K-species). It is now proposed that these same categories respectively approximate to the C-, R-, and S-strategies of evolutionary adaptation, hypothesized to have been adopted by planktonic algae.

What properties of these algae determine the evident differences in their ecologies? Given the rather remote phylogenies of the taxonomic groups represented in the freshwater phytoplankton, there is an obvious tendency to ascribe to each of them generalized adaptive traits. Certainly, the high dependence of the nonmotile, silicified diatoms on turbulence for suspension and for the replenishment of dissolved Si (Sommer, Chapter 6 this volume) separates them as essentially R-strategists. Similarly, the slow growth of the larger

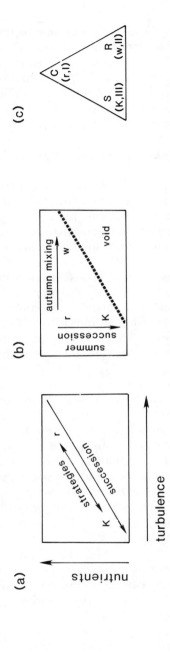

Fig. 10-1. (a) Representation of aquatic environments and the main environmental factors governing the distribution of r- and K-selected organisms, owing to Margalef (1978). (b) Representation, on similar axes, of periodic changes in phytoplankton as hypothesised by Reynolds et al. (1983b); w species are selectively favored in well-mixed water columns. (c) Rescaling and reorientation of the triangular area in (b) to analogize the r-, K-, and w-strategies of phytoplankton to Grime's (1979) primary strategic categories – competitors C, stress toleraters S, and disturbance-tolerant ruderals R – and to the periodic groupings (I,III,II) discerned by Reynolds (1984b).

dinoflagellates (*Peridinium, Ceratium* spp.) and their propensity to dominate nutrient-depleted environments (Pollingher, Chapter 4 this volume) suggests a more *K*-selected *S*-strategy. However, among the Cyanobacteria (Reynolds 1984b; Paerl, Chapter 7 this volume) and the Chlorophyta (Happey-Wood, Chapter 5 this volume), it is possible to distinguish species whose distributions conform unmistakably to typical *C*-patterns *(Synechococcus, Chlorella), S*-patterns *(Microcystis, Sphaerocystis),* and *R*-patterns (large *Oscillatoria,* desmid spp.). Ecological adaptation therefore evidently transcends taxonomic boundaries.

Many previous studies have differentiated among the growth rates of individual species on the basis of the size of cells or the size of colonial units (Laws 1975; Banse 1976; Malone 1980; Reynolds 1984a). Indeed, many tangible aspects of cell metabolism are actually or apparently related to cell morphology, including the efficiency of light interception and utilization (Harris 1978; Harris, Piccinin, & van Ryn 1983; Kirk 1983; Raven 1984; Tilzer 1984), the capacity to absorb and store essential nutrients (Eppley, Rogers, & McCarthy 1969; Sournia 1981, 1982; Sommer 1984), and the manner in which these processes respond to external temperature fluctuations (Tamiya et al. 1953; Foy, Gibson, & Smith 1976). Morphological properties of phytoplankton are also known to influence their potential rates of removal by grazing animals (Burns 1968; Porter 1977; Gliwicz 1980; Runge & Ohman 1982; Lehman & Sandgren 1985) and their susceptibilities to loss from suspension (Smayda 1970; Reynolds 1984a). Some or all of these effects may be modified in motile organisms, which have an additional potential ability to migrate to water depths offering more favorable opportunities for light and nutrient absorption (Raven and Richardson 1984). It is not surprising, then, that superficial correlations should have been discerned between spatial and temporal distributions of phytoplankton in lakes, on the one hand, and the dynamic population responses related to rates of cellular growth and loss (Allen & Koonce 1973; Allen, Bartell, & Koonce 1977; Lewis 1977; Sommer 1981) or to unit morphology (Reynolds 1984b), on the other. That the latter is ultimately related to cellular DNA content (Cavalier-Smith 1982) emphasizes the evolutionary aspect of the adaptations.

If the hypothesized link between the responses of planktonic algae to imposed environmental constraints and morphologically related properties is substantial, then it should be possible to discriminate morphological features that are consistently allied to one of the primary strategies. In other words, the supposed strategies should be distinguishable in terms of the functional mechanisms that enable organisms to survive under disturbed, stressed, or more favorable conditions, rather than on the pragmatic interpretation of distributions observed in nature.

Empirical support for this contention is sought in the subsequent presentation. The approach to the problem follows Grime's (1976) method of devising triangular ordinations of documented properties of planktonic algal species such that the more extreme values fall consistently toward one or another of the corners of the triangles, corresponding with *C*-, *S*-, or *R*-strategies (Fig. 10-1c) appropriate to the aquatic environments distinguished in Fig. 10-1b.

"PACKAGING": MORPHOLOGICAL VARIATION AMONG THE PLANKTONIC ALGAE

Although the unit volumes of freshwater planktonic algae span seven or eight orders of magnitude, from <10 μm^3 (single cells of *Synechococcus* and *Ankyra* spp.) to some 10^8 μm^3 (larger mucilaginous colonies of *Microcystis*), the ratios of their surface areas to their volumes (SA/V) vary more conservatively within only three orders (0.01 to $\sim$4 μm^{-1}; Reynolds 1984a). If the mucilaginous colonies are excluded, typical SA/V ratios fall in the range 0.2 to 4 μm^{-1}. Lewis (1976) argued that this apparent conservatism is metabolically essential. Among spherical cells, the critical ratio is attained as a direct consequence of small absolute size; any supposed advantage of increased size accruing to larger cells is compensated by a relative expansion in surface area that is achieved through attenuation in one or more planes or through the possession of protruding structures. Lewis (1976) represented this principle by plotting the maximum linear dimensions (MLD) of selected planktonic algae against their approximate SA/V ratios. Following the same method, Reynolds (1984b) plotted a different data set, based on observations made on algae in British waters. These data were separable into categories variously representing the species dominating vernal isothermally mixed water columns, those of the early stages of the summer stratification and those characterizing the later, often nutrient-stressed stages of the stagnation period. Some of the same data, together with several additional entries, are plotted against redrawn axes in Fig. 10-2a. The original three groupings are preserved, but each is now centered toward the apex representing the appropriate growth strategy (see insets, Fig. 10-2b–d).

These plots effectively separate those species with low SA/V ratios (Fig. 10-2c) from those whose small size determines that SA/V is relatively large (>0.3 μm^{-1}, Fig. 10-2b), and from those larger forms that achieve a similar SA/V by virtue of morphological distortion (Fig. 10-2d). That these same species should be similarly distinguishable also

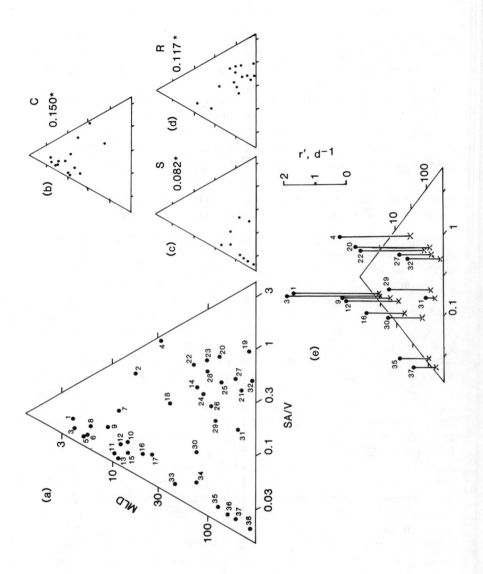

Fig. 10-2. (a) Triangular ordinations of morphological properties of selected phytoplankton algae: the mean maximum linear dimension (MLD) is plotted against the mean surface area-to-volume ratio (SA/V) of individual units (cells, filaments, or colonies). (b),(c),(d) show the same data but apportioned between primarily C-, S-, and R-strategists. A "nearest-neighbor" statistic ϕ ($= md^2$, where m is the density of points per plot and d is the mean distance between each point and its nearest neighbor) is appended; significant underdispersion of points is indicated by values <0.25. (e), reported maximal growth rates in culture (r') are projected on to the same triangular ordination, which has been tilted. Morphometric data are from Table 3 of Reynolds (1984a) with the exception of no. 6 (unpublished data of Kennedy and Sandgren reproduced with permission), no. 13 (Oliver, Kinnear, & Ganf 1981), 23 (author's approximations from illustrations in FBA Fritsch Collection) and no. 34 (Pollingher and Berman 1975). Data on light- and nutrient-saturated growth rates at ~20° are drawn from Table 16 of Reynolds (1984a). The species are numbered, in order of ascending unit volume, as follows: 1: *Synechococcus* sp.; 2: *Ankyra judayi* (G.M.Sm.) Fott; 3: *Chlorella pyrenoidosa* Chick; 4: *Ankistrodesmus braunii* (Näg.) Collins; 5: *Kephyrion* sp.; 6: *Chlamydomonas reinhardii* Dangeard; 7: *Rhodomonas minuta* Skuja var. *nannoplanktica* Skuja; 8: *Chrysochromulina parvula* Lackey; 9: *Monodus* sp.; 10: *Chromulina* sp.; 11: *Chrysococcus* sp.; 12: *Stephanodiscus hantzschii* Grunow; 13: *Chlorococcum* sp.; 14: *Scenedesmus quadricauda* (Turp.) Bréb. (4-cell coenobium); 15: *Cyclotella meneghiniana* Kütz.; 16: *Cryptomonas ovata* Ehrenb.; 17: *Stephanodiscus astraea* (Ehrenb.) Grunow; 18: *Mallomonas caudata* Iwanoff; 19: *Closterium aciculare* T. West; 20: *Asterionella formosa* Hassal (8-cell colony); 21: *Melosira italica* (Ehrenb.) Kütz. subsp. *subarctica* Müll. (10-cell filament); 22: *Fragilaria crotonensis* Kitton (10-cell colony); 23: *Diatoma elongatum* (Lyngby) Agardh. (5-cell colony); 24: *Staurastrum pingue* Teiling; 25: *Dinobryon* sp. (10-cell colony); 26: *Synedra ulna* (Nitzsch) Ehrenb.; 27: *Tabellaria flocculosa* (Roth) Kütz. var. *asterionelloides* (Roth) Knuds. (8-cell colony); 28: *Pediastrum boryanum* (Turp.) Meneghin (32-cell coenobium); 29: *Aphanizomenon flos-aquae* Ralfs *ex* Born. *et* Flah. (flake); 30: *Anabaena circinalis* Rabenh. *ex* Born. *et* Flah. (20-cell trichome with bounding mucilage); 31: *Ceratium hirundinella* O. F. Müller; 32: *Oscillatoria agardhii* Gom. (trichome 300 μm in length); 33: *Sphaerocystis schroeteri* Chodat (mucilaginous colony, 46 μm in diameter); 34: *Peridinium cinctum* (Mull.) Ehrenb, fa. *westii* (Lemm.) Lef.; 35: *Eudorina unicocca* G.Sm. (colony, 130 μm in diameter); 36: *Uroglena lindii* Bourrelly; 37: *Microcystis aeruginosa* Kütz. emend. Elenkin (mucilaginous colony, 200 μm in diameter); 38: *Volvox aureus* Ehrenb. (coenobium, including enclosed hollow space, 450 μm in diameter.)

on the basis of their seasonal distributions is unlikely to be mere coincidence. Rather, it is probable that the dynamic responses of the algae to environmental variations in temperature, vertical mixing and its effect on the exposure of cells to the underwater light field, as well as their susceptibilities to nutrient stress and to loss processes (sinking, grazing), are essentially conditioned by morphological consequences of their "packaging" (Reynolds 1984b). It would follow, therefore, that there should be corresponding distinctions among the physiological and metabolic properties of individual species, the demonstration of which is attempted here.

PHYSIOLOGICAL AND METABOLIC ADAPTATIONS

Optimum Growth Rates

The ultimate expression of metabolic efficiency lies in the rate of growth. When all environmental requirements are saturated, the limiting condition is the rate at which the materials for new cell production can be assimilated, assembled, and partitioned into the next generation. Such idealized growth conditions are probably found only in laboratory cultures grown in nutrient-replete artificial media and under continuous, saturating illumination. In Fig. 10-2e, a selection of maximum-recorded growth rates (r'_{max}) of species in culture as derived from the literature and, as far as possible applying to typical vegetative forms maintained at comparable temperatures (20°–25°C), is projected onto the (tilted) triangular ordination of MLD versus SA/V (cf. Fig. 10-2a). This representation emphasizes the correlation of increasing r'_{max} with decreasing unit size and, especially, with increasing SA/V ratio, as detected in earlier studies (Banse 1976; Reynolds 1984a). Presumably, these correlations reflect the relative metabolic efficiency with which such cells are able to absorb and assimilate the requisite raw materials (Eppley et al. 1969; Malone 1980; Raven 1982; Sournia 1982). These are, moreover, the properties expected to characterize species with opportunistic, C-strategies for growth and dominance under favorable environmental conditions.

Light Harvesting

In natural waters, growth conditions fall short of those that can be provided in the laboratory. Even in waters of high clarity and low light attenuance that offer a good depth range in which light-saturated photosynthetic rates can be attained, there remains the alternation between day and night. At increased levels of turbidity and vertical extinction, the absolute depth to which saturating light intensities can penetrate is correspondingly truncated. If the water column is simultaneously mixed to a greater depth, then cells may be transported to depths where photosynthesis is light limited or is even insufficient to compensate the dark respiration rate (i.e., the mixing depth exceeds the euphotic depth). Populations so mixed are likely to experience rapid fluctuations in irradiance and an effective reduction in total daily light dose (Farmer & Takahashi 1982). Under conditions of suboptimal irradiance, the ability of cells to maximize their opportunities for light absorption may assume critical importance.

Owing to the varying capacities of planktonic algae to adapt to low light intensities, either by increasing the sizes or the numbers of their photosynthetic units (Falkowski 1980; Falkowski & Owens 1980) and/ or by accumulating accessory pigments that spread the absorption spectrum of the light traps (Harris 1978; Falkowski 1984), it is often difficult to discern the contribution of cell morphology to the potential intracellular arrangement of the photosynthetic apparatus. Moreover, despite the large number of determinations of photosynthetic rates of natural phytoplankton reported in the literature (for a review, see Harris 1978), there is a dearth of information at the species level (Talling 1984). Some indications are available, however, from the light-limited portions of in situ profiles of photosynthetic behavior invade on natural populations dominated by a single species.

Two components influence light-limited photosynthetic efficiency (Raven 1984; Tilzer 1984): the quantum yield ϕ (the amount of carbon fixed per unit of light energy absorbed [units: mol C·(mol photon)$^{-1}$]) and the specific absorption coefficient of the photosynthetic pigment, k_c ([units: m^2·(mg chl a)$^{-1}$]). The product $\alpha = \phi \cdot k_c$ [units: mol C fixed·(mol photon)$^{-1}$·(mg chl a)$^{-1}$·m^2] is then equivalent to the slope of photosynthetic rate P plotted against the intensity of irradiance I over the range from zero light to the onset of light saturation (I_k). Both ϕ and k_c are variables. Other factors notwithstanding, 8 mole photons are required to fix 1 mole carbon (Raven 1984); that is, the maximum value of ϕ is 0.125 mol C·(mol photon)$^{-1}$. Close approximations to this value have been determined for natural suspensions of marine nannoplankton (Kirk 1983) and of *Synechococcus*-dominated lake plankton (Bindloss 1974). It is probably rather lower in larger-celled algae, and the yield is also affected by other biochemical pathways competing for the ATP and NADPH produced (Raven 1984). Bannister's (1974; see also Bannister and Weideman 1984) approximation of $\phi = 0.06$ to 0.08 mol C·(mol photon)$^{-1}$ may be more generally applicable. The specific absorption coefficients have been recorded more frequently in the literature, being said to range between 0.004 and 0.020 $\mu^2 \cdot$(mg chl a)$^{-1}$ (Talling 1960; Harris 1978), with the higher values applying to the more morphologically distorted species such as *Asterionella* and *Scenedesmus*. Potentially, then, variation among specific photosynthetic efficiencies range over an order of magnitude [α = 0.24 to 2.5 mmol C$\cdot$(mol photon)$^{-1}\cdot$(mg chl a)$^{-1}\cdot$m^2].

In light-saturated environments where the limiting step in cell production is *not* light (Reynolds et al. 1985), the principal selective advantage is likely to accrue from a high quantum yield of photosynthesis per se. For cells suspended in optically deep, well-mixed columns or stratified deep in the light gradient, the selective advantage might move toward species with higher efficiencies of light utilization. Although larger or low-light adapted cells containing relatively high concentrations of photosynthetic pigments can absorb absolutely more photons from a given rate of flux than can small cells, photon absorption per unit of cell carbon clearly benefits from a larger projected cell area, that is, when k_c is increased (Raven 1984).

These distinctions are illustrated by the data for individual algae represented in the plot (Fig. 10-3a) of photosynthetic efficiency (α) against the chlorophyll-specific coefficient of light absorption (k_c). In particular, the high photosynthetic efficiency of *Synechococcus,* a C-species with a high quantum yield, is distinguished from the less efficient, larger *S*-species (of *Ceratium* and *Peridinium*) and from the more attenuated cells and filaments of planktonic *R*-species, whose higher k_c values offer an increased cell-specific capacity for efficient light harvesting despite rapid fluctuations in perceived irradiance levels.

Nutrient Uptake and Assimilation

Autotrophic growth also depends on cells obtaining sufficient nutrients, including carbon dioxide, to sustain the assembly of new biomass. Rates of growth therefore depend on the rates of absorption and assimilation of those nutrients most likely to fall into short supply. The complex kinetic relationships between growth and availability of particular limiting nutrients have been extensively investigated in recent years, and there is already a large literature on the subject (for recent reviews, see Rhee 1982; Tilman et al. 1982; Droop 1983, Turpin, Chapter 8 this volume). Generally, the nutrient-limited growth rates of cells (r') can be expressed as a fraction of the nutrient-saturated growth rate (r'_{max}) according to a Monod equation:

$$r' = \frac{r'_{max}S}{K_s + S}$$

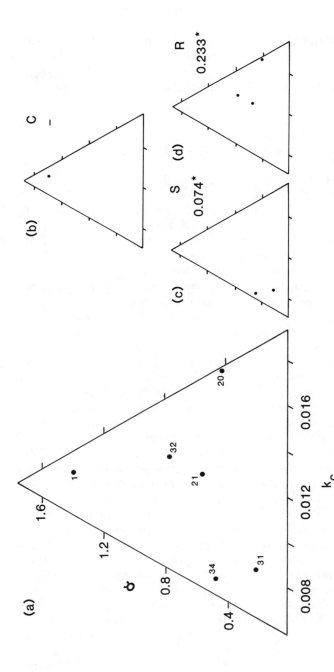

Fig. 10-3. (a) Triangular ordination of light-utilization properties of some planktonic algae. Photosynthetic efficiency, in mmols C fixed per mg chlorophyll a per mol photon flux received, is plotted against the chlorophyll-specific coefficient of light absorption (k_c m^2 per mg chlorophyll a). (b),(c),(d) show the same data for C-, S-, and R-species, respectively, with nearest-neighbor statistic. Data for *Synechococcus* (1) from Bindloss (1974); for *Asterionella* (20) from Reynolds (1984a); for *Melosira* (21) and *Oscillatoria* (32) from Jewson (1976); for *Ceratium* (31) from Harris, Heaney, & Talling (1979); for *Peridinium* (34) from Dubinsky & Berman (1976, 1979).

where S is the nutrient concentration and K_s represents the concentration of that nutrient at which the rate of growth is one-half of the maximum rate (i.e., $r'_{max}/2$). In fact, growth rate responds directly to the intracellular concentration of the nutrient, according to Droop's (1974) cell-quota model:

$$r' = \frac{r'_{max} (q - q_0)}{K_s + (q - q_0)}$$

where q, the "cell quota," is the amount of limiting nutrient available to each cell and q_0 is the absolute minimum at which no further growth can be sustained ($r' = 0$). At steady state, these formulations reputedly yield equivalent rates of growth for a given species (Goldman 1977; Burmaster 1979), owing to the equivalence between the cell-specific rates of uptake (V_m) and consumption of the limiting nutrient. However, r'_{max}, K_s, and V_m vary interspecifically to the extent that, under a given steady state, one species is always likely to grow faster than ("outcompete") others (e.g., Tilman and Kilham 1976).

The functional basis for these interspecific differences has yet to be investigated in any depth, although it is likely to be relevant to the comparative dynamics of potential competitors that apparently coexist in natural, nonequilibrium environments. It may be proposed that the rate of nutrient uptake in cells is related, in part, to the area of absorptive surface and the localized gradients that are maintained by the active uptake mechanisms operating at the cell periphery. The maximum rate of absorption per unit cell mass is thus likely to be a correlative of the effective SA/V ratio of the protoplast (Sournia 1982), although movement of the cell, relative to the adjacent medium (for instance, by sinking or swimming), has been shown to be advantageous in maintaining favorable uptake gradients (Munk and Riley 1952; Pasciak and Gavis 1974; Canelli and Fuhs 1976). Once inside the cell, the nutrients must diffuse or be transported to the sites of metabolic assembly. In larger cells, transport and assembly of nutrients seem more likely to represent a rate-limiting step than in smaller cells, but larger cells may also have a greater capacity to divert the nutrient influx into intermediate storage condensates, such as polyphosphate granules or protein bodies. These "internal stores" might then be consumed during growth when the external concentration of the relevant nutrient falls to low levels and be "topped-up" ("luxury-uptake") when external supplies are freely available. Moreover, over evolutionary time scales, natural selection determines adaptation of K_s to a value commensurate with high, low, or fluctuating concentrations of limiting nutrients (Crowley 1975).

Accordingly, the experiments devised by Sommer (1984) distinguished several broad strategies among the freshwater phytoplankton for contending with variable supplies of phosphorus. Sommer (1984) differentiated among *velocity-adapted* species, in which high maximum rates of nutrient uptake were matched by rapid cellular growth; *storage-adapted* species, in which the ratio between the uptake rate and a relatively lower growth rate permitted nutrient reserves to be accumulated intracellularly; and *affinity-adapted* species having a low half-saturation constant for growth that was rapidly saturated by increased phosphorus availability. The first group was proposed to be superior to the others in nutrient-rich conditions, the last in nutrient-depleted conditions, and the storage-adapted species were more suited to fluctuating environments.

This idea has been developed in the construction of Fig. 10-4a. Using published data on the rates of phosphorus uptake in phosphorus-starved cells (indicative of $V_{m\ max}$), the half-saturation constants of P-uptake (K_m) and r'_{max}, the rate of light- and nutrient-saturated growth in culture (at or near 20°C), ratios have been derived ($V_{m\ max}/K_m$, $V_{m\ max}/r'_{max}$) that, respectively, give some empirical expression of *affinity* of species for phosphorus (it is equivalent to the initial slope of uptake rate versus phosphate concentration) and of *storage* potential (the faster the uptake rate relative to the growth rate then the more is available for retention). The dimensions of the axes have been chosen to correspond to the previous triangular ordinations. In fact, the distribution of the *S*-species (Fig. 10-4c) shows a significant tendency toward being relatively storage-adapted (being grouped well to the left of the figure at high ratios of V_m/r'), whereas that of *C*-species (Fig. 10-4b) are grouped more to the right, toward velocity adaptation. The supposed *R*-strategists seem more variable in this context. None of the groupings appears to have a narrow distribution with respect to the affinity axis, suggesting that affinity adaptation is influenced less by cell morphology than, as argued by Crowley (1975), by evolutionary adjustment relative to ambient concentrations of substrate in a particular environment.

Temperature Regulation of Growth Rates

Seasonal changes in the abundance of individual species of phytoplankton, especially in temperate lakes, as well as general compositional differences among lakes at different latitudes, have often been assumed to be related to differential temperature optima and differing specific sensitivities of growth rate to temperature variation, yet there have been few systematic experimental studies on which to base any

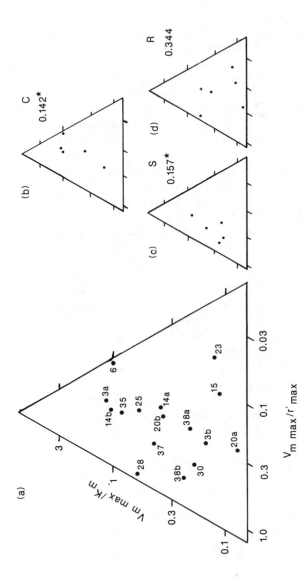

Fig. 10-4. Triangular ordination of nutrient uptake and processing properties of some planktonic algae. The ratio between the maximum rate of phosphate uptake ($V_{m\,max}$ in μmol $\times 10^{-9}$ per cell per minute) and the half-saturation coefficient of phosphorus uptake (K_m, in μmol P$\cdot$L^{-1}) is plotted against the ratio between $V_{m\,max}$ and the maximum rate of cellular growth at 20° (r_{max}, per day). Figures 10-4b,c,d show the same data for C-, S-, and R-species, respectively, with the nearest-neighbor statistic. Data for *Chlorella* (3a) from Jeanjean (1969) and (3b) Nyholm (1977); for *Chlamydomonas* (6) and *Eudorina* (35) from Kennedy & Sandgren (unpublished, reproduced with permission); for *Scenedesmus* (14a) from Rhee (1973) and (14b) Nalewajko & Lean (1978); for *Cyclotella meneghiniana* (15) from Tilman & Kilham (1976); for *Asterionella* (20a) from Tilman & Kilham (1976) and (20b) Holm and Armstrong (1981); for *Diatoma* (23) from Kilham, Kott & Tilman (1977); for *Dinobryon* (25) from Lehman (1976a); for *Pediastrum duplex* (28) from Lehman, Botkin, & Likens (1975); for *Anabaena flosaquae* (30) from Nalewajko and Lean (1978); for *Microcystis* (37) from Holm & Armstrong (1981); for *Volvox aureus* (38a) from Senft Hunchberger & Roberts (1981) and (38b) Kennedy & Sandgren, unpublished.

tangible hypotheses. Interpretations based on serial observations of seasonal specific growth patterns are complicated by such interacting variables as daylength, light penetration, and thermal stratification, which are not readily dissociated (Lund 1965). Temporal lags in seasonal warming and cooling relative to changes in the underwater light climate have permitted preliminary classification of specific tolerances to low light and temperature (Findenegg 1947). Laboratory experiments have confirmed that species respond differently to temperature (Lund 1949; Cloern 1977; Konopka and Brock 1978; Morgan and Kalff 1979) and in their sensitivities to simultaneous adjustments in light intensity and photoperiod (Foy et al. 1976). These studies serve to emphasize that resultant growth rates depend on several constituent processes, each of which responds to fluctuations in external temperature. Light-saturated photosynthetic rates and dark respiration rates alter by factors (Q_{10}) of 2.0–2.3 (see Harris 1978) per 10°C rise or fall in temperature between approximately 4° and 25°C. Intracellular transport and assimilation of photosynthate and other nutrients into proteins and new cell material are equally or more temperature-dependent ($Q_{10} > 2.3$), especially when the assembly sites are relatively remote from the sources of uptake or formation. Of these processes, only photosynthesis is essentially light dependent. Thus, among smaller or narrow cells, growth rate is likely to relate more closely to net photosynthetic rate over a range of temperatures than among larger cells, whose growth at low temperatures is liable to limitation by intracellular translocation rates. In environments simultaneously characterized by low temperatures and low daily light (exacerbated by full column mixing), such as are encountered in temperate lakes during winter and early spring, the growth rate of small cells might be more likely limited by light and that of large cells and colonies by temperature. Those smaller or narrow cells having a high capacity for light harvesting at low light intensities would then seem to be the best suited to these conditions (Reynolds 1984b).

The relative sensitivities of selected planktonic algae to light-dose and temperature fluctuations are represented in Fig. 10-5a. The axes describe possible ranges in the threshold of the mean exposure of cells, suspended in natural water columns, to saturating light intensities (in hours per day) and in the Q_{10} values of temperature limitation of light-saturated growth rate. The data plotted are drawn exclusively from Reynolds's (1986) synopses of the maximal growth rates (r') observed over several years in large experimental enclosures ("Lund tubes") and which were related to an index of the daily dose of saturating irradiance (daylength times the ratio of Secchi-disk extinction depth to mixed depth) separately calculated for two or three defined tempera-

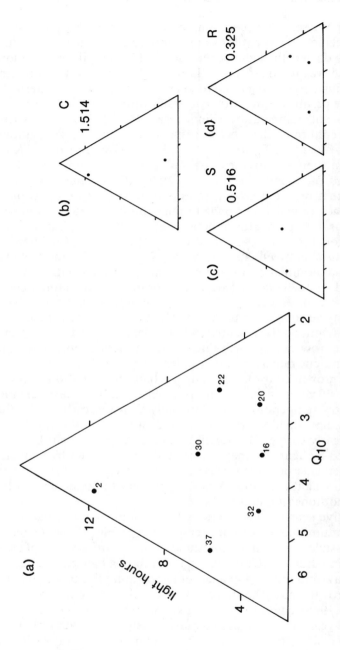

Fig. 10-5. (a) Triangular ordination of the sensitivities of in situ growth rate of some planktonic algae to low-light doses and low temperatures. The light threshold for net growth, in hours per day, is plotted against the Q_{10} of growth rates at different temperatures extrapolated to 24 h · day^{-1}. Data from Reynolds (1986). The same data for C-, S-, and R-species are shown in (b),(c),(d).

ture ranges. The intercept of each fitted curve defined the threshold of light dose; the extrapolations of each curve to 24 $h \cdot d^{-1}$ defined the maximal value of r' at the given temperature, from which the Q_{10} response to temperature was then calculated.

Sufficient data for these derivations were available only for seven species. Nevertheless, Fig. 10-5a attests to the substantial differences between the deduced light-dose threshold of *Ankyra* (no. 2), a frequent summer nannoplanktonic representative, and of *Asterionella* (no. 20), the diatom typically dominating the vernal increase, as well as differences between the Q_{10} values of *Ankyra* and *Asterionella* and that of the larger, colonial *Microcystis* (no. 37), the growth of which is confined to the summer period. The other species shown occupy intermediate positions on the plot. *Anabaena* and *Fragilaria,* respectively, conform to the expectations of planktonic *S*- and *R*-species (Fig. 10-5c,d). However, the trends are confounded somewhat by the capacity of the *Cryptomonas* sp. for adaptation to low-light intensities and by the high apparent Q_{10}-value derived for the *Oscillatoria.*

PROLONGATION OF SUSPENSION

A successful strategy for maximizing the opportunities for growth depends, in part, on the maintenance of the reproductive stock within the trophogenic zone. For a population to increase, its rate of growth must exceed the sum of the rates of its removal, whether due to "permanent" settling out of the trophogenic layer, or to exploitation by animals (as food), or by parasitic organisms (as hosts), or to other sources of mortality. Not only are the removal mechanisms, or "loss processes," known to exert important controls on the dynamics and phasing of net increase of phytoplankton, but they also alter through time in their relative magnitude and they act differentially on simultaneously coexisting species (Kalff & Knoechel 1978; Crumpton & Wetzel 1982; Reynolds 1984a; Lehman & Sandgren 1985). In different ways, the development of direct morphological or behavioral adaptations to resist the rapid elimination of cells and the investment in growth rates sufficient to ensure that the rates of loss are exceeded for substantial periods both represent alternative strategic mechanisms for the prolongation of net population increase.

The mechanisms for prolonging suspension per se are, in reality, rather more complex than the early popular view of a universal requirement of phytoplankton to limit their sinking rates. Most freshwater species of phytoplankton are continuously or frequently denser than water and, in consequence, sink in relation to the water immediately adjacent. This remains true even when the water itself is in tur-

bulent motion, wherein planktonic organisms are constantly random-
ized throughout the mixed layer, but remain liable to be lost from
suspension across its lower, nonturbulent boundary. Within the con-
fines of the mixed layer, the prime requirement is to reduce the opportunity for disentrainment from the turbulent eddies. The shallower the
mixed layer, however, the greater is the rate of sinking loss of heavier,
nonmotile species and the greater becomes the reliance on slow sinking
rates or on the ability of organisms to make compensatory movements
to maintain or recover their vertical station. Under conditions of
simultaneous nutrient stress, morphological streamlining (for rapid
vertical movement) coupled with a capacity for efficient swimming or
adjustment of buoyancy (among cyanobacteria) is likely to be at a pre-
mium. Thus, it is possible to differentiate among the phytoplankton
on the basis of their maximum rates of vertical movement in situ and
according to whether such movements are controlled or consequential
upon their high densities per se. As shown in Fig. 10-6, such distinctions are broadly consistent with the conceptual representation of
planktonic C-, S-, and R-strategies. Small-celled, fast-growing C-spe-
cies can be expected to respond well in clear-water environments providing favorable insolation. The S-species represented may perform
relatively better when turbulent mixing is weak and the light and
nutrient resources tend to become vertically segregated. The ability of
S-species to adjust vertical position considerably enhances the opportunities to exploit the nutrient resources available in the water column
(Raven & Richardson 1984). The larger, nonmotile R-species, especially diatoms, always rely on an absolute critical depth of vertical
mixing (Reynolds 1983, 1984b), although interspecific differences in
performance may be sensitive to relative light penetration and nutrient
availability in the mixed layer (Smol, Brown, & McIntosh 1984).

Biotic influences upon the residence times of planktonic species
have been evaluated in some detail by Reynolds (1984a). Save for grazing by animals, these processes are not primarily related to cell morphology, nor do they necessarily impinge upon the concept of adaptive
strategies. Many factors contribute to the rates of removal by herbivorous animals, including the species composition and abundance of
the pelagic fauna, their specific energetic requirements, and their food
preferences (see review in Reynolds 1984a). At high concentrations of
potential food ($>0.2 \mu g \ C \cdot mL^{-1}$), however, the optimal foraging strategy is arguably filter-feeding (Lehman 1976b), which is perhaps most
successfully employed by species of Cladocera (especially daphnids). The size range of algal particles on which they feed is determined
by the spacing of the filtering setules ($\sim 1 \ \mu m$) and the width of the
opening of the carapace valves (varying interspecifically from <35

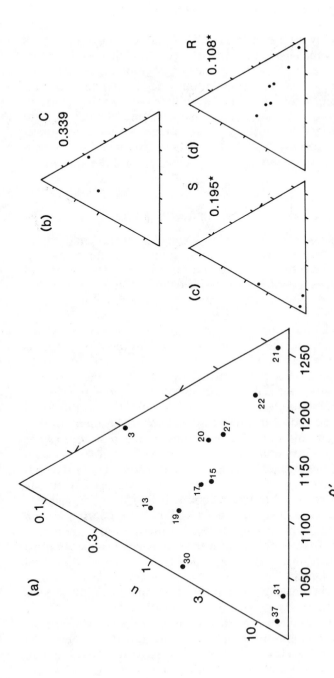

Fig. 10-6. (a) Triangular ordinations of the sinking and buoyant properties of some planktonic algae. Maximum vertical velocities of live, unaggregated cells or colonies, whether passively sinking or actively swimming (u, in m·d⁻¹) are plotted against measurements of unit density (ρ, in kg·m⁻³). The same data are shown in (b),(c),(d) for C-, S-, and R-species, respectively. Data for *Chlorella* (3), *Chlorococcum* (13), and *C. meneghiniana* (15) from Oliver et al. (1981); for *S. astraea* (17), *Asterionella* (20), *Melosira* (21) *Fragilaria* (22), and *Tabellaria* (27) from Reynolds (1984a); for *Anabaena circinalis* (30) from Reynolds (1984a); for *Ceratium* (31) from Heaney and Furnass (1980); and for *Microcystis* (37) from Reynolds et al. (1981).

μm to <60 μm; Gliwicz 1980; Reynolds 1984a). Colonial algae with gelatinous sheaths may be undigestable, even though they be of an ingestible size (Porter 1977). The species of phytoplankton therefore liable to depletion by *Daphnia* populations and the dynamic control by populations generating daily filtration rates equivalent to or in excess of the net growth rates of the algae consumed (≥ 0.5 d^{-1}) are generally characterized by unit sizes in the range 1–60 μm. Comparison with Fig. 10-2a suggests that the majority of phytoplankton species likely to be so affected are the *C*-species located toward the apex of the triangle.

REGENERATIVE STRATEGIES

The ability of species to survive periods when environmental conditions are unfavorable to growth and to maintain potential "inocula" poised to exploit the return of more favorable conditions constitutes another important aspect of the strategies of planktonic algae. As has been indicated, the criteria determining the suitability of environments for different species do not necessarily coincide and their specific response mechanisms are strikingly variable. The onset of stable thermal stratification and truncation of the depth of the mixed layer is usually stimulatory to slow sinking and motile species and, if accompanied by significant surface warming, to larger *S*-species also. The same conditions permit the acceleration of disentrainment and sinking loss of the nonmotile diatoms. Moreover, should the stratification abruptly follow a period of deep mixing and low-light adaptation of the cells, then those disentrained near the surface may experience excessive insolation, physiological impairment and a sharp increase in settling velocity (Reynolds & Wiseman 1982; Reynolds 1983). The latter contributes to an enhanced rate of sinking loss, enabling a relatively greater fraction to "escape" to lower depths, before they are seriously or irreversibly damaged. This enhanced sinking may be viewed as a survival mechanism contributing to the maintenance of a potential inoculum of cells in the water column to exploit a subsequent return of more mixed conditions. Conversely, conditions of increased vertical mixing, perhaps attendant on autumnal cooling and decreasing insolation, will be perceived as being adverse to many *C*- and *S*-species (Figs. 10-5, 10-6).

The growth of all species is eventually subject to deficiencies in the supply of one or more major nutrients. The more sensitive species will be those with the smallest internal stores; the least so will be those with enhanced capacities for nutrient acquisition and storage (see Figs. 10-4, 10-6). Limitations in the carbon dioxide supply (Talling 1976) also

have a differential selective effect among particular species of phytoplankton, especially against certain chrysophytes (Reynolds 1986;
Sandgren, Chapter 2 this volume), and the deficiencies in the availability of combined nitrogen are presumed to benefit nitrogen-fixing
species of cyanobacteria (e.g., Rhee 1982; Paerl, Chapter 7 this volume). Falling concentrations of dissolved silicic acid impose severe
limitations on the growth of diatoms, even when other environmental
conditions favor it. Near exhaustion of the available silicon remains
the major controlling factor in the seasonal growth among larger and
deeper lakes where the minimum extent of vertical mixing critical to
the maintenance of suspension is rarely approached (Reynolds 1984a).

In such cases, where further growth becomes limited or prevented,
populations become liable either to direct mortality or to losses by
other agencies. For many species, the ability to respond quickly to the
return of conditions promoting growth appears to rely heavily upon
the survival of small stocks of vegetative cells and the capacity of these
refuge populations to generate rapid rates of increase. Thus, the length
of the period of unfavorable conditions may prove critical to the
potential for a positive response and, hence, to whether given species
make any substantial contribution to the subsequent composition of
the community.

Alternatively, many species respond to the onset of adverse conditions by producing distinctive "resting" propagules. These include the
cysts of dinoflagellates (e.g., of *Ceratium* spp.; see Chapman, Dodge,
& Heaney 1982), the akinetes produced by many species of Nostocales
(*Anabaena, Aphanizomenon, Gloeotrichia;* Roelofs & Oglesby 1970;
Wildman, Loescher, & Winger 1975; Rother & Fay 1977; Cmiech,
Reynolds, & Leedale 1984), the sexually formed zygotes of larger Volvocales (*Eudorina:* Reynolds et al. 1982), the distinctive cysts of planktonic Chrysophyceae (Sandgren, Chapter 2 this volume), the densely
contracted protoplasts of the centric diatoms *Melosira* and *Stephanodiscus* (Lund 1954; Reynolds 1973), and the physiologically distinct
overwintering colonies of *Microcystis* (Reynolds et al. 1981; Cáceres &
Reynolds 1984). Although many of these structures are characterized
by thick external walls and abundant storage condensates and some
have been shown to retain their vitality for very long periods of time
(see, for instance, Livingstone & Jaworski 1980), the use of terms like
spores and *resting cysts* may give an exaggerated view of their long-
term survival value for which maintenance consumption of stored
reserves must be accommodated. The primary function of these propagules may well be directed toward the short-term preservation of a
substantial proportion of the assembled biomass and to removing it as
rapidly as possible to the bed of the water body. Akinete production in

Anabaena, for example, is usually initiated in response to metabolic imbalances occurring during surface-bloom formation (Rother & Fay 1979; Cmiech et al. 1984) and is thus analogous to settling-rate acceleration among planktonic diatoms (see earlier sections). Moreover, substantial germination can follow shortly afterward rather than the bulk of the akinete population persisting as long-term resting propagules (Rother & Fay 1977).

Mass-germination of propagules provides a second adaptive advantage in restoring a potentially large inoculum of biomass to the water column when favorable growth conditions return. In terms of population dynamics, this ability represents an alternative strategy to rapid growth of a refuge population for reestablishing the vegetative phase, the size of the initial inoculum compensating for a relatively slow rate of growth. Reynolds (1984a) pointed out that an apparent inverse proportionality exists between the relative production of propagules by given planktonic species and the typical rates of growth they were able to sustain in their natural environments. The same data are reworked and plotted in Fig. 10-7 against the maximal specific growth rates attained under field conditions; the axes are arranged in the standard triangular layout. This format emphasizes that the production of discrete, regenerative propagules is largely confined to the more stress-tolerant *S*-species and to certain *R*-species, and, thus, conceivably represents a further facet of the differential survival strategies among the freshwater phytoplankton.

STRATEGIES IN ACTION : PATTERNS OF PHYTOPLANKTON DISTRIBUTION

Moving toward the final stages of this synthesis, let us now consider how the various strategies adopted by the planktonic algae are manifest at the community level, with particular reference to the aspects that influence the composition of the plankton – species representation, growth, dominance, and coexistence – and the ways in which changes in composition occur through time. There is a certain circularity of the argument here that is unavoidable: having first identified strategic traits among phytoplankton species from their distributions in space and time, it is only to be expected that there will then be a close correlation among the distribution of species ascribed to one or another of the strategic categories. On the other hand, it is pertinent to relate the morphological and physiological features hypothesized to characterize *C*-, *S*-, and *R*-strategists to the broad environmental variables that selectively differentiate among them.

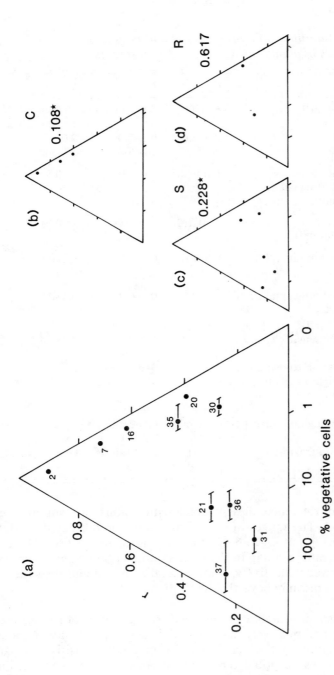

Fig. 10-7. Regenerative strategies in some phytoplankton: incidences of the fraction of vegetative populations forming resting propagules plotted against the maximum in situ increase rates. Replotted in triangular format from Fig. 76 of Reynolds (1984a), with additional unpublished data for *Melosira italica* in Blelham Enclosure C during 1983.

Table 10-2. *Summary of morphometric and physiological characteristics of planktonic C-, R-, and S-strategists.*

	C	R	S
Unit volume [μm^3]	5–5000	500–100,000	10,000–10,000,000
SA/V [μm^{-1}]	0.3–3.0	0.3–2.0	0.03–0.3
MLD [μm]	3–80	10–300	30–500
Protosynthetic efficiency α[mmoles C·(mg chl a)· m^{-1}(mol photon)$^{-1}$·m^2]	Up to 1.5	0.4 1.0	< 0.5
Projected area k_c[m^2(mg chl a)$^{-1}$]	< 0.012	> 0.011	< 0.013
Maximum growth rate (20°C) r'_{max} [d^{-1}]	0.8–1.8	0.8–1.8	0.2–0.9
Cellular phosphorus uptake rate relative to r'_{max} [$\mu M \times 10^{-9}$ cell·$^{-1}$·d^{-1}]	0.2–0.5	0.1–0.7	0.4–0.9
Temperature sensitivity of r', Q_{10}	2.2–3.2	1.9–3.4	2.4–4.4
Threshold dose of exposure to saturating light intensity [$h \cdot d^{-1}$]	3–8	3–6	> 5
Motility	Variable	Mostly −	+
Minimum sinking rate [$m \cdot d^1$]	Generally ≪ 0.6	Generally 0.2 −1	0
Susceptibility to grazing, ψ	→1	Generally < 0.6	Generally < 0.3

For ease of reference, these characteristic features are summarized in Table 10-2. The ranges of empirical evaluation are based on data presented in Figs. 10-2–10-6 or otherwise noted in the text. There is often considerable overlap in the ranges quoted against any given feature, but, viewed collectively, some generalizations about the suites of adaptive features may nevertheless be discerned.

C-strategists are small, distinguished by high ratios of surface area to volume and relatively high metabolic activity over a wide range of temperatures but which is usually markedly sensitive to light dose; they are susceptible to removal by grazers but less so to settling through stable layers.

S-strategists are large, with low SA/V ratios, a relatively low metabolic activity that is more sensitive to temperature limitation than to reductions in light dose; low rates of growth in situ are compensated by enhanced resistance to sinking and grazing losses, high nutrient-storage capacity, and the potential ability to augment growing populations with recruitment from stocks of resting propagules.

R-strategists are of intermediate to large size but with morphologies that preserve a high SA/V ratio, high metabolic activity, and potentially rapid growth rates. Most can be maintained at reduced temperatures and daily light doses. They generally depend on high coefficients of turbulent mixing to offset losses to passive sinking but (mostly) exceed the size ranges of particles readily ingested by filter-feeding zooplankton.

On the assumption that planktonic algae will grow wherever and whenever they have the opportunity to do so (Reynolds 1984b), it may be presumed that the evolutionary strategies that they have adopted will not necessarily confine them to given ranges of environmental variability or to those that will simultaneously exclude species that have adopted alternative strategies. Rather, there are circumstances in which the growth and recruitment of particular strategists will be preferentially selected, and, in consequence, these species are more likely to build up the largest fraction of the sustainable biomass (that is, to become dominant) so long as the same environmental circumstances persist. This may be achieved through the ability to maintain relatively faster rates of net increases at such times than their supposed competitors or to decrease more slowly or, where appropriate, to recruit a larger inoculum of propagules at the commencement of the growth phase.

Some of the environmental variables that may be supposed to differentiate selectively among the primary strategies of phytoplankton are listed in Table 10-3, together with symbolic representations of the nature (either positive or negative) and extent (one or two symbols) of the responses of C-, S-, and R-strategists. As indicated, the most favorable environments, characterized by high clarity, long daily photoperiods, and an abundant supply of inorganic nutrients (including carbon) are the province of C-strategists. If the water is also warm ($\geq 15°$) and well mixed ($z_m > 3$ m but substantially $< z_{eu}$), good growth of both the S- and R-strategists will also be supported but, at least initially, at inferior rates to C-species, which remain most likely to dominate. Conversely, if the water is cold and nonturbulent (which conditions may occur under snow-free ice cover on small, high-latitude lakes), C-strategist species are the most likely to respond positively.

Table 10-3. *Environmental factors likely to discriminate positively*
(+) or negatively (−) among planktonic C-, S-, and R-strategists.

	C	S	R
Physical factors			
Low temperature ($\theta < 5°C$)	+	−−	+
Stability ($N^2 > 500 \times 10^{-6}$, s^{-2})	++	++	−−
Shallow absolute mixed depth ($z_m < 3$ m)	++	++	−
High turbidity of mixed layer ($z_m/z_{eu} > 1.5$)	−−	−	+
Chemical factors			
Low nutrient concentration ([P] < 0.3 μg· L^{-1})	−−	+	−
Biotic factors			
CFR > 0.6 d^{-1}	−	++	+

Descriptions of the late spring and summer phytoplankton developing
under ice cover in small lakes of the Arctic and Antarctic, largely dominated by small motile chrysophytes, chlorophytes, and dinoflagellates,
generally conform to this view (e.g., Kalff et al. 1975; Moore 1979;
Light, Ellis-Evans, & Priddle 1981).

The capacity of allegedly favorable environments to support rapid
phytoplankton growth is often altered through time as a consequence
of that growth. The euphotic depth is reduced, the concentrations of
dissolved nutrients decline, and the increased availability of potential
foods can promote the growth, reproduction, and, eventually, the
increase in the rate of consumption by the zooplankton. Moreover, set-
tlement of live cells and, especially, of faecal pellets produced by the
planktonic animals feeding upon them contribute to a process of vertical nutrient transport and environmental segregation in the water
column, tending toward the establishment of a discrete, well-illuminated but nutrient-deficient upper layer overlying progressively
deeper, darker but richer layers. Such environments offer fewer micro-
habitats capable of supporting *C*-strategists, the net growth rate of
which becomes depressed at the same time as being subjected to
greater rates of loss of live cells. Selection now moves in favor of species that are resistant to sinking and grazing losses, especially those
that are able to regulate their vertical position to derive the maximum
benefit from the segregated resources. Such traits are well developed
among the stress-tolerant, *S*-strategists, especially *Ceratium* and

Microcystis spp., which characteristically dominate the summer plankton of temperature eutrophic lakes (Reynolds 1984b) and which are able to maintain positive rates of increase through a high degree of vertical self-regulation (Heaney & Talling 1980; Ganf & Oliver 1982).

Alternatively, the environmental capacity to support phytoplankton growth can be altered externally ("allogenically") by the superimposition of intense, wind-driven vertical mixing. Turbulent flow and attendant eddy diffusivity entrains and continuously randomizes the mixed-layer distributions of phytoplankton, nonliving particles, and environmental resources, save light, which continues to sustain photosynthetic production of organisms only while they are transported through the euphotic zone. Motility becomes less advantageous than adaptations toward maximizing entrainment and, when the mixed depth extends well beyond the euphotic limit, toward efficient cellular light absorption. Here the selection directly operates against many *C*-strategists and in favor of more ruderal (*R*) species. Nevertheless, *R*-species that grow in optically deeper water columns (*Oscillatoria, Melosira, Asterionella* spp.) may be distinguished from those that are relatively restricted to clearer, more oligotrophic epilimnia (*Cyclotella, Tabellaria,* and many desmid spp.), whereas low temperatures appear to be more generally tolerated by the high SA/V diatoms (Smol et al. 1984; Reynolds 1984a).

Major floristic differences are recognized between the composition of phytoplankton in eutrophic lakes subject to high nutrient loadings and in oligotrophic waters chronically deficient in one or more major nutrients (Jarnefelt 1952; Rawson 1956; Hutchinson 1967). The factors that select in favor of the distinctive species assemblages indicative of oligotrophic lakes, variously involving certain centric diatoms (notably *Rhizosolenia* and *Cyclotella* spp.), chrysophyte flagellates (*Dinobryon, Mallomonas, Synura, Uroglena* spp.), desmids (e.g., *Staurastrum, Staurodesmus, Micrasterias*), colonial green algae drawn from the Chlorococcales (*Coenococcus, Sphaerocystis*) and Tetrasporales *(Gemellicystis)* and dinoflagellates (*Peridinium, Ceratium* spp.), but against the diatoms *(Fragilaria, Diatoma, Stephanodiscus),* Volvocales *(Eudorina, Volvox),* and cyanobacteria *(Anabaena, Aphanizomenon, Microcystis)* of eutrophic lakes, have been the subject of a number of experimental studies (Moss 1973; Lund 1981; Reynolds 1986). Basin morphometry, hydrological flushing, bicarbonate alkalinity and water clarity are contributing variables, but the role of limiting nutrients (especially of phosphorus) has been usually judged to be crucial. Oligotrophic species, together with those eutrophic species surviving summer periods of nutrient scarcity, must therefore be efficient in obtaining and assimilating nutrients. Experimental evidence (Sommer

1983, 1984; see also the review of Tilman et al. 1982) is still inconclusive on this point, and it is difficult to relate known half-saturation constants (K_s) to the morphological properties of the species studied.

It is nevertheless apparent that the proposed classification of phytoplankton according to their morphologies does not discriminate between essentially "oligotrophic" and "eutrophic" species; supposedly oligotrophic species are partitioned among the three primary strategies in much the same way as the more eutrophic species, albeit preferentially toward the S and R apices (in Figs. 10-2, 10-3, 10-5–10-7). Nutrient requirements apart, the oligotrophic and eutrophic species seem likely to respond to the environmental variables listed in Table 10-3 in analogous ways. In a previous analysis of the growth-rate responses of phytoplankton maintained for up to a year in limnetic enclosures but subject to contrasted nutrient loadings, Reynolds (1984a) was able to define the ranges of, for example, water temperature, mixed depth, and relative light penetration as well as nutrient concentration within which the particular species had grown. These results are replotted in Figs. 10-8 and 10-9 to illustrate the same responses to interactions among the variables that might select among the organisms on the basis of their strategic adaptations. Each of the triangular plots in Fig. 10-8 traces incidences of net population increase of the given species in terms of the concentration of dissolved reactive phosphorus and the stability (N^2, the Brunt–Väsälä frequency) of the upper 6m of the enclosure then obtaining. Several species are apparently restricted in relation to N^2, but none is confined to a narrow range of [SRP]. Nevertheless, there are differences in the minimal concentrations tolerated (cf. *Eudorina* and *Uroglena*). In Fig. 10-9 the same incidences are plotted against temperature and the ratio between the depths of column mixing and Secchi-disk disappearance (larger values of this ratio being indicative of greater optical depth of the mixed layer).

Viewed in conjunction, these plots reveal some apparent specific preferences and tolerances. In the series of R-strategists, *Staurastrum*, *Fragilaria*, *Asterionella*, and *Oscillatoria*, growth shows an increasing tolerance of mixed, unstable conditions and enhanced optical depth ($z_m z_s \rightarrow 4$); *Staurastrum* growth occurred mainly at intermediate z_m/z_s ratios (≤ 3) and at the higher temperatures ($\geq 8c$) observed. The S-strategists, *Microcystis*, *Anabaena*, and *Ceratium*, grew best in warmer ($> 14°C$), more stable conditions, when $z_m/z_s < 2$. *Sphaerocystis* and *Eudorina* growth tolerated slightly lower temperatures ($> 8°C$) and some mixing, provided the water remained clear ($z_m/z_s < 2$). *Uroglena* also grew above 8° and tolerated more extensive mixing than other S-strategists. *Ankyra* (C-strategist) showed greater preference for warm, well-illuminated layers. The *Cryptomonas* showed the widest toler-

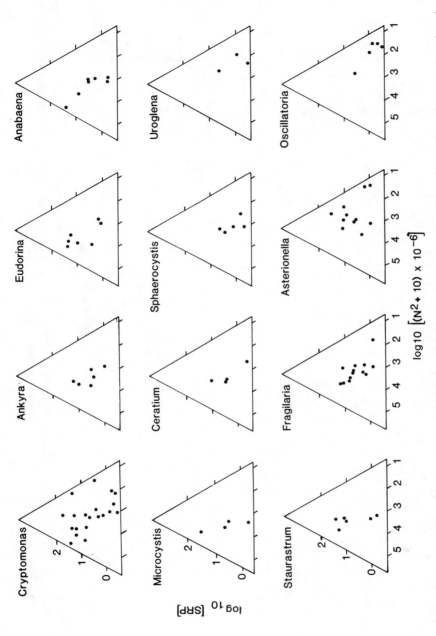

Fig. 10-8. The scatter of points representing incidence of specific increase of the named species in limnetic enclosures plotted against pairs of variables representing phosphorus concentration ($\log_{10}$ [SRP], in $\mu g \cdot L^{-1}$) and water column stability [$\log_{10}(N^2 + 10)$, s^{-2}]. Replotted from the data used in the construction of Figs. 71–75 of Reynolds (1984a).

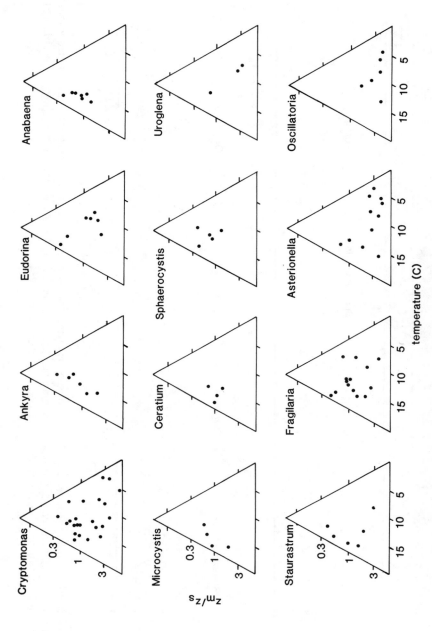

Fig. 10-9. Scatter of points representing incidences of specific increase of the species in limnetic enclosures plotted against pairs of variables representing water column disturbance (the ratio of mixed depth, z_m, to Secchi-disk depth, z_s) and temperature. Replotted from the data used in the construction of Figs. 71–75 of Reynolds (1984a).

ances with respect to fluctuations in temperature, mixed depth, and turbidity.

That these distinctive associations between the phasing of growth of particular species and quantifiable aspects of the physical environment are not merely coincidental but rather contribute to the explanation of discontinuous phytoplankton distribution is further supported by the fact that the plots combine data for different years and in what amount to contrasted limnetic environments. Similar specific growth responses have been observed at other times in the Blelham enclosures, even when the alternations between phases of mixing and restratification have been artificially truncated (see Reynolds et al. 1983b; Reynolds, Wiseman, & Clarke 1984). Comparing analyses of typical and atypical phasing of seasonal growth of phytoplankton in Crose Mere, a small eutrophic lake in the English Midlands that supports a number of species common to the Blelham sequences, Reynolds & Reynolds (1985) found strongly similar between-year responses to the physical variables, even though these were themselves phased differently.

It is an interesting feature of these observations that the responses should have been consistently individualistic and with little evidence of biological interactions among the species, as might be expected had they been in direct competition. It is apparent that for a long time before the species composition of natural communities can come to approach the equilibrium condition predicted from the outcome of resource-based competition, it will continue to be influenced primarily by the relative abundances of the individual species present and their respective abilities to maintain their effective rates of change, whether positive or negative. Moreover, because these aspects are more immediately sensitive to fluctuations in the ambient physical environment, so the progress toward equilibrium is redirected each time the physical conditions are altered. At any intermediate stage, therefore, the community may comprise species simultaneously either waxing or waning in an apparent nonequilibrium state of coexistence. This view amounts to a restatement, in terms of the population dynamics of the component species, of the contemporaneous disequilibrium hypothesis that Richerson, Armstrong, and Goldman (1970) advanced to explain the "paradoxical" (Hutchinson 1967) violation of Hardin's (1960) competitive exclusion principle by planktonic communities.

TEMPORAL CHANGE IN COMMUNITY COMPOSITION

The ranges of many of the growth-determining environmental factors fluctuate seasonally: variations in daylength, insolation, temperature, hydraulic resistance to wind mixing, and rainfall, itself influencing flushing- and nutrient-loading rates, all contribute (in order of decreasing predictability) to seasonal changes in the underwater environments of oligotrophic and eutrophic lakes. It is the extent of this seasonal variability, expressed through the medium of population dynamics, that largely drives the conspicuous periodic variations in the abundance, species composition, and dominance of the phytoplankton. Individual species should be selected on the basis of their strategic adaptations such that, at any given time, either *C*-, *S*-, or *R*-species would be selectively favored to attain dominance. Hydraulic stability operates in favor of *C*- and *S*-species and against most *R*-species, but low temperatures select against *S*-strategists; deep mixing and grazing work against *C*-species but in favor of *R*-species (Table 10-3). Thus, although the seasonal progressions of species dominance vary between lakes and between years, it is theoretically possible to construct a preliminary qualitative model that describes the directions of change observed in actual lakes. In Fig. 10-10, Grime's (1979) triangular *C-S-R* matrix (Fig. 10-1c) is superimposed upon Margalef's (1978) rectangular turbulence-versus-nutrients representation of (marine) aquatic environments in order to locate the areas within which *C*-, *S*-, or *R*-strategists might be preferentially selected. Temporal progressions in community dominance can then be tracked in relation to environmental variability. Some examples are inserted to demonstrate the principle.

Generalized annual cycles in temperate, stratifying oligotrophic and eutrophic lakes are respectively represented by the tracks labeled (a) and (b) in Fig. 10-10. These trace the effects of vernal growth on relative light penetration in the well-mixed water column, the depletion of nutrients, the onset of summer stratification and of the restoration of full mixing and nutrient concentrations in winter. At appropriate stages in the oligotrophic cycle (a), species that are either more disturbance tolerant (*R*-strategists) or more stress tolerant (*S*-strategists) are selected. The sequence adequately describes a range of annual cycles of phytoplankton dominance, from those in Carinthian subalpine lakes (Findenegg 1943), where *Cyclotella–Rhizosolenia* diatom associations are replaced later in the year by dinoflagellate-dominated assemblages that become increasingly influenced by desmids, to those of a more mesotrophic status such as Windermere, U.K. (see Reynolds 1980), where vernal diatom dominance (*Asterionella-Melosira* spp.) is sequentially replaced by *Sphaerocystis-Gemellicystis*/chrysophyte associations, then by *Ceratium-Gomphosphaeria* and with increased late-summer mixing, by a diatom-desmid *(Tabellaria-Fragilaria-Staurastrum)* assemblage.

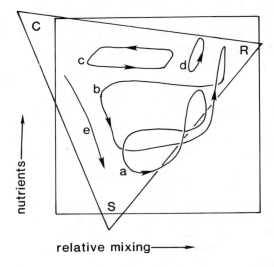

Fig. 10-10. The *C-S-R* triangle superimposed upon Margalef's (1978) representation of the aquatic environment, showing tracks of hypothetical, temporal variations in the relative magnitude of stress and disturbance factors and the species groups likely to be selected. (a) and (b) represent seasonal sequences in oligotrophic and eutrophic lakes respectively; (c) and (d) are more constant environments; (e) shows the general course of autogenic succession in undisturbed environments.

The cycles in eutrophic lakes (b in Fig. 10-10) are essentially similar, but the greater concentrations of nutrients available permit more opportunities for *C*-strategists to thrive. This track represents the seasonal cycles of the phytoplankton in many small eutrophic lakes, exemplified by Sjon Erken, Sweden (Nauwerck 1963), Astotin Lake, Canada (Lin 1972), Sagami-Ko, Japan (Saito 1978), and Crose Mere, U.K. (Reynolds 1980). These feature sequential phases of dominance by diatoms (especially of *Asterionella* and *Stephanodiscus*), cryptomonads, Volvocales (e.g., *Eudorina, Pandorina*) or chrysophytes (Nostocales, *Microcystis,* and/or *Ceratium*) before increased mixing again favors diatoms.

The cycles do not have to be tracked at an even rate, nor are they necessarily completed on an annual basis. Among tropical lakes where the stability of stratification is governed as much by seasonal fluctuations in weather (wind, rainfall) as by changes in daylength and insolation, similar cyclical tracks might be inserted in Fig. 10-10 but are traversed in periods markedly less than one year. Phytoplankton growth and dominance nevertheless continue to track the environmental fluctuations. Lewis's (1978) detailed description of the seasonal changes in the plankton of Lake Lanao, Philippines, could be reasonably represented by the (b) track in Fig. 10-10, but it would be completed within four to six months. Even in temperate lakes and reservoirs subject to extreme fluctuations in mixed depth on time scales of 5–50 days, alternation between phases of increase and dominance by R-species (e.g., *Stephanodiscus, Synedra* spp.) and by C-S groupings (e.g., cryptomonads, *Chlamydomonas, Oocystis,* Volvocales, and *Aphanizomenon*) are clearly distinguishable (Haffner, Harris, & Jarai 1980; Ferguson & Harper 1982). Again, in nutrient-rich lakes where the alternations result in the lake being either more constantly mixed or stably stratified, so the dominant species would be (respectively) more frequently R-species (as in Embalse Rapel, Chile; Cabrera et al. 1977) or C-species (as in Montezuma Well, U.S.A.; Boucher, Blinn, & Johnson 1984). These possibilities are represented in Fig. 10-10 by the track labeled (c).

Examples of lakes that are more or less constantly mixed seem to be generally dominated by R-species (diatoms and/or *Oscillatoria* spp.; see Gibson et al. 1971; Berger 1984), as shown by track (d). Examples of seasonal sequences in lakes that remain stably stratified for months on end are less easy to find, although evidence to be gleaned from descriptions of the vertical distribution of plankton communities during summer episodes in sheltered continental lakes of the types encountered in central North America (e.g., Eberley 1959, 1964; Brook, Baker, & Klemer 1971) and in the rain forest of tropical Brazil (Reynolds, Tundisi, & Hino 1983a), points to the selective role of vertical environmental segregation in driving autogenic successional sequences that might be represented by the track labeled (e) in Fig. 10-10.

Though far from exhaustive, these spatial representations attest to the consistency with which the growth and dominance of particular groups of organisms can be associated with particular interactions between stress and disturbance factors as they have been applied to freshwaters. Moreover, reversible changes in the relative representation of C-, S-, and R-strategists are stimulated by significant shifts in the locus of factor interaction that occur over a range of temporal scales. The strength of the deduction, however, lies in the fact that the distinctive responses of the C-, S-, and R-species can also be related to morphological and physiological properties of the organisms themselves (Table 10-2). The hypothesis that the diversity of form among freshwater phytoplankton is the product of the alternative adaptive strategies that they have evolved to contend with different dimensions of environmental variability and that, accordingly, those adaptations will be selectively advantageous in given habitats where and when the appropriate conditions obtain is not disproved. Though general in its assessments of species groupings and of limnetic environments, the model (Fig. 10-10) could be usefully developed to give more precise definition to the ecological ranges of given organisms and to the probable composition of phytoplankton communities in given freshwater systems.

CONCLUSIONS

The purpose of this article has been to distinguish among the biological properties of different species of phytoplankton, categories of features that might be ascribable to Grime's (1979) concept of the primary adaptive strategies of living organisms. In spite of a dearth of suitable information about the biological properties of individual species, the argument has been developed that there is substantial evidence of coherence among the axes variously describing the ranges of size, relative surface area, light interception properties, photosynthetic efficiencies, nutrient uptake kinetics, rate of growth, sensitivity to light and temperature limitation, susceptibility to sinking and grazing, and the efficiency of propagule production. This coherence enables the assembly of suites of adaptive features, which are preferentially selected in aquatic environments that are variously deficient in the supply of certain nutrients, in daily insolation levels (abetted by turbulent circulation beyond the euphotic limits), or in environments that are more nearly saturated with respect to the immediate requirements for growth. The successful species in such environments are, respectively, the stress-tolerant S-strategists, the ruderal R-strategists, and the fastest-growing C-strategists. Low temperature selects against most S-species, shallow absolute mixed depth against most R-species, and grazing, deep mixing, and nutrient stress against many C-species. Within the primary strategies, some species may be distinguished as being (say) stress-tolerant ruderals (SR-species: e.g., *Oscillatoria*) or "competitive" stress tolerators (CS-species: e.g., *Eudorina*). Environmental variability distinguishes among the responses of individual strategists to influence conspicuous spatial and temporal variations in phytoplankton community composition, generally selecting for the appropriate primary strategies.

These conclusions are not only wholly consistent with Grime's (1979) appreciation of terrestrial plant strategies and the processes governing vegetation development, thereby consolidating the general concept of evolutionary strategies, but they would indicate that the fresh-water phytoplankton provides a strong case in point. Moreover, the short (days to months) temporal scales of environmental variation to which the dynamics of individual species and the structure of plank-tonic communities each respond are amenable to more exhaustive eco-logical study and experimentation than are those of terrestrial ecosys-tems. In time, the differential adaptations and autecologies of the freshwater phytoplankton may come to afford the clearest insights into the operation of evolutionary strategies.

REFERENCES

Allen, T. F. H., Bartell, S. M., and Koonce, J. F. (1977). Multiple stable con-figurations in ordination of phytoplankton community change rates. *Ecology,* 58, 1076–84.

Allen, T. F. H. and Koonce, J. F. (1973). Multivariate approaches to algal stra-tegems and tactics in systems analysis of photoplankton. *Ecology,* 54, 1234–47.

Bannister, T. T. (1974). Production equations in terms of chlorophyll concen-tration, quantum yield and upper limit to production. *Limnol. Ocean-ogr.,* 19, 1–12.

Bannister, T. T. and Weideman, A. D. (1984). The maximum quantum yield of phytoplankton photosynthesis *in situ. J. Plankton Res.,* 6, 275–94.

Banse, K. (1976). Rates of growth, respiration and photosynthesis of unicel-lular algae as related to cell size – a review. *J. Phycol.,* 12, 135–40.

Berger, C. (1984). Consistent blooming of *Oscillatoria agardhii* Gom. in shal-low hypertrophic lakes. *Verhandlungen der internationale Vereinigung für theoretische unde angewandte Limnologie,* 22, 910–16.

Bindloss, M. E. (1974). Primary productivity of phytoplankton in Loch Level, Kinross. *Proc. Roy. Soc. Edinburgh B,* 74, 157–81.

Boucher, P., Blinn, D. W., and Johnson, D. B. (1984). Phytoplankton ecology in an unusually stable environment (Montezuma Well, Arizona, U.S.A.). *Hydrobiologia,* 199, 149–60.

Brook, A. J., Baker, A. L., and Klemer, A. R. (1971). The use of turbidimetry in studies of the population dynamics of phytoplankton populations, with special reference to *Oscillatoria agardhii* var. *isothrix. Mitteilungen der internationale Vereinigung für theoretische und angewandte Limnol-ogie,* 19, 244–52.

Burmaster, D. (1979). The continuous culture of phytoplankton: mathematical equivalence among three steady-state models. *Am. Natur.,* 113, 123–34.

Burns, C. W. (1968). Direct observations of mechanisms regulating feeding behaviour of *Daphnia* in lakewater. *Internationale Revue der gesamten Hydrobiologie und Hydrographie,* 53, 83–100.

Cabrera, S., Montecino, V., Vila, I., Bahamonde, N., Bahamondes, I., Barends,

I., Rodriguez, R., Ruiz, R., and Soto, D. (1977). Caracteristicas limnologicas del Embalse Rapel, Chile Central. *Seminario sobre medio ambiente y represas,* 1, 40–61.

Cáceres, O. and Reynolds, C. S. (1984). Some effects of artificially-enhanced anoxia on the growth of *Microcystis aeruginosa* Kütz. emend. Elenkin, with special reference to the initiation of its annual growth cycle in lakes. *Archiv für Hydrobiologie,* 99, 379–97.

Canelli, E., and Fuhs, G. H. (1976). Effect of sinking rate on two diatoms (*Thalassiosira* spp.) on uptake from low concentrations of phosphate. *J. Phycol.,* 12, 93–9.

Cavalier-Smith, T. (1982). Skeletal DNA and the evolution of genome size. *Ann. Rev. Biophys. Bioeng.,* 11, 273–302.

Chapman, D. V., Dodge, J. D., and Heaney, S. I. (1982). Cyst formation in the freshwater dinoflagellate *Ceratium hirundinella* (Dinophyceae). *J. Phycol.,* 18, 121–9.

Cloern, J. E. (1977). Effects of light-intensity and temperature on *Cryptomonas ovata* (Cryptophyceae) growth and nutrient uptake. *J. Phycol.,* 13, 389–95.

Cmiech, H. A., Reynolds, C. S., and Leedale, G. F. (1984). Seasonal periodicity, heterocyst differentiation and sporulation of planktonic Cyanophyceae in a shallow lake, with special reference to *Anabaena solitaria. Br. Phycol. J.,* 19, 245–57.

Crowley, P. H. (1975). Natural selection and the Michaelis constant. *J. Theoret. Biol.,* 50, 461–75.

Crumpton, W. G. and Wetzel, R. G. (1982). Effects of differential growth and mortality in the seasonal succession of phytoplankton populations in Lawrence Lake, Michigan. *Ecology,* 63, 1729–39.

Droop, M. R. (1974). The nutrient status of algal cells grown in continuous culture. *J. Mar. Biol. Assoc. U.K.,* 54, 825–55.

(1983). 25 years of algal growth kinetics. A personal view. *Bot. Mar.,* 26, 99–112.

Dubinsky, Z. and Berman, T. (1976). Light utilization efficiencies of phytoplankton in Lake Kinneret (Sea of Galilee). *Limnol. and Oceanogr.,* 21, 226–30.

(1979). Seasonal changes in the spectral composition of downwelling irradiance in Lake Kinneret (Israel). *Limnol. and Oceanogr.,* 24, 652–63.

Eberley, W. R. (1959). The metalimnetic oxygen maximum in Myers Lake. *Investigations of Indiana Lakes and Streams,* 5, 1–46.

(1964). Primary production in the metalimnion of McLish Lake (Northern Indiana), an extreme plus-heterograde lake, *Verhandlungen der Internationale Vereinigung für theoretische und angewandte Limnologie,* 15, 394–401.

Eppley, R. W., Rogers, J. N., and McCarthy, J. J. (1969). Half-saturation constants for uptake of nitrate and ammonium by marine phytoplankton. *Limnol. and Oceanogr.,* 14, 912–20.

Falkowski, P. G. (1980). Light-shade adaptation in marine phytoplankton. In:

Primary Productivity in the Sea, ed. P. G. Falkowski, pp. 99–119. New York: Plenum.

(1984). Physiological responses of phytoplankton to natural light regimes. *J. Plankton Res.,* 6, 295–307.

Falkowski, P. G. and Owens, T. G. (1980). Light-shade adaptation. Two strategies in marine phytoplankton. *Plant Physiology, Lancaster,* 66, 592–5.

Farmer, D. M. and Takahashi, M. (1982). Effects of vertical mixing on photosynthetic responses. *Japan. J. Limnol.,* 43, 173–81.

Ferguson, A. J. D. and Harper, D. M. (1982). Rutland water phytoplankton: the development of an asset or a nuisance? *Hydrobiologia,* 88, 117–33.

Findenegg, I. (1943). Untersuchungen über die Ökologie und die Produktionsverhältnisse des Planktons in Karnter Seengebeite. *Internationale Revue der gesamten Hydrobiologie und Hydrographie,* 43, 368–429.

(1947). Über die Lichtanspruche planktischer Susswasseralgen. *Sitzungsberichte der Akademie der Wissenschaften in Wien,* 155, 159–71.

(1966). Factors controlling primary productivity especially with regard to water replenishment, stratification and mixing. *Mem. Ist. Ital. Idrobiologia,* 18(Suppl.), 105–19.

Foy, R. H., Gibson, C. E., and Smith, R. E. (1976). The influence of day length, light intensity and temperature on the growth rates of planktonic bluegreen algae, *Br. Phycol. J.,* 11, 151–63.

Ganf, G. G. and Oliver, R. L. (1982). Vertical separation of light and available nutrients as a factor causing replacement of green algae by blue-green algae in the plankton of a stratified lake. *J. Ecol.,* 70, 829–44.

Gibson, C. E., Wood, R. B., Dickson, E. L., and Jewson, D. H. (1971). The succession of phytoplankton in L. Neagh, 1968–1970. *Mitteilungen der internationale Vereinigung für theoretische und angewandte Limnologie,* 19, 146–60.

Gliwicz, Z. M. (1980). Filtering rates, food size selection and feed rates in cladocerans – another aspect of interspecific competition in filter-feeding zooplankton. In: *Evolution and Ecology of Zooplankton Communities,* ed. W. C. Kerfoot, pp. 282–91. Hanover, NH: University of New England.

Goldman, J. C. (1977). Steady-state growth of phytoplankton in continuous culture: comparison of internal and external nutrient concentrations. *J. Phycol.,* 13, 251–8.

Grime, J. P. (1979). *Plant Strategies and Vegetation Processes.* New York: Wiley.

Haffner, G. D., Harris, G. P., and Jarai, M. K. (1980). Physical variability and phytoplankton communities. III. Vertical structure in phytoplankton populations. *Archiv für Hydrobiologie,* 89, 363–81.

Hardin, G. (1960). The competitive exclusion principle. *Science,* 131, 1292–7.

Harris, G. P. (1978). Photosynthesis, productivity and growth: the physiological ecology of phytoplankton. *Ergebnisse der Limnologie,* 10, 1–171.

(1980). Temporal and spatial cycles in phytoplankton ecology. Mechanisms, methods, models and management. *Can. J. Fish. Aquat. Sci., 37,* 877–900.

(1983). Mixed-layer physics and phytoplankton populations: studies in equilibrium and non-equilibrium ecology. *Prog. Phycol. Res.,* 2, 1–52.

Harris, G. P., Heaney, S. I., and Talling, J. F. (1979). Physiological and environmental constraints in the ecology of the planktonic dinoflagellate *Ceratium hirundinella. Freshwat. Biol.,* 9, 413–28.

Harris, G. P., Piccinin, B. B., and van Ryn, J. (1983). Physical variability and phytoplankton communities. V. Cell size, niche diversification and the role of competition. *Archiv Hydrobiol.* 98, 215–39.

Heaney, S. I., and Furnass, T. I. (1980). Laboratory models of diel vertical migration in the dinoflagellate *Ceratium hirundinella. Freshwat. Biol.* 10, 163–70.

Heaney, S. I. and Talling, J. F. (1980). Dynamic aspects of dinoflagellate distribution patterns in a small productive lake. *J. Ecol.,* 68, 75–94.

Holm, N. P. and Armstrong, D. E. (1981) Role of nutrient limitation and competition in controlling the populations of *Asterionella formosa* and *Microcystis aeruginosa* in semicontinuous culture. *Limnol. Oceanogr.,* 26, 622–34.

Hutchinson, G. E. (1967). *A Treatise on Limnology,* Vol. IV. New York: Wiley-Interscience.

Ivanovici, A. M. and Wiebe, W. J. (1981). Towards a working "definition" of "stress": a review and critique. In: *Stress Effects on Natural Ecosystems,* eds G. W. Barrett and R. Rosenbury, pp. 13–27. New York: Wiley.

Jarnefelt, H. (1952). Plankton als Indikator der Trophiegruppen der Seen. *Ann. Acad. Sci. Fenn. Ser. A.,* 18, 1–29.

Jeanjean, R. (1969). Influence de la carence en phosphor sur les vitesses d'absorption du phosphate par les Chlorelles. *Bull. Soc. Française Physiol. Vegetale, 15,* 159–71.

Jewson, D. H. (1976). The interaction of components controlling net phytoplankton photosynthesis in a well-mixed lake (Lough Neagh, Northern Ireland). *Freshwat. Biol.,* 6, 551–76.

Kalff, J., Kling, H. J., Holmgren, S. H., and Welch, H. E. (1975). Phytoplankton growth and biomass cycles in an unpolluted and in a polluted polar lake. *Verh. Int. Verein. Limnol.* 19, 487–95.

Kalff, J. and Knoechel, R. (1978). Phytoplankton and their dynamics in oligotrophic and eutrophic lakes. *Ann. Rev. Ecol. Syst.,* 9, 475–95.

Kilham, P. and Kilham, S. S. (1980). The evolutionary ecology of phytoplankton. In: *The Physiological Ecology of Phytoplankton,* ed. I. Morris, pp. 571–97. Oxford: Blackwell.

Kilham, S. S., Kott, C. L., and Tilman, D. (1977). Phosphate and silicate kinetics for the Lake Michigan diatom *Diatoma elongatum. J. Great Lakes Res.,* 3, 93–9.

Kirk, J. T. O. (1983). *Light and Photosynthesis in Aquatic Ecosystems.* Cambridge: Cambridge University Press.

Konopka, A. E. and Brock, T. D. (1978). Effect of temperature on blue-green algae (Cyanobacteria) in Lake Mendota. *Appl. Environ. Microbiol.,* 36, 572–6.

Laws, E. A. (1975). The importance of respiration losses in controlling the size distribution of marine phytoplankton. *Ecology,* 56, 419–26.

Lehman, J. T. (1976a). Ecological and nutritional studies on *Dinobryon* Ehrenb.: seasonal periodicity and the phosphate toxicity problem. *Limnol. Oceanogr.,* 21, 646–58.

(1976b). The filter-feeder as an optimal forager, and the predicted shapes of feeding curves. *Limnol. Oceanogr.,* 21, 501–16.

Lehman, J. T., Botkin, D. B., and Likens, G. E. (1975). The assumptions and rationales of a computer model of phytoplankton population dynamics. *Limnol. Oceanogr.,* 20, 343–64.

Lehman, J. T. and Sandgren, C. D. (1985). Species-specific rates of growth and grazing loss among freshwater algae. *Limnol. Oceanogr.,* 30, 34–46.

Lewis, W. M. (1976). Surface/volume ratio: implications for phytoplankton morphology. *Science,* 192, 885–7.

(1977). Net growth rate through time as an indicator of ecological similarity among phytoplankton species. *Ecology,* 58, 149–57.

(1978). Dynamics and succession of the phytoplankton in a tropical lake: Lake Lanao, Philippines. *J. Ecol.,* 66, 849–80.

(1983). A revised classification of lakes based on mixing. *Can. J. Fish. Aquat. Sci.,* 40, 1779–87.

Light, J. J., Ellis-Evans, J. C., and Priddle, J. (1981). Phytoplankton ecology in an Antarctic lake. *Freshwat. Biol.,* 11, 11–26.

Lin, C. K. (1972). Phytoplankton succession in a eutrophic lake with special reference to blue-green algal blooms. *Hydrobiologia,* 39, 321–34.

Livingstone, D. and Jaworski, G. H. M. (1980). The viability of akinetes of blue-green algae recovered from the sediments of Rostherne Mere. *Br. Phycol. J.,* 15, 357–64.

Lund, J. W. G. (1949). Studies on *Asterionella.* I. The origin and nature of the cells producing seasonal maxima. *J. Ecol.,* 37, 389–419.

(1954). The seasonal cycle of the plankton diatom *Melosira italica* subsp. *subarctica* O. Mull. *J. Ecol.,* 42, 151–79.

(1965). The ecology of the freshwater phytoplankton. *Biol. Rev. Cambridge Phil. Soc.,* 40, 231–93.

(1981). Investigations on phytoplankton with special reference to water usage. *Occasional Publ. Freshwat. Biol. Assoc.,* No. 13.

MacArthur, R. H. and Wilson, E. O. (1967). *The Theory of Island Biogeography.* Princeton: Princeton University Press.

Malone, T. C. (1980). Algal size. In: *The Physiological Ecology of Phytoplankton,* ed. I. Morris, pp. 433- 63. Oxford: Blackwell.

Margalef, R. (1978). Life-forms of phytoplankton as survival alternatives in an unstable environment. *Oceanologica Acta,* 1, 493–509.

Moore, J. W. (1979). Factors influencing the diversity, species composition and abundance of phytoplankton in twenty-one Arctic and sub-Arctic lakes. *Int. Revue ges. Hydrobiol.,* 64, 485–9.

Morgan, K. C., and Kalff, J. (1979). Effect of light and temperature interactions on the growth of *Cryptomonas erosa* (Cryptophyceae). *J. Phycol.,* 15, 127–34.

Moss, B. (1973). The influence of environmental factors on the distribution of freshwater algae: an experimental study. IV. Growth of test species in natural lake water, and conclusion. *J. Ecol.*, 61, 193–211.

Munk, W. H. and Riley, G. A. (1952). Absorption of nutrients by aquatic plants. *J. Mar. Res.*, 11, 215–40.

Nalewajko, C., and Lean, D. R. S. (1978). Phosphorus kinetics – algal growth relationships in batch cultures. *Mitt. Int. Verein. Limnol.*, 21, 184–92.

Nauwerck, A. (1963). Die Beziehungen zwischen Zooplankton und Phytoplankton in See Erken. *Symb. Bot. Upsal.*, 17, 1–163.

Nyholm, N. (1977). Kinetics of phosphate limited algal growth. *Biotechnology and Bioengineering*, 19, 467–92.

Oliver, R. L., Kinnear, A. J., and Ganf, G. G. (1981). Measurement of cell density of three freshwater phytoplankters by density gradient centrifugation. *Limnol. Oceanogr.*, 26, 285–94.

Pollingher, U. and Berman, T. (1975). Temporal and spatial patterns of dinoflagellate blooms in Lake Kinneret, Israel (1969–1974). *Verh. Int. Verein. Limnol.*, 19, 1370–82.

Porter, K. G. (1977). The plant-animal interface in freshwater ecosystems. *Am. Sci.*, 65, 159–70.

Raven, J. A. (1982). The energetics of freshwater algae: energy requirements for biosynthesis and volume regulation. *New Phytologist*, 92, 1–20.

(1984). A cost-benefit analysis of photon absorption by photosynthetic unicells. *New Phytologist*, 98, 593–625.

Raven, J. A. and Richardson, K. (1984). Dinophyte flagella: a cost-benefit analysis. *New Phytologist*, 98, 259–76.

Rawson, D. S. (1956). Algal indicators of trophic lake types. *Limnol. Oceanogr.*, 1, 18–25.

Reynolds, C. S. (1973). The seasonal periodicity of planktonic diatoms in a shallow eutrophic lake. *Freshwat. Biol.*, 3, 89–110.

(1980). Phytoplankton assemblages and their periodicity in stratifying lake systems. *Holarctic Ecology*, 3, 141–59.

(1983). A physiological interpretation of the dynamic responses of a population of a planktonic diatom to physical variability of the environment. *New Phytologist*, 95, 41–53.

(1984a) *The Ecology of Freshwater Phytoplankton*. Cambridge: Cambridge University Press.

(1984b). Phytoplankton periodicity: the interactions of form, function and environmental variability. *Freshwat. Biol.*, 14, 111–42.

(1986). Experimental manipulations of the phytoplankton periodicity in large, limnetic enclosures in Blelham Tarn, English Lake District. *Hydrobiologia*, 138, 43–64.

Reynolds, C. S., Harris, G. P., and Gouldney, D. N. (1985). Comparison of carbon-specific growth rates and rates of cellular increase of phytoplankton in large limnetic enclosures. *J. Plank. Res.*, 7, 791–820.

Reynolds, C. S., Jaworski, G. H. M., Cmiech, H. A., and Leedale, G. F. (1981). On the annual cycle of the blue-green alga *Microcystis aeruginosa* Kutz. emend. Elenkin. *Phil. Trans. Roy. Soc. London B*, 293, 419–77.

Reynolds, C. S. and Reynolds, J. B. (1985). The atypical seasonality of phytoplankton in Crose Mere, 1972: an independent test of the hypothesis that variability in the physical environment regulates community dynamics and structure. *Br. Phycol. J.*, 20, 227–42.

Reynolds, C. S., Thompson, J. M., Ferguson, A. J. D., and Wiseman, S. W. (1982). Loss processes in the population dynamics of phytoplankton maintained in closed systems. *J. Plank. Res.*, 4, 561–600.

Reynolds, C. S., Tundisi, J. G., and Hino, K. (1983a). Observations on a metalimnetic *Lyngbya* population in a stably stratified tropical lake (Lagoa Carioca, Eastern Brasil). *Arch. Hydrobiol.*, 97, 7–17.

Reynolds, C. S. and Wiseman, S. W. (1982). Sinking losses of phytoplankton maintained in closed limnetic system. *J. Plank. Res.*, 4, 489–522.

Reynolds, C. S., Wiseman, S. W., and Clarke, M. J. O. (1984). Growth- and loss-rate responses of phytoplankton to intermittent artificial mixing and their application to the control of planktonic algal biomass. *J. Appl. Ecol.*, 21, 11–39.

Reynolds, C. S., Wiseman, S. W., Godfrey, B. M., and Butterwick, C. (1983b). Some effects of artificial mixing on the dynamics of phytoplankton populations in large limnetic enclosures. *J. Plank. Res.*, 5, 203–34.

Rhee, G.-Y. (1973). A continuous culture study of phosphate uptake, growth rate and polyphosphate in *Scenedesmus* sp. *J. Phycol.*, 9, 495–506.

(1982). Effect of environmental factors on phytoplankton growth. In: *Advances in Microbial Ecology*, vol. 6, ed. K. C. Marshall, pp. 33–74. New York: Plenum.

Richerson, P., Armstrong, R., and Goldman, C. R. (1970). Contemporaneous disequilibrium, a new hypothesis to explain the paradox of the plankton. *Proc. Nat. Acad. Sci.*, 67, 1710–14.

Roelofs, T. D. and Oglesby, R. T. (1970). Ecological observations on the planktonic cyanophyte *Gloeotrichia echinulata*. *Limnol., Oceanogr.*, 15, 244–9.

Rother, J. A. and Fay, P. (1977). Sporulation and the development of planktonic blue-green algae in two Salopian meres. *Proc. Roy. Soc. London B*, 196, 317–32.

(1979). Some physiological-biochemical characteristics of planktonic blue-green algae during bloom-formation in three Salopian meres. *Freshwat. Biol.*, 9, 369–79.

Runge, J. A., and Ohman., M. D. (1982). Size fractionation of phytoplankton as an estimate of food available to herbivores. *Limnol. Oceanogr.*, 28, 570–6.

Saito, S. (1978). Seasonal succession of phytoplankton in Sagami Reservoir from 1973 to 1977. *Japan. J. Limnol.*, 39, 147–55.

Senft, W. H., Hunchberger, R. A., & Roberts, K. E. (1981). Temperature dependence of growth and phosphorus uptake in two species of *Volvox* (Volvocales, Chlorophyta). *J. Phycol.*, 17, 323–9.

Smayda, T. J. (1970). The suspension and sinking of phytoplankton in the sea. *Oceanogr. Mar. Biol. Ann. Rev.*, 8, 353–414.

Smith, I. R. (1985). The influence of events on population growth. *Ann. Rep. Inst. Terrestrial Ecol.,* 1984, 33–4.

Smol, J. P., Brown, S. R., and McIntosh, H. J. (1984). A hypothetical relationship between differential algal sedimentation and diatom succession. *Verh. Int. Verein. Limnol.,* 22, 1361–5.

Sommer, U. (1981). The role of *r*- and *K*-selection in the succession of phytoplankton in Lake Constance. *Acta Oecologia,* 2, 327–42.

(1983). Nutrient competition between phytoplankton species in multispecies chemostat experiments. *Arch. Hydrobiol.,* 96, 399–416.

(1984). The paradox of the plankton: fluctuations of phosphorus availability maintain diversity of phytoplankton in flow-through cultures. *Limnol. Oceanogr.,* 29, 633–6.

Sournia, A. (1981). The morphological bases of competition and succession. *Can. Bull. Fish. Aquat. Sci.,* 210, 339–46.

(1982). Form and function in marine phytoplankton. *Biol. Rev. Cambridge Phil. Soc.,* 57, 347–94.

Southwood, T. R. E. (1977). Habitat, the templet for ecological strategies? *J. Animal Ecol.,* 46, 337–65.

Talling, J. F. (1960). Self-shading effects in natural populations of a planktonic diatom. *Wetter Leben,* 12, 235–42.

(1971). The underwater light climate as a controlling factor in the production ecology of freshwater phytoplankton. *Mitt. Int. Verein. Limnol.,* 19, 214–43.

(1976). The depletion of carbon dioxide from lake water by phytoplankton. *J. Ecol.,* 64, 79–121.

(1984). Past and contemporary trends and attitudes in work on primary productivity. *J. Plank. Res.,* 6, 203–17.

Tamiya, H., Iwamura, T., Shibata, K., Hase, E., and Nihei, T. (1953). Correlation between photosynthesis and light-independent metabolism in the growth of *Chlorella. Biochim. Biophys. Acta,* 12, 23–40.

Tilman, D., and Kilham, S. S. (1976). Phosphate and silicate growth and uptake kinetics of the diatoms *Asterionella formosa* and *Cyclotella meneghiniana* in batch and semicontinuous culture. *J. Phycol.,* 12, 375–83.

Tilman, D., Kilham, S. S., and Kilham, P. (1982). Phytoplankton community ecology: the role of limiting nutrients. *Ann. Rev. Ecol. System.,* 13, 349–72.

Tilzer, M. M. (1984). A quantum yield as a fundamental parameter controlling vertical photosynthetic profiles of phytoplankton in Lake Constance. *Arch. Hydrobiol. (Suppl.),* 69, 169–98.

Walsby, A. E. and Reynolds, C. S. (1980). Sinking and floating. In: *The Physiological Ecology of Phytoplankton,* ed. I. Morris, pp. 371–412. Oxford: Blackwell.

Wildman, R., Loescher, J. H., and Winger, C. L. (1975). Development and germination of akinetes of *Aphanizomenon flos-aquae. J. Phycol.,* 11, 96–104.

INDEX

acid phosphatases, 16
acid tolerance: of phytoplankton, 50, 272
akinetes (of Cyanobacteria), 264, 297,
    302–4, 411–12
alkaline phosphatase, 16, 113, 330–1
allelopathy, 216, *see also* competition
alpine lakes: phytophankton of, 20, 32,
    109, 183, 195–6
ammonium: metabolism of, 16, 286;
    uptake of, 319, 329
arctic lakes: phytoplankton of, 20, 30–1,
    32, 109, 182–3, 416
assimilation, *see* nutrients, carbon
auxospores (of diatoms), 246–7
auxotrophy, 15

Bacillariophyceae, *see* diatoms
bacteria, 151; ingestion of by algae, 16–
    17, 114; ingestion of by zooplankton,
    300; interaction of with cyanobacterial
    metabolism, 291–7, resource
    partitioning by, 337
bicarbonate, 295, 321; utilization of by
    chrysophytes, 48–50
biogeographic distribution of
    phytoplankton, 412–21; of
    Chlorophyta, 181–8; of chrysophytes,
    18–28; of cryptomonads, 108–10, 120–
    1; of cyanobacteria, 262–4, 272, 277–8,
    278–9; of diatoms, 228; of
    dinoflagellates, 134
biomass of phytoplankton: relationship
    to phosphorus loading, 27, 334;
    relationship to zooplankton biomass,
    67, 69; seasonal cycles of, 28, 32, 33,
    34, 110
blooms, 182; of cryptomonads, 108; of
    cyanobacteria, 261, 271, 296–7; of
    dinoflagellates, 139
blue-green algae, *see* cyanobacteria

bouyancy regulation: in cyanobacteria,
    263, 273, 408

C-strategist phytoplankton, 391–2, 393,
    400, 403, 407, 408, 410, 412–26
Ca:Mg ratio, 52
calcium: as biogeographic factor, 52; *see
    also* cationic ratios
carbon: assimilation of by
    phytoplankton, 157, 195, 216, 324,
    328; as a biogeographic factor, 48–50,
    278–9, 411; as a growth limiting factor,
    48–50, 295, 389, 411; to phosphorus
    ratio of algal biomass, 342; uptake
    kinetics of 319
carbon dioxide, 295, 334; as carbon
    source for chryosphytes 48–50, 411
carbon nutrition, *see* nutrition, *listings
    under various algal groups*
carbonic anhydrase, 48–50
cationic ratios: as biogeographic and
    growth limiting factor, 52–3
cell contents, of phytoplankton, 11, 107,
    135, 261; *see also listings under various
    algal groups*
cell coverings: of phytoplankton cells, 12,
    106, 135, 228
cell density, 238–9, 408
cell division, 73, 148, 157, 162, 378;
    phasing and synchrony of, 115, 148,
    249
cell generation (doubling) time, *see*
    growth rates
cell morphology, 395–8; as a metabolic
    scaling factor, 394; *see also* cell shape,
    cell size, morphological diversity of
    phytoplankton, shape of
    phytoplankton, size of phytoplankton
cell motility, *see* bouyancy regulation,
    flagella

cell nutrient quotas, 61, 231–2, 389; of phosphorus, 62–4, 402–3; as related to nutrient uptake, assimilation and growth, 321, 324–9, 340–2, 378–82, 402–3; as related to resource competition, 342–5; of silicon for diatoms, 229

cell shape, 136–8, 378, 384, 395; *see also* colonial growth forms, shape

cell size, ranges of, 136–8, 177, 227, 395; *see also* size of phytoplankton

cell volumes: of chrysophytes, 62–3; of phytoplankton, 395

centric diatoms, *see* diatoms

chelating agents, 16, 280–1, 295; as factor in biogeographic and seasonal distribution, 60–1, 281; as growth limiting factor, 55–61

chloride: as biogeographic factor for phytoplankton, 53

Chlorococcales (*see also* Chlorophyta), 205–10; carbon nutrition of, 216; growth rates of, 194, 205, 208–9, 395; growth strategies of, 208–10; light requirements of, 208–9; photosynthetic efficiency of, 399–400; population dynamics of, 209; phosphorus physiology of, 403–4; seasonal periodicity, 205, 421–5; sensitivity of to hydraulic mixing, 205–8, 408–9; survival strategies of 218–19; vertical distribution of, 189–90

Chlorophyceae, *see* Chlorophyta

chlorophyll: cell contents of, 157, 216, 399–400

Chlorophyta (green algae), 175–226; allelopathic effects of, 216; biogeographical distribution of, 176, 181–8; cell organization of, 175, 176; growth rates of, 194, 205, 208–9, 212–13, 395; horizontal patchiness of, 176–7; morphological diversity of, 190–4; parasitism of, 217; phosphorus physiology of, 403–4; photosynthetic efficiency of, 399–400; reproduction and resting stages of, 181, 214, 411; seasonal periodicity of, 181–8; sinking and bouyancy of, 408–9; survival strategies of, 214–20; taxonomic position of, 176; use of bicarbonate by, 48; vertical distribution of, 181, 186, 188–90; *see also* Chlorococcales, desmids, filamentous green algae, micro–green algae

chrysomonads, *see* chrysophytes

Chrysophyta or Chrysophyceae, *see* chrysophytes

chrysophytes, 9–104; acid phosphatases of, 16; biogeographic distribution of, 18–28, 86–7; carbon dioxide utilization by, 47–50, 411; cell organization of, 11; effect of N:P supply ratios on, 55–9; as a food source for zooplankton, 67–73; growth in relation to diatom blooms, 60; growth rates as function of temperature, 42–5; growth strategies of, 35, 86; maximum growth rates of, 62–4, 395; metalimnetic populations of, 14, 15, 35–6; morphological diversity of, 11–13; nutrition of, 15–18, 36; phosphorus physiology of, 62–7, 86, 403–4; phosphorus toxicity problem of, 53; physiological tolerances of, 42–54; population dynamics of, 30–5, 41; reproduction of, 73–83; resting stages of, *see* statospores; seasonal periodicity of, 28–36, 87; sensitivity of to hydraulic mixing, 36; taxonomic position of, 11; vertical migration of, 14–15

chytrids: parasitic attacks on algae by, 41, 119–20, 152–3, 213

ciliates: food selection by, 119, 152, 375

cladocerans, planktonic, 370; feeding efficiency of, 249, 382–4, 408; filtering mechanism of, 377; filtering (clearance) rates of, 67, 372–3; food selection by, 67–8, 118–19, 298–9, 408–10; nutrient regeneration by, 2, 68, 119, 186, 375–7; reproduction of, 371; *see also* filter feeding, grazing, zooplankton

clear phase: in phytoplankton seasonal periodicity, 34, 373

coenobial colonies, 380

colonial growth forms, 13, 263–4

colony formation: as adaptive strategy, 301–2, 395–8; significance of in grazer avoidance, 68–9, 301–2, 370, 375, 410

color, *see* water color

competition, 179–81, 194, 252, 276, 392, 421; and interactions between resource continua, 349–51; interference, 68, 215; interspecific, 153–6, 194; for light, 208–9; for nutrients, 61–7, 113, 151, 208–9, 233–8, 286, 316–18, 334, 402, 411; and resource variability (patchiness), 236–7, 352–8; role of specific chelators in, 60–1, 280–1, 295, 330–1; for single resource, 335–9; for several potentially limiting resources, 339–49

competitive exclusion principle, 391, 421

conductivity: as biogeographic factor, 22–3, 51–2, 271